Real and Complex Singularities

Photo courtesy of Walter Fukuhara.

CONTEMPORARY MATHEMATICS

569

Real and Complex Singularities

XI International Workshop on
Real and Complex Singularities
July 26–30, 2010
Instituto de Ciências Matemáticas e de Computação
Universidade de São Paulo
São Carlos, SP, Brazil

Victor Goryunov
Kevin Houston
Roberta Wik-Atique
Editors

American Mathematical Society
Providence, Rhode Island

2010 *Mathematics Subject Classification.* Primary 58Kxx, 57Rxx, 57Qxx, 32Sxx, 14Pxx, 37Cxx.

Library of Congress Cataloging-in-Publication Data

International Workshop on Real and Complex Singularities (11th : 2010 : Universidade de São Paulo) Real and complex singularities : XI International Workshop on Real and Complex Singularities, July 26-30, 2010, Universidade de São Paulo, São Carlos, SP Brazil / Victor Goryunov, Kevin Houston, Roberta Wik-Atique, editors.

p. cm. — (Contemporary mathematics ; v. 569)

Includes bibliographical references.

ISBN 978-0-8218-5359-7 (alk. paper)

1. Singularities (Mathematics)–Congresses. I. Goryunov, Victor, 1955- II. Houston, Kevin, 1968- III. Wik-Atique, Roberta, 1964- IV. Title.

QA614.58.1527 2010
514′.746—dc23   2011052531

---

10 9 8 7 6 5 4 3 2 1   17 16 15 14 13 12

This volume is dedicated to David Mond
on the occasion of his 60th birthday

Photo courtesy of Eleonore Faber.

Contents

Preface

This volume is a collection of papers presented at the 11th Workshop on Real and Complex Singularities, São Carlos, Brazil, July 26-30, 2010. The meeting was a part of a highly successful series of biennial conferences organized by the Singularity Theory group at São Carlos, University of São Paulo, Brazil. It is the longest running sequence of international workshops in singularities, which have been providing an exceptional opportunity for both young researchers and recognised leaders working in the field to meet together in a very productive scientific atmosphere. In 2010 a total of 160 participants from 13 countries (Brazil, Canada, Denmark, France, Germany, Japan, Mexico, Norway, Poland, Portugal, Spain, UK, USA) came to the Workshop.

The meeting in 2010 had a special dedication – to David Mond's 60th birthday. David Mond, professor at the University of Warwick, United Kingdom, is one of the leading experts in the area and has inspired many young mathematicians. Many years ago he fell in love with South America. Nowadays David has very strong connections with the Singularity group at Sao Carlos.

The main subject of singularity theory is the geometry and topology of spaces and maps defined by polynomials or analytic equations which are not regular. The theory uses techniques from several branches of mathematics and contributes to the development of rather distant fields, including algebraic geometry, knot theory, optics, computer vision, and many others. The possibility of application in a large number of different areas is one of the reasons of the success of singularity theory.

This book reflects the high level of the conference. It discusses the most recent results and applications of singularity theory and shows promising directions for future research in the field. Therefore, it will be an excellent reference for experienced researchers and an ideal introduction for younger, such as PhD students and post-docs, not only for current results but also for the variety of methods and techniques used in singularity theory.

The volume covers pure singularity theory (invariants, classification theory, Milnor fibres) and applications to other areas (singularities in topology and differential geometry, algebraic geometry and bifurcation theory). In particular, the reader will find here papers on plane curve singularities, metric theory of singularities, symplectic singularities, cobordisms of maps, Goursat distributions, sections of analytic varieties, Vassiliev type invariants, projections of hypersurfaces, properties of the Jacobian ideal.

We thank members of the Scientific Committee: Ragnar Buchweitz, Francisco Castro Jiménez, James Damon, Washington L. Marar, Walter Neumann, Juan José Nuño Ballesteros, Marcio Soares and Duco van Straten. We are also thankful to members of the Organizing Committee: Alexandre Fernandes, Nivaldo Grulha Jr,

Regilene Oliveira, Marcelo Saia, João Tomazella and Catiana Casonatto. We would like to express our gratitude to all others who work hard for the success of the meeting.

The workshop was funded by Brazilian funding bodies Fapesp, CNPq, CAPES, USP and INCTMat and the Japanese funding body JSPS, whose support we gratefully acknowledge.

We thank the referees for their diligent work in refereeing all the papers in this volume. We thank the staff members of the American Mathematical Society involved with the preparation of this book, and all those who have contributed in whatever way to these proceedings.

Victor Goryunov
Kevin Houston
Roberta Wik-Atique

David Mond

To celebrate the 60th birthday of David Mond the 2010 Real and Complex Singularities conference, a meeting held biennually in São Carlos, Brazil, was dedicated to him. David has been an organizer and proceedings editor for earlier Real and Complex Singularities conferences. Furthermore, having two of his PhD students based in São Carlos means that David has strong connections with the singularity group there and that the conference was a great opportunity to honour him.

David had not originally intended to become an academic. After completing his undergraduate degree in Mathematics and Philosophy at St. Catherine's College, Oxford, 1971, he was inspired by the work of Dutch furniture maker Gerrit Rietveld and initially pursued the ambition of making furniture himself. In 1973 with an Oxford friend he visited South America intending to travel through Venezuela and Colombia to Peru. Their journey came to a halt in Bogotá, Colombia, where David remained in the end for eight years.

It was in Colombia that David returned to mathematics, taking a Master's degree in Mathematics in the Universidad Nacional de Colombia, and later working as a mathematics instructor at Universidad de los Andes, and Universidad Nacional, both in Bogotá. His first five, essentially expository, papers appeared in Colombian mathematical journals and hence were written in Spanish – a language in which, along with Portuguese, he is fluent. After finishing the MSc in Mathematics at the Universidad Nacional, he moved in 1979 to the UK to begin a PhD at the University of Liverpool under the supervision of C.T.C. Wall FRS. His PhD, submitted in 1982, was entitled *The Classification of Germs of Maps from Surfaces to 3-space, with Applications to the Differential Geometry of Immersions.* From the late 1960s there had been much progress in the classification of singularities, particularly for right equivalence following V.I. Arnold's classifications. What was novel and significant about David's thesis was that it concerned the $\mathcal{A}$-equivalence (also known as right-left equivalence) for germs from $\mathbb{R}^2$ to $\mathbb{R}^3$. The stable maps in this case were classified by Whitney back in the 50s, David was classifying simple maps and those with $\mathcal{A}_e$-codimension less than or equal to 4. In the late 70s and early 80s such a classification required doing by hand long, tedious calculations involving tangent spaces, although David insists that it is often through long tedious calculations that one builds up an understanding of what is going on. Aspects of the classification were applied in differential geometry, for example to study the tangent developable of a space curve.

After his PhD, David had a brief research visit at the IHES in France and returned to Colombia to work again at the Universidad Nacional. In 1985 he moved to the University of Warwick in the UK. Of particular note during this time is the publication *Some remarks on the geometry and classification of germs of maps from surfaces to 3-space* in Topology 26 (1987) 361-383. It is here that David starts a thread that passes through much of his work by studying complex analytic maps, in this case investigating how invariants of map germs from $\mathbb{C}^2$ to $\mathbb{C}^3$ control the determinacy of the germ. It is in this paper that the (still unresolved) Mond conjecture first begins to take form: the $\mathcal{A}_e$-codimension of a map germ from $(\mathbb{C}^n, 0)$ to $(\mathbb{C}^{n+1}, 0)$ is less than or equal to the rank of the vanishing homology for the image of a local stabilization of the map, with equality if the map is quasihomogeneous. Various Mond students - including this author - have attempted to prove this statement during their PhD studies.

Also at this time David had his first PhD student, Washington 'Ton' Marar from Brazil, with whom he further developed the ideas in the Topology paper to show the multiple point spaces for finitely $\mathcal{A}$-determined corank 1 maps from $\mathbb{C}^n$ to $\mathbb{C}^p$, $n < p$, were in fact isolated complete intersection singularities (or zero-dimensional). His interest in classification continued with his second student, Diana Ratcliffe, producing a thesis on classifying multi-germs from $\mathbb{R}^2$ and $\mathbb{R}^3$. She later became his post-doctoral assistant and produced a computer program that relieved the drudgery from the extensive calculations involved in classifications. David remarked in one seminar that her program could do a calculation in seconds that he had spent weeks doing during his postgraduate days. A visit to Holland at this time led to a collaboration with Ruud Pellikaan, and later to a series of papers with Duco van Straten.

Other students followed in the early 90s with Thomas Cooper and this author both completing in 1994. By this time David had collaborated with other notable Singularity Theorists: Terry Gaffney, on the germs of maps from the plane to the plane; Jim Damon (in their paper for Inventiones Mathematicae) on a result that related $\mathcal{A}_e$−codimension and the rank of the vanishing homology for maps with $n \geq p$ (which added evidence to the Mond conjecture); Victor Goryunov, on a powerful spectral sequence that allowed one to calculate the rational cohomology of the image of a finite map.

David continued his association with São Carlos, co-supervising Roberta Wik-Atique with Maria Ruas, again on classification of singularities. However, following the importance of free divisors in the paper co-authored with Damon, he began to study these objects in more detail. A hypersurface D is a free divisor if the module of ambient vector fields which are tangent to D at its smooth points is freely generated. This property has significant consequences for the topology of the singularity, and, as the Inventiones paper showed, for the topology of its non-linear sections, because it implies the conservation of a certain singular multiplicity. This interest eventually produced a string of papers singly and co-authored with people such as Francisco Castro-Jiménez, Luis Narváez-Macarro, Martin Holland, Francisco Calderón Moreno, Ragnar-Olaf Buchweitz, Michel Granger, Alicia Nieto-Reyes, Mathias Schulze, Ignacio de Gregorio (another PhD student and later postdoctoral assistant) and Christian Sevenheck. In this string of papers, the important concept of linear free divisor is introduced. (This is a free divisor where the basis is formed by linear vector fields.) The importance partly stems from the connection to quiver representations as shown in the paper with Buchweitz. They also have connection with Frobenius manifolds since for a large class of examples, the Gauss-Manin system associated to a generic linear section of the Milnor fibre of a linear free divisor is a Frobenius manifold.

Recent students have been Paul Cadman (2010) and Ayşe Altintas (2011). The latter was one of David's students present at the conference in his honour. The conference was a great success and during the outing to a nearby lake the attendees were treated to a performance of David playing the flute. At the conference dinner friends, colleagues and ex-students took turns to pay tribute and sing the praises of a man they are proud to call their friend and teacher.

Kevin Houston
Leeds, UK, 2011.

Contemporary Mathematics
Volume **569**, 2012
http://dx.doi.org/10.1090/conm/569/11241

Milnor Fibrations and the Concept of d-regularity for Analytic Map Germs

José Luis Cisneros-Molina, José Seade, and Jawad Snoussi

ABSTRACT. In this expository article we review Milnor's proof of his celebrated fibration theorem and the way how it extends to the case of real analytic map germs. We highlight the similarities and the differences between the complex analytic and the real analytic settings, and we indicate the key geometric aspects in Milnor's proof that make it work in the complex case, unlike the theorem in the real setting which is more stringent. We explain too how, from the viewpoint of Milnor type fibrations, the difference between the real and the complex settings naturally leads to the concept of d-regularity.

Introduction

Milnor's fibration theorem for holomorphic maps is a key-stone in singularity theory. This theorem says that given a holomorphic map-germ $f\colon (\mathbb{C}^n, \underline{0}) \to (\mathbb{C}, 0)$ with a critical point at the origin $\underline{0} \in \mathbb{C}^n$, one has two locally trivial fibrations associated to it, and these fibrations are equivalent. The first is Milnor's fibration:

$$\phi := \frac{f}{|f|} : \mathbb{S}_\varepsilon \setminus K \longrightarrow \mathbb{S}^1 , \tag{1}$$

where K is the link of f at $\underline{0}$, that is $K = f^{-1}(0) \cap \mathbb{S}_\varepsilon$ with $\mathbb{S}_\varepsilon$ being a sufficiently small sphere around $\underline{0}$.

The second is nowadays called the Milnor-Lê fibration; this is a fibration on a "Milnor tube" $N(\varepsilon, \delta) = \mathbb{B}_\varepsilon \cap f^{-1}(\partial \mathbb{D}_\delta)$ where $\mathbb{B}_\varepsilon$ is a ball bounded by $\mathbb{S}_\varepsilon$, $\partial \mathbb{D}_\delta$ is the sphere bounding the ball of radius δ centred at the origin of $\mathbb{C}$. When $0 < \delta \ll \varepsilon \ll 1$ the map f induces a fibration

$$f\colon N(\varepsilon, \delta) \to \partial \mathbb{D}_\delta . \tag{2}$$

When $\underline{0}$ is an isolated critical point of f, the map ϕ actually defines an open-book decomposition of the sphere, with binding K.

2010 *Mathematics Subject Classification.* Primary 32S05, 32S55, 57Q45; Secondary 58K65.
Key words and phrases. Real and complex singularities, Milnor fibration, canonical pencil, d-regularity.
This research was partially supported by CONACYT (Mexico) grants G-36357-E, J-49048-F, U-55084, by DGAPA-UNAM-PAPIIT (Mexico) grants IN105806, IN102208, IN108111, by the CNRS, France and by ECOS–ANUIES grant M06-M02, France–Mexico.

The first author is a Regular Associate of the Abdus Salam International Centre for Theoretical Physics, Trieste, Italy.

It is natural to ask whether there is anything similar for real analytic map-germs $(\mathbb{R}^n, \underline{0}) \xrightarrow{f} (\mathbb{R}^p, 0)$ with $n \geq p$. This question goes back to Milnor's book, and actually to his unpublished preprint [**20**], where he proves that if f has an isolated critical point, then one always has a fibration in a tube, of type (2), now over a small $(p-1)$-sphere in $\mathbb{R}^p$, and this tube can always be inflated to the sphere, defining an equivalent fibration $\mathbb{S}^{n-1} \setminus K \xrightarrow{\psi} \mathbb{S}^{p-1}$, which restricted to a neighbourhood of K is the obvious projection $f/|f|$. Yet, this theorem has several weaknesses, as for instance the following (the first two of these were pointed out by Milnor in his book):

i) *The condition of having an isolated critical point is very stringent.* When this is satisfied we say that the corresponding map-germ satisfies *the Milnor condition.* In [**19, 9**] the authors classify the pairs (n, p) for which there exist analytic map-germs satisfying Milnor's condition. There are many other related works by various authors, as for instance N. A'Campo [**1, 2**], B. Perron [**25**], L. Rudolph [**30, 15**] and many others. In particular one has that when $p = 2$, even though the Milnor condition is indeed stringent, there are plenty of map-germs satisfying it, for all $n \geq 2$. More generally, one can study Milnor fibrations associated to the larger class of real analytic mappings with an isolated critical value. This was begun in [**27**] and continued in [**11, 12**].

ii) *We cannot always take the map ψ to be globally the canonical projection $\phi = f/\|f\|$.* When the equality holds we say that the map germ satisfies *the strong Milnor condition.* Examples of singularities with the strong Milnor condition have been given, for instance, in [**17, 31, 29, 10, 28, 23**].

iii) *If one has that the two types of fibrations exist, one on a Milnor tube and another on a sphere minus the link, a priori we do not know whether or not the two fibrations are necessarily equivalent.* This point was first noticed by R. N. Araujo dos Santos (see for instance [**3**]), and we fully answered it in [**12**].

We refer to our recent article [**13**] for a survey on the topic of Milnor fibrations for real singularities, where we explore carefully the above points and give an overview of the developments about them that one has in the literature.

Here we focus on explaining the main ideas and results in [**11, 12**], which give an essentially complete answer to the problem of existence and equivalence of the two types of Milnor fibrations for maps with an isolated critical value.

The key-idea is a certain regularity condition that we introduced in [**11, 12**], called d-regularity, which involves the family of varieties associated to every map-germ $(\mathbb{R}^n, \underline{0}) \xrightarrow{f} (\mathbb{R}^p, 0)$ as follows: for each line $\mathcal{L}$ through the origin in $\mathbb{R}^p$ set:

$$X_{\mathcal{L}} := \{x \in \mathbb{R}^n \mid f(x) \in \mathcal{L}\}.$$

The family $\{X_{\mathcal{L}}\}$ is *the canonical pencil* of f. When f has an isolated critical value, these analytic varieties are all smooth away from V, the axis of the pencil. The d-regularity property means that there exists $\varepsilon > 0$ such that every sphere of radius $\leq \varepsilon$ intersects transversally each manifold $X_{\mathcal{L}} \setminus V$.

We show that d-regularity is precisely the condition required in order to inflate the fibration on a Milnor tube conserving the value $\frac{f}{\|f\|}$, until we get the Milnor fibration on the sphere with projection map $\frac{f}{\|f\|}$. The theorem says:

THEOREM. *Let $f : (\mathbb{R}^n, \underline{0}) \to (\mathbb{R}^p, 0)$, $n > p$, be a real analytic map-germ, which is surjective over a neighbourhood of* 0 *and with* $\dim f^{-1}(0) > 0$. *Assume* *further that f has the Thom property, so one has a Milnor-Lê fibration*

$$f\colon N(\varepsilon,\delta) \longrightarrow \partial\mathbb{D}_\delta \cong \mathbb{S}^{p-1}\,.$$

Then one has also a locally trivial Milnor fibration:

$$\phi = \frac{f}{\|f\|} : \mathbb{S}_\varepsilon \setminus (\mathbb{S}_\varepsilon \cap f^{-1}(0)) \longrightarrow \mathbb{S}^{p-1},$$

if and only if f is d-regular, and in this case these two fibrations are equivalent.

The concept of d-regularity is actually implicit in Milnor's proof of his fibration theorem. This concept is also used in [**11**] to give an alternative way for looking at Milnor's fibration theorem for holomorphic mappings and its generalisation by Lê Dũng Tráng to functions on singular spaces. This concept also appears implicitly in [**27**, Lemma 5.7] to prove the equivalence of the two types of Milnor fibrations, as in (1) and (2), for functions of the type $f\bar{g}$. This regularity condition also appears implicitly in [**5**] for map-germs with an isolated critical point.

The name "d-regularity" comes from the facts that this concept is naturally associated to a function "distance to the critical point", and also that every map-germ (with an isolated singularity) which is (c)-regular in K. Bekka's sense, is d-regular in our sense (see [**28**] for details and [**6**] for definitions).

The purpose of this expository article is to give an introduction to d-regularity and explain the theorem above. We begin the article with a section on real and complex examples where it is easy to describe the two types of Milnor fibrations, and where d-regularity appears implicitly to give the equivalence of the two types of fibrations. Then we go back to the holomorphic setting, because it is actually from Milnor's work that this concept springs. We review Milnor's original proof of the fibration theorem from a perspective that leads towards the concept of d-regularity. Finally, in Section 3 we explain how Thom's a_f condition relates to these fibration theorems, we speak about d-regularity and explain how this concept allows to prove that whenever the two types of fibration exist, they are equivalent.

1. An example: weighted homogeneous singularities

In this section we look at Milnor fibrations for real and complex weighted homo geneous singularities. We start with a quick look at the celebrated Pham-Brieskorn singularities, which provide the paradigm of Milnor fibrations. Throughout this work, given a map germ $\mathbb{R}^n \to \mathbb{R}^p$, we denote by $\underline{0}$ the origin in the source, and by 0 that in the target.

1.1. Milnor fibrations for the Pham-Brieskorn singularities. Consider a Pham-Brieskorn polynomial $f : \mathbb{C}^n \to \mathbb{C}$:

$$(z_1, ..., z_n) \overset{f}{\mapsto} z_1^{a_1} + ... + z_n^{a_n}, \qquad a_i \geq 2\,.$$

It is clear that the origin $\underline{0} \in \mathbb{C}^n$ is its only critical point, so $V := f^{-1}(0)$ is a hypersurface with an isolated singularity at $\underline{0}$.

Notice that this polynomial is weighted homogeneous. In fact, let d be the lowest common multiple of the a_i and for each $i = 1, \cdots, n$ set $d_i = d/a_i$. Then for every non-zero complex number $\lambda \in \mathbb{C}^*$ one has a $\mathbb{C}^*$-action on $\mathbb{C}^n$ determined by

$$\lambda \cdot (z_1, \cdots, z_n) \mapsto (\lambda^{d_1} z_1, \cdots, \lambda^{d_n} z_n)\,.$$

Observe one has:

$$f(\lambda^{d_1} z_1, \cdots, \lambda^{d_n} z_n) = \lambda^d f(z_1, \cdots, z_n)\,.$$

We thus get a holomorphic action of $\mathbb{C}^*$ which has the origin $\underline{0}$ as its only fixed point and V is an invariant set, union of orbits. It is clear that this $\mathbb{C}^*$-action has the following four properties:

• **Property 1.** If we restrict the action to the positive real numbers $t \in \mathbb{R}^+$, then each such orbit is a curve that converges to $\underline{0}$ as t tends to 0, and it goes to infinity as t tends to ∞. Furthermore, a direct computation shows that for the points in each such orbit, the distance to $\underline{0}$ is a function with no critical point, so each orbit is transversal to all spheres centred at $\underline{0}$. Hence the variety V intersects transversally every $(2n-1)$-sphere $\mathbb{S}_r$ around the origin and the intersection $K_r := V \cap \mathbb{S}_r$ is a real codimension 2 smooth submanifold of the sphere $\mathbb{S}_r$. Also, this real flow determines a 1-parameter group of diffeomorphisms and given arbitrary spheres $\mathbb{S}_r, \mathbb{S}_{r'}$ centred at $\underline{0}$, we get a diffeomorphism from $\mathbb{S}_r$ into $\mathbb{S}_{r'}$ taking K_r into $K_{r'}$. Thence the diffeomorphism type of K_r is independent of the choice of the sphere $\mathbb{S}_r$. We denote this manifold simply by K and call it *the link of the singularity* (see [**21, 14**]).

• **Property 2.** The above real analytic flow also has the property that for points in $\mathbb{C}^n \setminus V$, the argument of the complex number $f(z)$ is constant on each orbit, *i.e.*, $f(z)/|f(z)| = f(tz)/|f(tz)|$ for all $t \in \mathbb{R}^+$.

• **Property 3.** If we restrict the $\mathbb{C}^*$-action to the complex numbers of unit length we get an $\mathbb{S}^1$-action that leaves invariant every sphere around $\underline{0}$, and for each $e^{i\theta}$ and each $z = (z_1, \cdots, z_n)$ we have $f(e^{i\theta} \cdot (z_1, \cdots, z_n)) = e^{i\theta d} \cdot f(z_1, \cdots, z_n)$. Hence for each $\zeta \in \mathbb{C}$ we have that the multiplication by $e^{i\theta}$ in $\mathbb{C}^n$ transports the fibre $f^{-1}(\zeta)$ into the fibre $f^{-1}(e^{i\theta d} \cdot \zeta)$. Therefore, if for each real number $\delta > 0$ we consider the "tube" $N(\delta) = f^{-1}(\partial \mathbb{D}_\delta)$, where $\partial \mathbb{D}_\delta \cong \mathbb{S}^1$ is the boundary of the disc in $\mathbb{C}$ of radius δ, centred at 0, then $\mathbb{S}^1$ acts on $N(\delta)$. A direct computation shows that the orbits of this action are transverse to the fibers of f. So the action determines a locally trivial fibration (a fibre bundle):

$$f\colon N(\delta) \to \partial \mathbb{D}_\delta\,.$$

Together with Property 2, this implies that the fibres $f^{-1}(\lambda)$ are all diffeomorphic.

Now observe that for each line $\mathcal{L}_\theta$ through the origin in $\mathbb{C}$, we may consider the set

$$X_\theta := \{z \in \mathbb{C}^n \mid f(z) \in \mathcal{L}_\theta\}\,.$$

Then each X_θ is the zero-set of f followed by an orthogonal projection. Hence each X_θ is a real analytic hypersurface with an isolated singularity at $\underline{0}$, their union fills the entire space $\mathbb{C}^n$ and their intersection is V. By property 1, each of these hypersurfaces is transversal to all the spheres, and by Property 3, the $\mathbb{S}^1$-action permutes these hypersurfaces: multiplication by e^{it} carries each X_θ into another one of these hypersurfaces. In other words one has:

• **Property 4.** These varieties define a pencil in $\mathbb{C}^n$, a sort of open-book where the binding is now the singular variety V, and each of these varieties is transverse to every sphere around $\underline{0}$. If we remove V from $\mathbb{C}^n$, for every ball around $\underline{0}$ we get a canonical projection

$$\Phi = \frac{f}{|f|} : \mathbb{B}^{2n} \setminus V \longrightarrow \mathbb{S}^1\,, \tag{3}$$

which actually is a fibre bundle; the fibre over a point $e^{i\theta}$ is the corresponding connected component of $X_\theta \setminus V$. (The other component is the fibre over $e^{-i\theta}$.)

We now focus our attention near the origin, say restricted to the unit ball $\mathbb{B}^{2n}$ in $\mathbb{C}^n$. Since each X_θ meets transversally the sphere $\mathbb{S}^{2n-1} = \partial\mathbb{B}^{2n}$, the intersection is a smooth codimension 1 submanifold of the sphere, containing the link $K = V \cap \mathbb{S}^{2n-1}$. And since the orbits of the $\mathbb{S}^1$-action preserve the sphere $\mathbb{S}^{2n-1}$, it follows that the restriction of Φ to $\mathbb{S}^{2n-1}$ defines a locally trivial fibration:

$$\phi = \frac{f}{|f|} : \mathbb{S}^{2n-1} \setminus K \longrightarrow \mathbb{S}^1 \tag{4}$$

This is the classical *Milnor fibration* for the map f. It is worth saying that for these maps, the Pham-Brieskorn polynomials, the fact that (4) is a fibre bundle was first proved by Pham and used by Brieskorn to prove important results about the topology of the corresponding links (for more about this, see [**32**, Chapter 1] and the bibliography in it).

REMARK 1.1. Since f has a unique critical point at $\underline{0}$, the implicit function theorem implies that at each point $x \in V^* := V \setminus \{\underline{0}\}$ we can find local coordinates so that f is locally a linear projection $\mathbb{C}^n \to \mathbb{C}$. Hence the fibers of f are locally (in a neighbourhood of x) parallel complex discs of complex dimension $n-1$. This implies the following statement, that we do not need here, but which is the key for generalising this discussion to the case of (real and complex) analytic mappings with non-isolated critical points: *every smooth map-germ with an isolated critical point has the Thom a_f-property.* We refer to 3.1 for a precise definition; for more about this important concept see for instance [**11**] and the bibliography in it.

In our case, that of the Pham-Brieskorn singularities, Remark 1.1 tells us that, since V is transverse to the unit sphere $\mathbb{S}^{2n-1}$, the fibre of f passing through each point in the sphere sufficiently near V, is also transverse to the sphere. Since $K = V \cap \mathbb{S}^{2n-1}$ is a compact set, it follows that there exists a real number $\delta > 0$ such that all fibres $f^{-1}(\lambda)$ with $|\lambda| = \delta$, are transverse to $\mathbb{S}^{2n-1}$. In other words, the tube $N(\delta)$ in Property 3 is everywhere transverse to $\mathbb{S}^{2n-1}$. Therefore, setting $N(1,\delta) := N(\delta) \cap \mathbb{B}^{2n}$, where the 1 means that the ball has radius 1, we have that the fibre bundle described by Property 3 determines a fibre bundle:

$$f \colon N(1,\delta) \to \partial\mathbb{D}_\delta \cong \mathbb{S}^1 . \tag{5}$$

This is the second classical version of Milnor's fibration for the map f. Following the modern literature, we call this the *Milnor-Lê fibration of* f.

Notice that by Property 2, each $\mathbb{R}^+$-orbit is everywhere transverse to the tube $N(\delta)$ and transverse to the sphere $\mathbb{S}^{2n-1}$, and the complex numbers $f(z)$ have constant argument along each orbit. Thence the integral lines of this action determine a diffeomorphism between $N(1,d)$ and $\mathbb{S}^{2n-1}$ minus the part of the sphere contained inside the open solid tube $f^{-1}(\overset{\circ}{\mathbb{D}}_\delta)$. This determines the classical equivalence between the Milnor fibration in the sphere (4) and the Milnor-Lê fibration in the tube (2).

We now remark that everything we said above works in exactly the same way for all weighted homogeneous complex singularities, *i.e.*, for all complex polynomials f for which there is a $\mathbb{C}^*$ action on $\mathbb{C}^n$ as above, for some positive integers $\{d; d_1, \cdots, d_n\}$,

$$\lambda \cdot (z_1, \cdots, z_n) = (\lambda_1^{d_1} z_1, \cdots, \lambda_n^{d_n} z_n),$$

satisfying that for all $\lambda \in \mathbb{C}^*$ and for all $z \in \mathbb{C}^n$ one has:

$$f(\lambda \cdot z) = \lambda^d \cdot f(z) .$$

This follows because if we look carefully at the above discussion we see that Properties 1 to 4 are intrinsically given by the weights of the action and the fact that f brings out scalars to the power d. All the statements above concerning the fibration theorems follow from these properties. More precisely:

i) Property 1 implies there is a flow on $V \setminus \{\underline{0}\}$ given by the $\mathbb{R}^+$-action, which is transverse to all spheres around the origin $\underline{0}$. Since f has an isolated critical point at $\underline{0}$, this implies that given a sphere $\mathbb{S}_\varepsilon$, all fibers $f^{-1}(t)$ sufficiently near V are also transverse to $\mathbb{S}_\varepsilon$. Property 3 then implies that given a tube $N(\varepsilon, \delta) := f^{-1}(\mathbb{D}_\delta) \cap \mathbb{B}_\varepsilon)$, the flow defined by the $\mathbb{S}^1$-action permutes the fibers in this tube and we have a Milnor-Lê fibration

$$f : N(\varepsilon, \delta) \to \partial\mathbb{D}_\delta \cong \mathbb{S}^1 . \tag{6}$$

ii) Properties 2 and 4 imply that one has a decomposition of all of $\mathbb{C}^n$ as an "open-book" with binding the singular variety V and pages the real hypersurfaces $X_\theta \setminus V$.

iii) Then Property 2 implies that the X_θ are transverse to every sphere $\mathbb{S}_\varepsilon$, and therefore the above open book decomposition of $\mathbb{C}^n$ determines the Milnor fibration

$$\phi = \frac{f}{|f|} : \mathbb{S}_\varepsilon^{2n-1} \setminus K \longrightarrow \mathbb{S}^1 . \tag{7}$$

iv) Finally, properties 1 and 2 imply that the flow given by the $\mathbb{R}^+$-action determines an equivalence between the two fibrations of f, the Milnor fibration and the Milnor-Lê-fibration.

As we will see in the sequel, there are several interesting families of singularities which can be equipped with flows that satisfy properties analogous to (i)-(iv), and whenever we have these properties, we have the two types of fibrations and they are equivalent.

1.2. Polar weighted singularities. Let us consider now polynomial maps $f : \mathbb{R}^{2n} \to \mathbb{R}^2$, $n \geq 1$ that carry the origin $\underline{0} \in \mathbb{R}^{2n}$ into the origin $0 \in \mathbb{R}^2$. We identify the plane $\mathbb{R}^2$ with the complex line $\mathbb{C}$ and equip $\mathbb{C}^*$ with polar coordinates $\{r\, e^{i\theta} \,|\, r > 0;\, \theta \in [0, 2\pi)\}$; notice that as a manifold $\mathbb{C}^*$ is the cylinder $\mathbb{S}^1 \times \mathbb{R}^+$. We also identify $\mathbb{R}^{2n}$ with $\mathbb{C}^n$.

The following concept is introduced in [**10**].

DEFINITION 1.2. A map f as above is a *polar weighted homogeneous polynomial map* if there exists an action of $\mathbb{S}^1 \times \mathbb{R}^+$ on $\mathbb{C}^n$ of the form

$$(\lambda, r) \cdot (z_1, \ldots, z_n) = (\lambda^{d_1} r^{p_1} z_1, \ldots, \lambda^{d_n} r^{p_n} z_n) , \quad \lambda \in \mathbb{S}^1 , r \in \mathbb{R}^+ ,$$

where the d_j, p_j are positive integers such that $\gcd(d_1, \ldots, d_n) = 1 = \gcd(p_1, \ldots, p_n)$, and one has:

$$f((\lambda, r) \cdot (z_1, \ldots, z_n)) = \lambda^d r^p f(z_1, \ldots, z_n) ,$$

for some positive integers d, p.

Notice this includes the singularities given by weighted homogeneous complex polynomials; in that case the degrees d and p coincide.

The paradigm of polar weighted homogeneous maps is given by the *Twisted Pham-Brieskorn singularities* studied in [**29**] (see also Chapter VII in [**32**]). These are the singularities defined by maps of the following type:

$$(z_1, ..., z_n) \mapsto z_1^{a_1} \bar{z}_{\sigma(1)} + ... + z_n^{a_n} \bar{z}_{\sigma(n)} \quad , \; a_i \geq 2 \, ,$$

where σ is a permutation of the set $\{1, ..., n\}$. The explicit weights are easily computable from the exponents a_i and the permutation σ (see Lemmas 4.3 and 4.4 in Chapter VII of [**32**]). The name comes from their similarity with the classical Pham-Brieskorn polynomials and the fact proved in [**29, 24**], that if σ is the identity, then these singularities are equivalent to the usual Pham-Brieskorn singularities:

$$(z_1, ..., z_n) \mapsto z_1^{a_1 - 1} + ... + z_n^{a_n - 1} \, .$$

When σ is not the identity, these singularities are not equivalent to complex singularities, generally speaking (by [**26**]).

Using the *polar action* given in Definition 1.2 one can show (see [**10**]) that polar weighted homogeneous polynomials have $0 \in \mathbb{C}$ as the only critical value, and assuming that the dimension of $V = f^{-1}(0)$ is more than 0, the above Properties 1-4 continue to hold in this setting. Therefore every polar weighted singularity has the two types of Milnor fibrations that complex singularities have: one on the sphere given as in equation (4) and another in the tube as in equation (6). Furthermore, just as in the complex case, these two fibrations are equivalent, by exactly the same arguments as above.

In fact one has an associated pencil as in Property 4. The elements X_θ in this pencil are real algebraic hypersurfaces in $\mathbb{C}^n$ which are smooth away from $V := f^{-1}(0)$, they fill out the whole ambient space and meet exactly at V. The $\mathbb{R}^+$ action leaves invariant every element of the pencil, which therefore is transverse to all spheres around $\underline{0}$. And the orbits of the $\mathbb{S}^1$-action are tangent to all the spheres around $\underline{0}$ and permute the hypersurfaces X_θ. Therefore one has a global fibration as in equation (3), which restricts to the Milnor fibration (4) on each sphere, and by Property 2 this is equivalent to the fibration on a tube as in equation (6).

1.3. Milnor fibrations for weighted homogeneous real polynomial mappings. Consider now a weighted homogeneous polynomial map germ

$$f := (f_1, \cdots, f_p) : (\mathbb{R}^n, \underline{0}) \to (\mathbb{R}^p, 0) \, ,$$

with an isolated critical point at $\underline{0}$ and $\dim V > 0$, where $V = f^{-1}(0)$. We now have a weighted action of $\mathbb{R}$ on $\mathbb{R}^n$ of the type

$$t \cdot (x_1, \cdots, x_n) \mapsto (t^{d_1} x_1, \cdots, t^{d_n} x_n), \, d_i \geq 1$$

such that for each $i = 1, \cdots, p$ one has:

$$f_i(t \cdot (x_1, \cdots, x_n)) = t^d f_i(x_1, \cdots, x_n) \, .$$

Observe that in this setting we have a real analytic flow that clearly has Property 1 as well as the following property, analogous to Property 2:

- **Property 2′.** The line through the origin in $\mathbb{R}^p$ determined by a point $f(x)$ is the same for all x in the same $\mathbb{R}^+$-orbit.

Regarding Property 3, now there is no $\mathbb{S}^1$ action, so Property 3 makes no sense. Recall that this property was used only to show that the tube $N(\delta) = f^{-1}(\partial \mathring{\mathbb{D}}_\delta)$ was a fiber bundle over the boundary of the δ-disc, with projection map f. We claim (following [**21**, Chapter 10]) that the mere fact that f has an isolated critical point implies that it satisfies:

• **Property** $3'$. If we let $\widehat{N}(\varepsilon,\delta)$ be the tube $f^{-1}(\mathbb{D}_\delta \setminus 0) \cap \mathbb{B}_\varepsilon$, where $\mathbb{D}_\delta$ is now a ball in $\mathbb{R}^p$, centred at 0, of radius $\delta > 0$, then for $\varepsilon \gg \delta > 0$ sufficiently small one has that $\widehat{N}(\varepsilon,\delta)$ is everywhere transverse to the sphere $\mathbb{S}_\varepsilon$.

Then, just as in Milnor's work (see for instance [**20**, p. 6]), one has a locally trivial fibration:

$$f\colon \widehat{N}(\varepsilon,\delta) \longrightarrow \mathbb{D}_\delta \setminus 0,$$

whose restriction to $N(\varepsilon,\delta) = f^{-1}(\partial\mathbb{D}_\delta) \cap \mathbb{B}_\varepsilon$ is also a fibration over $\partial\mathbb{D}_\delta \cong \mathbb{S}^{p-1}$. This is just a version of Ehresmann's fibration theorem relative to the boundary of the tube.

It is now an exercise to show that properties 1 and $2'$ imply that we can carry the tube $N(\varepsilon,\delta)$ into the sphere $\mathbb{S}_\varepsilon$ and get a Milnor fibration on every sphere centred at $\underline{0}$:

$$\phi := \frac{f}{\|f\|}\colon \mathbb{S}_\varepsilon \setminus (V \cap \mathbb{S}_\varepsilon) \longrightarrow \mathbb{S}^{p-1}\,.$$

Regarding Property 4, for each line $\mathcal{L}$ through the origin in $\mathbb{R}^p$ consider the set:

$$X_{\mathcal{L}} = \{x = (x_1,\cdots,x_n) \,|\, f(x) \in \mathcal{L}\}\,.$$

Each of these is a real algebraic variety of dimension $n-p+1$, smooth away from V, and invariant under the $\mathbb{R}$-action. Hence they meet transversally every sphere around $\underline{0}$. It is now an exercise to show that Properties (1), ($2'$) and ($3'$) imply that one has Property (4) and there is a locally trivial fiber bundle:

$$\Phi := f/\|f\| : \mathbb{R}^n \setminus V \to \mathbb{S}^{p-1}\,,$$

whose restriction to the boundary sphere is the Milnor fibration.

So, in this latter example, Property 4 was a consequence of the previous properties (1), ($2'$) and ($3'$). Yet, we will see in Section 3 that given a real analytic map germ $\mathbb{R}^n \to \mathbb{R}^p$ with an isolated critical point and $\dim f^{-1}(0) > 0$, having an analogous "Property (4)" is a necessary and sufficient condition for having a Milnor fibration on the spheres, and in this case this is equivalent to the fibration on a Milnor tube, which all such maps have. "Property (4)" is precisely the d-regularity condition mentioned in the title of this article. This will be discussed in Section 3.

2. On the geometry in Milnor's proof of the fibration theorem

As we mentioned in the introduction, Milnor proved that given any complex polynomial $f\colon (\mathbb{C}^n,\underline{0}) \to (\mathbb{C},0)$, the map $\phi = \frac{f}{|f|}\colon \mathbb{S}^{2n-1}_\varepsilon \setminus K \to \mathbb{S}^1$ is a locally trivial fibration. Instead of using a $\mathbb{C}^*$-action restricted to $\mathbb{S}^1$ and to $\mathbb{R}^+$ as in Section 1, Milnor constructs two suitable vector fields which do the same job (essentially properties 1 to 4).

The aim of this section is to describe the geometry behind Milnor's original proof of his fibration theorem for holomorphic maps. We shall describe the geometric ideas without giving all the technical details for which we refer to Milnor's book [**21**, §4].

2.1. Complex and real gradients. Let $\langle\ ,\ \rangle$ be the standard Hermitian inner product on $\mathbb{C}^n$ given by

$$\langle v, w\rangle = \sum_{j=1}^{n} v_j \overline{w}_j, \quad \text{for all } v = (v_1, \dots, v_n), w = (w_1, \dots, w_n) \in \mathbb{C}^n.$$

The Hermitian vector space can also be thought of as a Euclidian vector space of dimension $2n$ over the real numbers, defining the Euclidean inner product $\langle\ ,\ \rangle_{\mathbb{R}}$ to be the real part of the Hermitian inner product. That is,

$$\langle v, w\rangle_{\mathbb{R}} = \Re\langle v, w\rangle, \quad \text{for all } v, w \in \mathbb{C}^n.$$

Let $f\colon \mathbb{C}^n \to \mathbb{C}$ be a complex analytic function and define its *gradient* by

$$\operatorname{grad} f = \left(\overline{\frac{\partial f}{\partial z_1}}, \dots, \overline{\frac{\partial f}{\partial z_n}}\right).$$

With this definition the chain rule for the derivative of f along a path $z = p(t)$ has the form

$$\frac{df}{dt}(p(t)) = \left\langle \frac{dp}{dt}, \operatorname{grad} f \right\rangle. \tag{8}$$

In other words, the *directional derivative* of f along a vector v at the point z is equal to the inner product $\langle v, \operatorname{grad} f\rangle$.

Let $f\colon (\mathbb{C}^n, \underline{0}) \to (\mathbb{C}, 0)$ be an analytic function with a critical value at 0 and let $V = f^{-1}(0)$. By the Bertini-Sard Theorem [**33**, Thm. (3.3)] there exists a ball $\mathbb{B}_\varepsilon$ of radius ε centred at $\underline{0}$ such that $0 \in \mathbb{C}$ is the only critical value of f restricted to $\mathbb{B}_\varepsilon$. Writing f as

$$f(z) = |f(z)| e^{i\theta(z)} \tag{9}$$

we can associate to f the real analytic functions

$$\begin{aligned} \theta\colon \mathbb{B}_\varepsilon \setminus V &\to \mathbb{R}, \\ \log|f|\colon \mathbb{B}_\varepsilon \setminus V &\to \mathbb{R}. \end{aligned}$$

where θ is locally well defined.

From (9) we have that these functions are related to the function $\log f$ as follows:

$$\log f = \log|f| + i\theta.$$

Thus,

$$\begin{aligned} \theta &= -i \log f + i \log|f|, \\ \log|f| &= \log f - i\theta. \end{aligned}$$

Since $i \log|f|$ and $i\theta$ are pure imaginary, we have that

$$\theta = \Re(-i \log f)\,, \tag{10}$$

$$\log|f| = \Re(\log f). \tag{11}$$

Differentiating (10) and (11) along a curve $z = p(t)$ we obtain

$$\begin{aligned}
\frac{d\theta(p(t))}{dt} &= \Re\Big(\frac{d(-i\log f)}{dt}(p(t))\Big) \\
&= \Re\Big\langle \frac{dp}{dt}, \operatorname{grad}(-i\log f(p(t))\Big\rangle \\
&= \Re\Big\langle \frac{dp}{dt}, i\operatorname{grad}\log f(p(t))\Big\rangle \quad (12)\\
&= \Big\langle \frac{dp}{dt}, i\operatorname{grad}\log f(p(t))\Big\rangle_{\mathbb{R}}, \quad (13)
\end{aligned}$$

and

$$\begin{aligned}
\frac{d\log|f(p(t))|}{dt} &= \Re\Big(\frac{d(\log f)}{dt}(p(t))\Big) \\
&= \Re\Big\langle \frac{dp}{dt}, \operatorname{grad}\log f(p(t))\Big\rangle \quad (14)\\
&= \Big\langle \frac{dp}{dt}, \operatorname{grad}\log f(p(t))\Big\rangle_{\mathbb{R}}. \quad (15)
\end{aligned}$$

Hence, by (8), (12) and (14) we have that

$$\frac{d\theta(p(t))}{dt} = \Re\Big(-i\Big\langle \frac{dp}{dt}, \operatorname{grad}\log f(p(t))\Big\rangle\Big) = \Im\Big(\frac{d\log f(p(t))}{dt}\Big), \tag{16}$$

$$\frac{d\log|f(p(t))|}{dt} = \Re\Big\langle \frac{dp}{dt}, \operatorname{grad}\log f(p(t))\Big\rangle = \Re\Big(\frac{d\log f(p(t))}{dt}\Big). \tag{17}$$

That is:

PROPOSITION 2.1. *The directional derivative of the function $\theta(z)$ (respectively of $\log|f(z)|$) in the direction $\frac{dp}{dt}$ is the imaginary (respectively the real) part of the directional derivative of the function $\log f$ in the direction $\frac{dp}{dt}$. Therefore (by (13) and (15)) their corresponding* real *gradients are given by*

$$\operatorname{grad}_{\mathbb{R}} \theta = i\operatorname{grad}\log f, \tag{18}$$

$$\operatorname{grad}_{\mathbb{R}} \log|f| = \operatorname{grad}\log f, \tag{19}$$

and these are of course normal to the respective level hypersurfaces.

For the case of the function $\theta(z)$, the level hypersurface corresponding to a constant angle θ_0 is given by

$$\begin{aligned}
\theta^{-1}(\theta_0) &= \{\, z \in \mathbb{B}_\varepsilon \setminus V \mid \theta(z) = \theta_0 \,\} \\
&= \{\, z \in \mathbb{B}_\varepsilon \setminus V \mid f(z) \in \mathcal{L}_{\theta_0}^{+} \,\} \\
&= f^{-1}(\mathcal{L}_{\theta_0}^{+}) \cap \mathbb{B}_\varepsilon,
\end{aligned}$$

where $\mathcal{L}_{\theta_0}^{+}$ is the open real ray in $\mathbb{C}$ with angle θ_0 with respect to the positive real axis. We shall denote

$$E_{\theta_0} = f^{-1}(\mathcal{L}_{\theta_0}^{+}) \cap \mathbb{B}_\varepsilon.$$

Since f is a submersion on $\mathbb{B}_\varepsilon$ outside V we have that E_{θ_0} is a codimension 1 submanifold of $\mathbb{C}^n$. Hence by (13) we have the following lemma:

LEMMA 2.2. *The vector $i\operatorname{grad}\log f(z)$ is normal to $E_{\theta(z)}$ at the point $z \in \mathbb{B}_\varepsilon \setminus V$.*

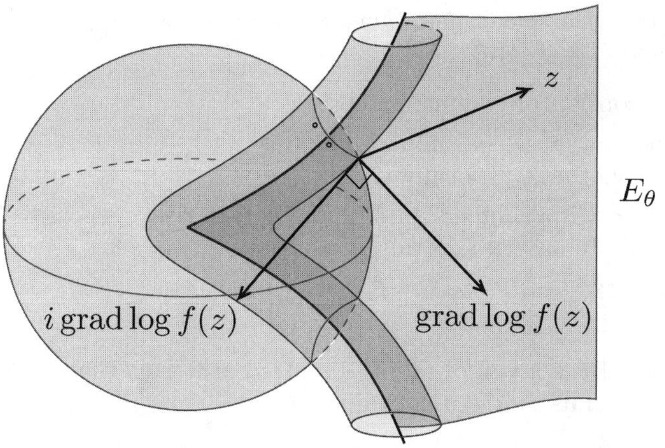

FIGURE 1. The vectors z, $\operatorname{grad}\log f(z)$ and $i\operatorname{grad}\log f(z)$.

For the case of the function $\log|f(z)|$, the level hypersurface corresponding to a positive constant value $\log\delta$ is given by

$$\begin{aligned}\log|f|^{-1}(\log\delta) &= \{\, z \in \mathbb{B}_\varepsilon \mid \log|f(z)| = \log\delta \,\}\\ &= \{\, z \in \mathbb{B}_\varepsilon \mid |f(z)| = \delta \,\}\\ &= f^{-1}(\partial\mathbb{D}_\delta) \cap \mathbb{B}_\varepsilon,\end{aligned}$$

where $\partial\mathbb{D}_\delta$ is the boundary of the disc of radius δ in $\mathbb{C}$. In other words, the level hypersurface of $\log|f(z)|$ corresponding to the value $\log\delta$ is the tube

$$N(\varepsilon,\delta) = f^{-1}(\partial\mathbb{D}_\delta) \cap \mathbb{B}_\varepsilon.$$

Again, since f is a submersion outside V we have that $N(\varepsilon,\delta)$ is a codimension 1 submanifold of $\mathbb{C}^n$. Hence by (15) we have the following lemma:

LEMMA 2.3. *The vector* $\operatorname{grad}\log f(z)$ *is normal to the tube* $N(\varepsilon, |f(z)|)$ *at the point* $z \in \mathbb{B}_\varepsilon \setminus V$.

Figure 1 ilustrates Lemmas 2.2 and 2.3 showing the vectors $\operatorname{grad}\log f(z)$ and $i\operatorname{grad}\log f(z)$ and the fact that they are orthogonal.

REMARK 2.4. Note that from the definition of the gradient we have that

$$\operatorname{grad}\log f(z) = \frac{\operatorname{grad} f(z)}{\bar{f}(z)} \tag{20}$$

so it is well defined everywhere, even though $\log f$ is only locally defined as a single-valued function.

2.2. Milnor Fibration. Now that we have described the real gradients of the real functions θ and $\log|f|$ in terms of $\log f$ the aim is to prove

MILNOR'S FIBRATION THEOREM. *Let* $f\colon (\mathbb{C}^n, \underline{0}) \to (\mathbb{C}, 0)$ *be a holomorphic map-germ. Then the map*

$$\phi := \frac{f}{|f|}\colon \mathbb{S}_\varepsilon \setminus K \longrightarrow \mathbb{S}^1$$

is the projection of a smooth fibre bundle, where $\mathbb{S}_\varepsilon$ is a sufficiently small sphere centred at $\underline{0}$ and $K = f^{-1}(0) \cap \mathbb{S}_\varepsilon$.

The proof consists of two main steps:

(1) To prove that ϕ is a submersion.
(2) To prove that ϕ is locally trivial.

To show that ϕ is a submersion, Milnor characterises its possible critical points [**21**, Lemma 4.1]. In order to do this, consider the map

$$\Phi := \frac{f}{|f|} \colon \mathbb{B}_\varepsilon \setminus V \to \mathbb{S}^1,$$

where as before, $\mathbb{B}_\varepsilon$ is a ball of radius ε centred at $\underline{0}$ such that the restriction of f to $\mathbb{B}_\varepsilon$ has $0 \in \mathbb{C}$ as its only critical value.

LEMMA 2.5. *The map Φ is a submersion.*

PROOF. The map Φ can be seen as the composition of the restriction of f to $\mathbb{B}_\varepsilon \setminus V$ and the projection $\pi \colon \mathbb{C} \setminus \{0\} \to \mathbb{S}^1$ given by $\pi(x) = \frac{x}{|x|}$. The lemma follows since both maps are submersions. □

REMARK 2.6. The fibres of Φ are precisely the E_θ since given $e^{i\theta} \in \mathbb{S}^1$

$$\begin{aligned} \Phi^{-1}(e^{i\theta}) &= f^{-1}\big(\pi^{-1}(e^{i\theta})\big) \cap \mathbb{B}_\varepsilon \\ &= f^{-1}\big(\mathcal{L}_\theta^+\big) \cap \mathbb{B}_\varepsilon \\ &= E_\theta. \end{aligned}$$

Therefore, given $z \in E_\theta$ we have that $\ker d_z\Phi = T_z E_\theta$.

Now we can give Milnor's characterisation of the critical points of the map $\phi := \frac{f}{|f|} \colon \mathbb{S}_\varepsilon \setminus K \to \mathbb{S}^1$.

LEMMA 2.7 ([**21**, Lemma 4.1]). *The critical points of the map $\phi \colon \mathbb{S}_\varepsilon \setminus K \to \mathbb{S}^1$ are precisely those points $z \in \mathbb{S}_\varepsilon \setminus K$ for which the vector $i \operatorname{grad} \log f(z)$ is a real multiple of the vector z.*

PROOF. The map ϕ is the restriction of the map Φ to $\mathbb{S}_\varepsilon \setminus K$. Thus, a point $z \in \mathbb{S}_\varepsilon \setminus K$ is a critical point of ϕ if and only if the tangent space to $\mathbb{S}_\varepsilon \setminus K$ at z coincides with the kernel of $d_z\Phi$, i.e., $T_z(\mathbb{S}_\varepsilon \setminus K) = \ker d_z\Phi = T_z E_\theta$, since $\mathbb{S}_\varepsilon \setminus K$ and E_θ have both codimension 1. But $T_z(\mathbb{S}_\varepsilon \setminus K) = T_z E_\theta$ if and only if the normal vector to E_θ at z is a real multiple of the normal vector to $\mathbb{S}_\varepsilon \setminus K$ at z. By Lemma 2.2 the normal vector to E_θ at z is $i \operatorname{grad} \log f(z)$ and the normal vector to $\mathbb{S}_\varepsilon \setminus K$ at z is z itself. □

REMARK 2.8. Observe that just as in Section 1, for each line $\mathcal{L}_\theta$ through the origin in $\mathbb{C}$ we may consider the set

$$X_\theta := \{z \in \mathbb{C}^n \mid f(z) \in \mathcal{L}_\theta\}.$$

Then each X_θ is a real analytic hypersurface with singular set equal to the singular set of V. Notice that $X_\theta \setminus V = E_\theta \cup E_{\theta+\pi}$, where the latter manifolds are those in Lemma 2.2. The family $\{X_\theta\}$ of all these real hypersurfaces is *the canonical pencil* of f (see Section 3). Then Lemma 2.7, together with Lemma 2.2, yields to the following proposition.

PROPOSITION 2.9. *The critical points of the map* ϕ *are the points in* $\mathbb{S}_\varepsilon \setminus K$ *where the elements* $X_\theta \setminus V$ *of the canonical pencil are tangent to the sphere* $\mathbb{S}_\varepsilon$.

Using Lemma 2.7, to show that $\phi := \frac{f}{|f|} : \mathbb{S}_\varepsilon \setminus K \to \mathbb{S}^1$ is a submersion for every sufficiently small ε, we only need to prove the following lemma.

LEMMA 2.10 ([**21**, Lemma 4.2]). *There exists* $\varepsilon_0 > 0$ *such that for every* $z \in \mathbb{C}^n \setminus V$ *with* $\|z\| \leq \varepsilon_0$, *the two vectors* z *and* $i \operatorname{grad} \log f(z)$ *are linearly independent over* $\mathbb{R}$.

REMARK 2.11. Notice that Lemma 2.10 is equivalent to saying that for every $z \in \mathbb{C}^n \setminus V$ with $\|z\| < \varepsilon_0$, the manifolds $\mathbb{S}_{\|z\|} \setminus V$ and $E_{\theta(z)}$ are transverse at z, where $\mathbb{S}_{\|z\|}$ is the sphere of radius $\|z\|$. In other words, in $\mathbb{B}_{\varepsilon_0}$ all the spheres $\mathbb{S}_\varepsilon \setminus V$ with $\varepsilon \leq \varepsilon_0$ and all the E_θ are transverse. In other words, this will show that every complex valued holomorphic map is d-regular.

To prove Lemma 2.10 Milnor proves a slightly stronger statement when f is a polynomial, which is the heart of the proof of his fibration theorem.

LEMMA 2.12 ([**21**, Lemma 4.3]). *Given any polynomial* f *which vanishes at the origin, there exists a number* $\varepsilon_0 > 0$ *so that, for all* $z \in \mathbb{C}^n \setminus V$ *with* $\|z\| \leq \varepsilon_0$, *the two vectors* z *and* $i \operatorname{grad} \log f(z)$ *are either linearly independent over the complex numbers or else*

$$\operatorname{grad} \log f(z) = \lambda z$$

where λ *is a non-zero complex number whose argument has absolute value less than say* $\frac{\pi}{4}$.

It is easy to see that Lemma 2.12 implies Lemma 2.10: in the first case, if the vectors z and $i \operatorname{grad} \log f(z)$ are linearly independent over the complex numbers, then they are linearly independent over the real numbers; in the second case, λ lies in the open quadrant of the complex plane which is centred about the positive real axis. Thus $\Re(\lambda) > 0$ and therefore λ cannot be pure imaginary. This implies that z and $i \operatorname{grad} \log f(z)$ cannot be linearly dependent over $\mathbb{R}$.

IDEA OF THE PROOF OF LEMMA 2.12. The proof is a nice application of the Curve Selection Lemma [**21**, Lemma 3.1] and it is done by contradiction: suppose that there were points $z \in \mathbb{C}^n \setminus V$ arbitrarily close to the origin with

$$\operatorname{grad} \log f(z) = \lambda z \neq 0, \tag{21}$$

and with $|\arg \lambda| > \frac{\pi}{4}$. In other words, assume that λ lies in the open half-plane $\Re\big((1+i)\lambda\big) < 0$ or the open half-plane $\Re\big((1-i)\lambda\big) < 0$.

Then consider the set W of points $z \in \mathbb{C}^n$ for which the vectors $\operatorname{grad} f$ and z are linearly dependent. We have that W is an algebraic set. Using (20) is easy to see that $z \in \mathbb{C}^n \setminus V$ is in W if and only if equation (21) holds for some complex number $\lambda(z)$. Let U_+ (respectively U_-) be the open set consisting of all z satisfying the real polynomial inequality $\Re\big((1+i)\lambda'(z)\big) < 0$ (respectively $\Re\big((1-i)\lambda'(z)\big) < 0$), where $\lambda'(z)$ is some real positive multiple of $\lambda(z)$ defined by a real polynomial function, and thus $\lambda(z)$ and $\lambda'(z)$ have the same argument. The original supposition implies that there exists points z arbitrarily close to the origin with $z \in W \cap (U_+ \cup U_-)$. Then the Curve Selection Lemma (see [**21**, Lemma 3.1] for the precise statement) asserts that there exists a real analytic curve $p\colon [0, \varepsilon) \to \mathbb{C}^n$ with $p(0) = 0$ and $p(t) \in W \cap (U_+ \cup U_-)$ for all $t > 0$. This proves Lemma 2.12 since the existence of such a curve contradicts Lemma 2.13. □

LEMMA 2.13 ([**21**, Lemma 4.4]). *Let* $p\colon [0,\varepsilon) \to \mathbb{C}^n$ *be a real analytic path with* $p(0) = 0$ *such that, for each* $t > 0$, *the number* $f\big(p(t)\big)$ *is non-zero and the vector* $\operatorname{grad}\log f\big(p(t)\big)$ *is a complex multiple* $\lambda(t)p(t)$. *Then the argument of the complex number* $\lambda(t)$ *tends to zero as* $t \to 0$.

IDEA OF THE PROOF. By Remark 2.4 we have that

$$\operatorname{grad} f(p(t)) = \lambda(t)p(t)\overline{f(p(t))}. \tag{22}$$

Consider the Taylor expansions of $p(t)$, $f(p(t))$ and $\operatorname{grad} f(p(t))$ denoting their corresponding non-zero leading coefficients by a, b and c, and their corresponding leading exponents by α, β and γ, which are integers with $\alpha \geq 1$, $\beta \geq 1$ and $\gamma \geq 0$. Substituting these Taylor expansions in (22) one can prove that $\lambda(t)$ has a Taylor expansion of the form

$$\lambda(t) = \lambda_0 t^{\gamma-\alpha-\beta}(1 + k_1 t + k_2 t^2 + \dots), \tag{23}$$

and the leading coefficients satisfy the equation

$$c = \lambda_0 a \bar{b}. \tag{24}$$

Substituting (24) in the power series expansion of the identity

$$\frac{df}{dt} = \langle \frac{dp}{dt}, \operatorname{grad} f \rangle,$$

and comparing leading coefficients we obtain

$$\beta = \alpha \|a\|^2 \bar{\lambda}_0,$$

which proves that λ_0 is a positive real number. Therefore

$$\begin{aligned}\lim_{t\to 0} \frac{\lambda(t)}{|\lambda(t)|} &= \frac{\lambda_0 t^{\gamma-\alpha-\beta}(1 + k_1 t + k_2 t^2 + \dots)}{\lambda_0 t^{\gamma-\alpha-\beta}\|1 + k_1 t + k_2 t^2 + \dots\|} \\ &= \lim_{t\to 0} \frac{1 + k_1 t + k_2 t^2 + \dots}{\|1 + k_1 t + k_2 t^2 + \dots\|} = 1.\end{aligned}$$

Hence $\arg \lambda(t) \to 0$ as $t \to 0$. □

Now by Lemma 2.7 and Lemma 2.10 we conclude the first main step of the proof of the Fibration Theorem.

COROLLARY 2.14 ([**21**, Cor. 4.5]). *If* $\varepsilon \leq \varepsilon_0$ *then the map*

$$\phi = \frac{f}{|f|} \colon \mathbb{S}_\varepsilon \setminus K \to \mathbb{S}^1$$

is a submersion.

REMARK 2.15. Recall that by Remark 2.6 the fibres of the map Φ are given by the E_θ. Since ϕ is the restriction of Φ to $\mathbb{S}_\varepsilon \setminus K$, we have that the fibres of ϕ are given by

$$F_\theta := E_\theta \cap \mathbb{S}_\varepsilon.$$

By Remark 2.11 this intersection is transverse and therefore F_θ is a smooth $(2n-2)$-dimensional manifold.

To prove that ϕ is the projection of a locally trivial fibration it is enough to give a *complete* vector field w on $\mathbb{S}_\varepsilon \setminus K$ which projects under ϕ to the unit vector field u tangent to $\mathbb{S}^1$ given by $u(e^{i\theta}) = ie^{i\theta}$, so it is transverse to the fibres of ϕ. Let $p(t)$ be an integral curve of the vector field w. The function $p(t)$ depends smoothly on both t and the initial value $z_0 = p(0)$. We denote this dependence by

$$p(t) = h_t(z_0).$$

Then each h_t is a diffeomorphism mapping $\mathbb{S}_\varepsilon \setminus K$ to itself, which sends the fibre F_θ onto the fibre $F_{\theta+t}$.

PROOF OF THE FIBRATION THEOREM [**21**, Thm. 4.8]. Given $e^{i\theta} \in \mathbb{S}^1$ let U be a small neighbourhood of $e^{i\theta}$. Then the correspondence

$$U \times F_\theta \to \phi^{-1}(U)$$
$$(e^{i(t+\theta)}, z) \mapsto h_t(z),$$

for $|t| <$ constant, and $z \in F_\theta$, is a diffeomorphism, proving the local triviality of ϕ. □

Since ϕ is a submersion, by Corollary 2.14, we can take w to be a lifting of the vector field u on $\mathbb{S}^1$ to $\mathbb{S}_\varepsilon \setminus K$. The main difficulty is to guarantee that the vector field w is *complete*, since $\mathbb{S}_\varepsilon \setminus K$ is non-compact we need to insure that $p(t)$ cannot tend towards K as t tends toward some finite limit t_0. This is equivalent to guarantee that $f\big(p(t)\big)$ cannot tend to zero or that $\log|f(p(t))|$ cannot tend to $-\infty$ as t tends to a finite value t_0. One way to do this is to make $\log|f(p(t))|$ to increase or decrease "slowly" by keeping its derivative small in absolute value (compare with [**21**, Lemma 4.7]). Suppose that

$$\left|\frac{d\log|f(p(t))|}{dt}\right| < 1,$$

then we would have

$$\begin{aligned}\Big|\log|f(p(t_0))| - \log|f(p(0))|\Big| &= \left|\int_0^{t_0} \frac{d\log|f(p(t))|}{dt}dt\right| \\ &\leq \int_0^{t_0} \left|\frac{d\log|f(p(t))|}{dt}dt\right| \\ &< \int_0^{t_0} dt = t_0.\end{aligned}$$

Then $\log|f(p(t))|$ cannot tend to $-\infty$ as t tends to any finite limit t_0. Hence by (14) we need the vector field w to satisfy

$$\Big|\Re\Big\langle w\big(p(t)\big), \operatorname{grad}\log f\big(p(t)\big)\Big\rangle\Big| < 1.$$

As we mentioned above, the other property that the vector field w needs to have is to project under ϕ to the unit tangent vector field u of $\mathbb{S}^1$. In other words, we need that the integral curves $p(t)$ of w project under ϕ to the path on $\mathbb{S}^1$ which winds around the unit circle in the positive direction with unit velocity, this is the same as

$$\theta\big(p(t)\big) = t + \text{constant}.$$

In turn, by (12) this is equivalent to have

$$\frac{d\theta(p(t))}{dt} = \Re\Big\langle \frac{dp}{dt}, i\,\mathrm{grad}\log f(p(t))\Big\rangle = 1.$$

In summary, we need a vector field w on $\mathbb{S}_\varepsilon \setminus K$ with the following properties:

(1) $\Re\langle w(z), z\rangle = 0$ (tangent to $\mathbb{S}_\varepsilon$),

(2) $\Big|\Re\Big\langle w(z), \mathrm{grad}\log f(z)\Big\rangle\Big| < 1$ (complete),

(3) $\Re\Big\langle w(z), i\,\mathrm{grad}\log f(z)\Big\rangle = 1$ (projects onto u).

Milnor solves the problem in a beautiful way, using Lemma 2.12 to construct a vector field w on $\mathbb{B}_{\varepsilon_0} \setminus V$ with these properties as follows.

LEMMA 2.16 ([**21**, Lemma 4.6]). *If $\varepsilon \le \varepsilon_0$ then there exists a smooth tangential vector field $v(z)$ on $\mathbb{B}_{\varepsilon_0} \setminus V$ such that $v(z)$ is tangent to $\mathbb{S}_\varepsilon \setminus K$ and for each $z \in \mathbb{S}_\varepsilon \setminus K$, the complex inner product*

$$\langle v(z), i\,\mathrm{grad}\log f(z)\rangle$$

is non-zero, and the absolute value of its argument is less than $\frac{\pi}{4}$.

The condition $\langle v(z), i\,\mathrm{grad}\log f(z)\rangle \neq 0$ guarantees that $v(z)$ is transverse to the fibres of ϕ. In fact, suppose that $\langle v(z), i\,\mathrm{grad}\log f(z)\rangle = 0$, in particular we have that $\Re\langle v(z), i\,\mathrm{grad}\log f(z)\rangle = 0$ so $v(z)$ would be tangent to $E_{\theta(z)}$ (see Lemma 2.2). Since $v(z)$ is also tangent to $\mathbb{S}_\varepsilon$ then we would have that $v(z)$ is tangent to $F_{\theta(Z)} = \mathbb{S}_\varepsilon \cap E_{\theta(z)}$, but we want $v(z)$ to be transverse to the fibres of ϕ.

We shall follow Milnor's proof literally and we shall explain the geometric meaning of each step.

PROOF. It suffices to construct such a vector field locally, in the neighbourhood of some given point $z \in \mathbb{C}^n \setminus V$.

Case 1: If the vectors z and $\mathrm{grad}\log f(z)$ are linearly independent over $\mathbb{C}$, then the linear equations

$$\langle v, z\rangle = 0, \tag{25}$$

$$\langle v, i\,\mathrm{grad}\log f(z)\rangle = 1, \tag{26}$$

have a simultaneous solution v. The first equation guarantees that $\Re\langle v, z\rangle = 0$, so that v is tangent to $\mathbb{S}_\varepsilon$ at z.

If the vectors z and $\mathrm{grad}\log f(z)$ are linearly independent over $\mathbb{C}$, then z does not lie in the complex line generated by $\mathrm{grad}\log f(z)$ (where it also lies $i\,\mathrm{grad}\log f(z)$). Also it implies that z and $\mathrm{grad}\log f(z)$ are linearly independent over $\mathbb{R}$ which by Lemma 2.3 implies that the sphere $\mathbb{S}_{\|z\|}$ and the tube $N(\varepsilon_0, |f(z)|)$ are transverse at z. The intersection $T_z\mathbb{S}_{\|z\|} \cap T_zN(\varepsilon_0, |f(z)|)$ is precisely the vector space orthogonal (with respect to the real Euclidean inner product) to the *real* plane generated by z and $\mathrm{grad}\log f(z)$. The intersection $T_z\mathbb{S}_{\|z\|} \cap T_zN(\varepsilon_0, |f(z)|)$ has real dimension $2n-2$ and notice that it *does not* coincide with $T_zF_{\theta(z)} = T_z\mathbb{S}_{\|z\|} \cap T_zE_{\theta(z)}$ because in that case, it would be also orthogonal to $i\,\mathrm{grad}\log f(z)$ contradicting the fact that z does not lie in the complex line generated by $\mathrm{grad}\log f(z)$.

The idea now is to take a vector $v(z)$ in $T_z\mathbb{S}_{\|z\|} \cap T_zN(\varepsilon_0, |f(z)|)$ such that it satisfies property (3). In this way, it satisfies property (1) by construction and also property (2) since the directional derivative of $\log|f(z)|$ along a vector tangent to $N(\varepsilon_0, |f(z)|)$ is zero because $N(\varepsilon_0, |f(z)|)$ is the level hypersurface of $\log|f(z)|$.

The solution to Milnor's linear equations (25) and (26) is a vector of this kind: (25) obviously implies property (1) and Milnor kills two birds with one stone with equation (26) obtaining properties (2) and (3) at the same time since

$$\Re(\langle v, \operatorname{grad} \log f(z)\rangle) = \Im(\langle v, i \operatorname{grad} \log f(z)\rangle) = 0,$$
$$\Re(\langle v, i \operatorname{grad} \log f(z)\rangle) = 1.$$

Case 2: If $\operatorname{grad} \log f(z)$ is equal to a multiple λz, then set $v = iz$. Clearly

$$\Re\langle iz, z\rangle = 0; \tag{27}$$

and by Lemma 2.12 the number

$$\langle iz, i \operatorname{grad} \log f(z)\rangle = \bar{\lambda}\|z\|^2 \neq 0 \tag{28}$$

has argument with absolute value less than $\frac{\pi}{4}$.

In this case $T_z \mathbb{S}_{\|z\|} \cap T_z N(\varepsilon_0, |f(z)|)$ coincides with $T_z F_{\theta(z)}$, so we cannot take v there. Taking iz instead (27) obviously implies property (1), on the other hand, iz is transverse to $F_{\theta(z)}$, if it was tangent to $F_{\theta(z)}$, then iz would be orthogonal to $i \operatorname{grad} \log f(z)$ which implies that iz is a real multiple of $\operatorname{grad} \log f(z)$ which is equivalent to z being a real multiple of $i \operatorname{grad} \log f(z)$ which is impossible by Lemma 2.10. In this case we do not get directly properties (2) and (3), but we obtain them after a normalisation as we shall see.

In either case one can chose a local tangential vector field $v(z)$ which takes the constructed value v at z. The condition

$$|\arg\langle v(z), i \operatorname{grad} \log f(z)\rangle| < \frac{\pi}{4}$$

will hold throughout a neighbourhood of z. Using a partition of unity we obtain a global vector field $v(z)$ having the same property. □

Normalising by setting

$$w(z) = \frac{v(z)}{\Re\langle v(z), i \operatorname{grad} \log f(z)\rangle}$$

we obtain a smooth vector field w on $\mathbb{B}_{\varepsilon_0} \setminus V$ which satisfies

1. Property (1), since v satisfies it and this does not change multiplying by a real scalar.
2. Property (3), since
$$\Re\langle w(z), i \operatorname{grad} \log f(z)\rangle = \frac{\Re\langle v(z), i \operatorname{grad} \log f(z)\rangle}{\Re\langle v(z), i \operatorname{grad} \log f(z)\rangle} = 1.$$
3. Property (2), since
$$\langle w(z), i \operatorname{grad} \log f(z)\rangle$$
has argument with absolute value less than $\frac{\pi}{4}$ and its real part is equal to 1.

In this way we obtain the vector field w used in the proof of the fibration theorem.

2.3. Milnor-Lê Fibration. Let $0 < \delta \ll \varepsilon \leq \varepsilon_0$ and consider the Milnor tube $N(\varepsilon,\delta) = f^{-1}(\partial\mathbb{D}_\delta)$, where $\mathbb{D}_\delta$ is the disc of radius δ in $\mathbb{C}$ with centre at 0. Then the restriction

$$f\colon N(\varepsilon,\delta) \to \partial\mathbb{D}_\delta$$

is a locally trivial fibration. In the case when f has an isolated critical point, this was proved by Milnor (see [**20**, Thm. 2],[**21**, Thm. 11.2]) using the fact that the fibres in $N(\varepsilon,\delta)$ are transverse to the sphere $\mathbb{S}_\varepsilon$ (compare with Remark 1.1 and Example 3.2) and a version of Ehresmann Fibration Theorem for manifolds with boundary. In the general case it was proved by Lê in [**18**, Thm. (1,1)], using a theorem of Hironaka [**16**, §5, Cor. 1] which asserts that every holomorphic function $f\colon \mathbb{C}^n \to \mathbb{C}$ satisfies the Thom a_f-property, and therefore the fibres in $N(\varepsilon,\delta)$ are transverse to the sphere $\mathbb{S}_\varepsilon$ (compare with Proposition 3.3 in Section 3).

In order to prove that the fibration on the Milnor tube is equivalent to the one on the sphere, we need a vector field v whose integral curves are transverse to the tube $N(\varepsilon,\delta)$ and to the sphere $\mathbb{S}_\varepsilon$ and such that $f(z)$ has constant argument along each orbit. That is, we want a smooth vector field with the Properties (1) and (2) in Section 1. Once we have such a vector field, given an integral curve $p(t)$ of v we have that $f(p(t))$ lies on the ray $\mathcal{L}_\theta$ with $\theta = \arg f(z)$. Let z be a point in $N(\varepsilon,\delta)$ and let $p(t)$ be the integral curve of v through z. Following $p(t)$ we travel "away" form the origin, transversally to the tubes and to the spheres until we reach a point z' on $\mathbb{S}_\varepsilon \setminus K$. Since $f(p(t))$ has constant argument along $p(t)$ we have that

$$\frac{f(z)}{|f(z)|} = \frac{f(z')}{|f(z')|}.$$

Thus, the correspondence $z \mapsto z'$ is a diffeomorphism which gives the equivalence between the Milnor Fibration and the Milnor-Lê Fibration.

Milnor constructed such a vector field in the following lemma. Originally Milnor used this vector field just to prove that the fibre in the tube is diffeomorphic to the fibre in the sphere, since he did not have the fibration on the tube in the general case.

LEMMA 2.17 ([**21**, Lemma 5.9]). *Let $\varepsilon \leq \varepsilon_0$. There exists a smooth vector field v on $\mathbb{B}_\varepsilon \setminus V$ so that the inner product*

$$\langle v(z), \operatorname{grad}\log f(z)\rangle$$

is real and positive, for all $z \in \mathbb{B}_\varepsilon \setminus V$, and so that the inner product $\langle v(z), z\rangle$ has positive real part.

Let $p(t)$ be an integral curve of v, that is $\frac{dp}{dt} = v(p(t))$. By (8) we have that

$$\frac{d\log f(p(t))}{dt} = \langle \frac{dp}{dt}, \operatorname{grad}\log f(p(t))\rangle.$$

The condition that $\langle \frac{dp}{dt}, \operatorname{grad}\log f(p(t))\rangle = c$ with c real and positive, by (17) and (16) implies that

$$\frac{d\log|f(p(t))|}{dt} = \Re\Big\langle \frac{dp}{dt}, \operatorname{grad}\log f(p(t))\Big\rangle = c > 0, \tag{29}$$

$$\frac{d\theta(p(t))}{dt} = \Im\Big\langle \frac{dp}{dt}, \operatorname{grad}\log f(p(t))\Big\rangle = 0. \tag{30}$$

From (29) we have that

$$\log|f(p(t))| = ct + k, \qquad \text{with } k \text{ constant},$$

hence

$$|f(p(t))| = e^{ct+k},$$

whose derivative with respect to t does not vanish. Therefore the integral lines $p(t)$ of v are transverse to the Milnor tubes. From (30) we have that

$$\theta(p(t)) = \theta_0, \qquad \text{for a constant value } \theta_0,$$

in other words, $f(p(t))$ has constant argument as required. On the other hand, the condition that $\Re\langle v(z), z\rangle > 0$ implies that

$$\frac{d\|p(t)\|^2}{dt} = 2\Re\langle \frac{dp}{dt}, p(t)\rangle > 0,$$

therefore the integral lines $p(t)$ are transverse to all the spheres.

PROOF. The proof is analogous to the proof of Lemma 2.16. It suffices to construct such a vector field locally, in the neighbourhood of some given point $z \in \mathbb{B}_\varepsilon \setminus V$.

Case 1: If the vectors z and $\operatorname{grad}\log f(z)$ are linearly independent over $\mathbb{C}$, then the linear equations

$$\langle v, \operatorname{grad}\log f(z)\rangle = 1,$$
$$\langle v, z\rangle = k, \qquad \text{with } k \in \mathbb{C} \text{ and } \Re(c) > 0,$$

have a simultaneous solution v.

Case 2: If $\operatorname{grad}\log f(z)$ is equal to a multiple λz, set $v = \operatorname{grad}\log f(z)$. We have that

$$\langle \operatorname{grad}\log f(z), \operatorname{grad}\log f(z)\rangle = \langle \lambda z, \lambda z\rangle = \|\lambda z\|^2 \in \mathbb{R}^+,$$

since $z \neq 0$ and $\lambda \neq 0$. On the other hand,

$$\langle \operatorname{grad}\log f(z), z\rangle = \langle \lambda z, z\rangle = \lambda\|z\|^2,$$

since $|\arg\lambda| < \frac{\pi}{4}$, we have that $|\lambda\|z\|^2| < \frac{\pi}{4}$, that is, its real part is positive. In either case one can choose a local tangential vector field $v(z)$ which takes the constructed value v at z. Using a partition of unity we obtain a global vector field with the desired properties. □

REMARK 2.18. The existence of the vector field constructed in Lemma 2.17 is equivalent to saying that in $\mathbb{B}_{\varepsilon_0}$ all the spheres $\mathbb{S}_\varepsilon \setminus V$ with $\varepsilon \leq \varepsilon_0$ and all the E_θ are transverse (compare with Remark 2.11).

3. d-regularity for real analytic map-germs

We now consider a real analytic map-germ

$$f\colon \mathbb{R}^n \to \mathbb{R}^p$$
$$x \mapsto (f_1(x), \dots, f_p(x))$$

with $n > p$ and set $V = f^{-1}(0)$.

First we want to know under what conditions one can have a "Milnor-Lê" fibration:

$$f\colon N(\varepsilon,\delta) \to \partial\mathbb{D}_\delta, \tag{31}$$

where $0 < \delta \ll \varepsilon \leq \varepsilon_0$ and $N(\varepsilon, \delta) = f^{-1}(\partial \mathbb{D}_\delta) \cap \mathbb{B}_\varepsilon$ is the so-called Milnor tube, $\mathbb{D}_\delta$ being the ball of radius δ in $\mathbb{R}^p$ and $\mathbb{B}_\varepsilon$ the ball of radius ε in $\mathbb{R}^n$, both with centre at the origin. We have just seen that this is always the case when f is holomorphic.

For simplicity we require all along this section that the map $N(\varepsilon, \delta) \xrightarrow{f} \partial \mathbb{D}_\delta$, is surjective. This means that f has to be surjective over some neighborhood of the origin in $\mathbb{R}^p$. In this case we will call it *locally surjective.*

Milnor proved in [**21**] that every real analytic map $f : \mathbb{R}^n \to \mathbb{R}^p$ with $n > p$ and with an isolated critical point induces a fibration on tubes

$$f \colon N(\varepsilon, \delta) = \mathbb{B}_\varepsilon \cap f^{-1}(\partial \mathbb{D}_\delta) \to \partial \mathbb{D}_\delta .$$

In order to do that, he used an Ehresmann fibration theorem for manifolds with boundary, which in turn requires the fibers of the induced map to be transverse to the sphere. This last condition is always fulfilled when f has an isolated critical point, and this is essentially Example 3.2 and Proposition 3.3.

D.T. Lê proved in [**18**] that every holomorphic function $f : X \to \mathbb{C}$ induces a topological locally trivial fibration on tubes $N(\varepsilon, \delta)$. In order to apply Milnor's argument he first pointed out that any holomorphic function satisfies Thom's a_f-property ([**16**]) and hence all the fibers in the tube are transverse to the sphere. So it became clear that if one needs to prove a fibration theorem in the tube it is sufficient to prove transversality of the fibers with the spheres. Let us explain the relation between this transversality and Thom's condition.

Let $f : (\mathbb{R}^n, \underline{0}) \to (\mathbb{R}^p, 0)$, $n > p$ be a real analytic map with an isolated critical value at 0. Endow $V = f^{-1}(0)$ with a stratification $V = \bigcup_\alpha V_\alpha$ such that in a neighborhood U of the origin the partition

$$U = (U \setminus V) \cup \bigcup_\alpha V_\alpha$$

is a Whitney regular stratification of U.

DEFINITION 3.1. The map f has the *Thom a_f-property* if there exists a stratification as above, satisfying that for every sequence of points $x_n \in U \setminus V$ converging to a point x in some $V_\alpha \subset V$, such that the sequence of spaces $T_{x_n} f^{-1}(f(x_n))$, tangent to the fibers of f, converges to some linear space T, one has $T_x V_\alpha \subset T$. That is, the space T contains the tangent space to V_α at x .

EXAMPLE 3.2. Every smooth map-germ $(\mathbb{R}^n, \underline{0}) \to (\mathbb{R}^p, 0)$, $n > p$, with an isolated critical point at $\underline{0}$ has the Thom a_f-property. This is an immediate consequence of the Implicit Function Theorem.

Assume now that the map-germ f has Thom's a_f-property. Assume also that $\underline{0}$ is a stratum and that the neighbourhood U is sufficiently small so that every other stratum contains $\underline{0}$ in its closure. We know that for sufficiently small ε, every stratum V_α, other than $\underline{0}$ itself, intersects transversally the sphere $\mathbb{S}_\varepsilon^{n-1}$. Hence, by Thom's a_f-property, for each point $x \in V_\alpha \cap \mathbb{S}_\varepsilon^{n-1}$ there exists a neighborhood U_x such that for every $y \in U_x \cap \mathbb{S}_\varepsilon^{n-1}$ the fiber $f^{-1}(f(y))$ intersects transversally the sphere $\mathbb{S}_\varepsilon^{n-1}$. Since the sphere is compact and the map f is continuous, there exists $0 < \delta \ll \varepsilon$ such that for every $\xi \in \mathbb{R}^p$ such that $\|\xi\| = \delta$ the fiber $f^{-1}(\xi)$ is transversal to the sphere $\mathbb{S}_\varepsilon^{n-1}$ at each point of intersection. In other words, sufficiently small spheres are transverse to fibers sufficiently close to $f^{-1}(0)$. When

$f^{-1}(0)$ is reduced to the origin, this intersection is vacuous and hence transversal. Thus we have the following proposition whose complete proof can be read in [**12**, Proposition 5.1]:

PROPOSITION 3.3. *Let $f : \mathbb{R}^n \to \mathbb{R}^p$ be a real analytic map with isolated critical value satisfying Thom property at 0. There exists $\varepsilon_0 > 0$ such that for any $0 < \varepsilon \leq \varepsilon_0$ there exists $0 < \delta \ll \varepsilon$ and for any $\xi \in \mathbb{R}^p$ with $\|\xi\| = \delta$ the fiber $f^{-1}(\xi)$ is transverse to the sphere $\mathbb{S}^{n-1}_\varepsilon$.*

Now we observe that $N(\varepsilon, \delta) = f^{-1}(\partial \mathbb{D}_\delta) \cap \mathbb{B}_\varepsilon$ is a compact manifold with boundary, which allows us to apply the Ehresmann fibration theorem. Of course the same remarks apply if we consider the solid tube $\widehat{N}(\varepsilon,\delta) = f^{-1}(\mathbb{D}_\delta \setminus \{0\}) \cap \mathbb{B}_\varepsilon$, just as in Property $3'$ in Section 1. In fact, around each point $x \in \mathbb{D}_\delta \setminus \{0\}$ we can chose a closed ball $\mathbb{B}_x \subset \mathbb{D}_\delta \setminus \{0\}$. The restriction of f to $\mathbb{B}_\varepsilon \cap f^{-1}(\mathbb{B}_x)$ is a proper map whose fibers are transverse to the boundary sphere, so it induces a locally trivial fibration.

So we have the following theorem, which is proved by an easy extension of Milnor's arguments in the last chapter of his book (see [**27**, Theorem 1.3], also [**12**, Remark 5.7]).

THEOREM 3.4. *Let $f : (\mathbb{R}^n, \underline{0}) \to (\mathbb{R}^p, 0))$ be an analytic map-germ with an isolated critical value at 0. Assume further that $n > p$, and for every $\varepsilon > 0$ sufficiently small one has that there exists $\delta > 0$ such that every fiber $f^{-1}(t)$ with $\|t\| \leq \delta$ meets transversally the sphere $\mathbb{S}_\varepsilon$. Then one has locally trivial fibrations:*

$$f : \widehat{N}(\varepsilon,\delta) \to \mathbb{D}_\delta \setminus \{0\} \quad \textit{and} \quad f : N(\varepsilon,\delta) \to \partial\mathbb{D}_\delta \cong \mathbb{S}^{p-1}.$$

Now, what about fibrations from the sphere minus the link into the sphere?

In fact it is easy to adapt Milnor's methods to show that the above tube $N(\varepsilon,\delta)$ can always be inflated to the sphere $\mathbb{S}_\varepsilon$ using appropriate vector fields. This is similar to what we did in Section 1 using properties (1)-(3). One gets a locally trivial fibration:

$$\bar{\phi} : \mathbb{S}^{n-1}_\varepsilon \setminus V \to \mathbb{S}^{p-1},$$

which restricted to a neighbourhood of the link $K = V \cap \mathbb{S}^{n-1}$ is given by the projection $\frac{f}{\|f\|}$ (this is Theorem 1.3 in [**27**]). If we write $\overset{\circ}{T}(\varepsilon,\delta) = \mathbb{S}_\varepsilon \cap f^{-1}(\overset{\circ}{\mathbb{D}}_\delta \setminus\{0\})$, then by Theorem 3.4, the restriction of f to $\overset{\circ}{T}(\varepsilon,\delta)$ is a locally trivial fibration. If we compose it with the radial projection we get a locally trivial fibration

$$\phi_T = \frac{f}{\|f\|} : \overset{\circ}{T}(\varepsilon,\delta) \to \mathbb{S}^{p-1}. \tag{32}$$

We are interested in finding conditions ensuring that the above map $\bar{\phi}$ can be taken as $\frac{f}{\|f\|}$ globally. That is, we want to know when

$$\phi = \frac{f}{\|f\|} : \mathbb{S}^{n-1}_\varepsilon \setminus V \to \mathbb{S}^{p-1}$$

is a locally trivial fibration, equivalent to the fibration in Theorem 3.4. Here is where we come to d-regularity.

We use again an idea from J. Milnor in [**20, 21**]. Consider the map $f\colon N(\varepsilon,\delta) \to \partial\mathbb{D}_\delta$ and compose it with the projection to the unit sphere in $\mathbb{R}^p$ to obtain the map:

$$\phi_N = \frac{f}{\|f\|} : N(\varepsilon,\delta) \to \mathbb{S}^{p-1}.$$

If we have a fibration in the tube then this last map ϕ_N is also a fibration. We want to transform its fibers into those of ϕ. In order to do that we will need a flow that moves a point in the tube diffeomophically into a point in the sphere preserving the value $\frac{f}{\|f\|}$. This flow can be given by a vector field transverse to the tubes and the spheres and tangent to the fibers of $\hat{\phi} = \frac{f}{\|f\|} : \mathbb{B}_\varepsilon \setminus V \to \mathbb{S}^{p-1}$. In particular the spheres need to be transverse to the fibers of $\hat{\phi}$. This is the *d*-regularity condition. Let us explain it carefully.

Consider the two-fold covering of $\mathbb{RP}^{p-1}$ given by

$$\pi : \mathbb{S}^{p-1} \to \mathbb{RP}^{p-1} .$$

We have the following commutative diagramm:

$$\begin{array}{ccc} \mathbb{B}_\varepsilon \setminus V & \xrightarrow{\hat{\phi} = \frac{f}{\|f\|}} & \mathbb{S}^{p-1} \\ & \searrow^{\psi} & \downarrow \pi \\ & & \mathbb{RP}^{p-1} \end{array}$$

where the map ψ is given by the homogeneous coordinates $(f_1 : \dots : f_p)$.

Let $l \in \mathbb{RP}^{p-1}$ and call θ^- and θ^+ the corresponding antipodal points in the sphere $\mathbb{S}^{p-1}$. The fiber $\psi^{-1}(l)$ is the union of the fibers $\phi^{-1}(\theta^-)$ and $\phi^{-1}(\theta^+)$. If we call $\mathcal{L}$ the line in $\mathbb{R}^p$ corresponding to the projective point l, then the fiber $\psi^{-1}(l)$ is the intersection $(f^{-1}(\mathcal{L}) \setminus V) \cap \mathbb{B}_\varepsilon$.

DEFINITION 3.5. Let $f = (f_1, \dots, f_p) : U \subset \mathbb{R}^n \to \mathbb{R}^p$ be a locally surjective real analytic map defined in a neighborhood of the origin and $\mathcal{L} \subset \mathbb{R}^p$ be a line containing the origin. We define $X_\mathcal{L}$ to be the inverse image of the line $\mathcal{L}$ by the map f. The family of the spaces $X_\mathcal{L}$ when $\mathcal{L}$ varies in $\mathbb{RP}^{p-1}$ is called the *canonical pencil* associated to f.

This family of real analytic spaces has been introduced in [**31, 29, 28**], and later in [**11, 12, 5**]. By abuse of notation, we will make no difference between a line $\mathcal{L}$ containing the origin in $\mathbb{R}^p$ and the projective point corresponding to it in $\mathbb{RP}^{p-1}$.

Note that for any $\mathcal{L}$ and $\mathcal{L}'$ in $\mathbb{RP}^{p-1}$ the corresponding pre-images satisfy

$$X_\mathcal{L} \cap X_{\mathcal{L}'} = V$$

and the whole pencil covers the neighborhood where the map f is defined, *i.e.*,

$$U = \bigcup_{\mathcal{L} \in \mathbb{RP}^{p-1}} X_\mathcal{L}.$$

From now on we will suppose that the map f has an isolated critical value at 0 and is locally surjective. Then each $X_\mathcal{L}$ has dimension $n - p + 1$. In this case $X_\mathcal{L}$ is non-singular outside V.

Each line $\mathcal{L}$ intersects the sphere $\mathbb{S}^{p-1}$ in two antipodal points θ^- and θ^+. We can then decompose the line $\mathcal{L}$ into two open half lines $\mathcal{L}^-$ and $\mathcal{L}^+$ containing respectively the points θ^- and θ^+, so that we can express

$$\mathcal{L} = \mathcal{L}^- \cup \{0\} \cup \mathcal{L}^+.$$

If we define E_{θ^-}, respectively E_{θ^+}, to be the inverse image $f^{-1}(\mathcal{L}^-)$, respectively $f^-(\mathcal{L}^+)$, then we can express each element of the canonical pencil as the following union:

$$X_{\mathcal{L}} = E_{\theta^-} \cup V \cup E_{\theta^+}. \tag{33}$$

Using this notation we can describe the fibers of the map $\phi : \mathbb{S}_\varepsilon^{n-1} \to \mathbb{S}^{p-1}$ as follows:

$$\phi^{-1}(\theta^-) = E_{\theta^-} \cap \mathbb{S}_\varepsilon^{n-1}$$

and

$$\phi^{-1}(\theta^+) = E_{\theta^+} \cap \mathbb{S}_\varepsilon^{n-1}$$

If we denote by K the link of f in $\mathbb{S}_\varepsilon^{n-1}$, then we can write:

$$X_{\mathcal{L}} \cap \mathbb{S}_\varepsilon^{n-1} = (E_{\theta^-} \cap \mathbb{S}_\varepsilon^{n-1}) \cup K \cup (E_{\theta^+} \cap \mathbb{S}_\varepsilon^{n-1}) = \phi^{-1}(\theta^-) \cup K \cup \phi^{-1}(\theta^+).$$

We can now give the precise definition of d-regularity.

DEFINITION 3.6. Let $f : U \to \mathbb{R}^p$ be a locally surjective real analytic map with an isolated critical value at 0. We say that f *is d-regular* if there exists $\varepsilon_0 > 0$ such that for every $\varepsilon \leq \varepsilon_0$ and for every line $\mathcal{L}$ through the origin in $\mathbb{R}^p$, the sphere $\mathbb{S}_\varepsilon^{n-1}$ and the manifold $X_{\mathcal{L}} \setminus V$ are transverse.

REMARK 3.7. Recalling that each $X_{\mathcal{L}}$ can be decomposed as in equation (33) above, the d-regularity condition states that the E_θ's are transverse to the spheres. Since the family of the E_θ's depends on $(p-1)$-parameters, the existence of such an ε_0 is not guaranteed when $p > 1$ (compare with the second example in 3.9 below). On the other hand, it is well known that we have transversality (in the stratified sense) of V with all sufficiently small spheres [**21, 8, 33**]. Hence if f is d-regular, then the whole family $X_{\mathcal{L}}$ is transverse to the spheres. One can show further that, since the domain of f is smooth at $\underline{0}$, the transversality of the $X_{\mathcal{L}}$ with the spheres implies that the pencil has a local uniform conical structure as in [**11**].

Now, how does one verify whether or not a map is d-regular?

One can use the following procedure, which is implicit in [**5**, p. 179]: Consider the $(((p(p-1)/2)+1) \times n)$-matrix M whose first $p(p-1)/2$ rows are given by

$$f_i \operatorname{grad} f_j - f_j \operatorname{grad} f_i$$

for $1 \leq i < j \leq n$, and the last row is the vector

$$(X_1, \dots X_n).$$

PROPOSITION 3.8. *The map $f = (f_1, \dots, f_p)$ is d-regular if and only if there exists no arbitrary close point to the origin in $\mathbb{R}^n$ outside V where the matrix M has rank strictly less than p.*

PROOF. Let $x \in U$ be a point such that $f(x) \neq 0$. The element of the canonical pencil $X_{\mathcal{L}}$ that contains x is the inverse image by f of the line generated by the vector $(f_1(x), \dots, f_p(x))$. So the space $X_{\mathcal{L}}$ is defined by

$$X_{\mathcal{L}} = \{y \in \mathbb{R}^n \setminus V, (f_1(x) : \dots : f_p(x)) = (f_1(y) : \dots : f_p(y))\}.$$

This space is then defined by the equations

$$f_i(x) f_j - f_j(x) f_i = 0,\ 1 \leq i < j \leq p.$$

The tangent space to $X_{\mathcal{L}}$ at x is the kernel of the matrix given by the rows

$$f_i(x) \operatorname{grad}_x f_j - f_j(x) \operatorname{grad}_x f_i,\ 1 \leq i < p.$$

The space $X_{\mathcal{L}}$ and the sphere $\mathbb{S}^{n-1}_{\|x\|}$ are not transverse if and only if the dimension of the intersection of their corresponding tangent spaces is bigger than $n-p$. This is equivalent to saying that the matrix M has rank strictly smaller than p. □

EXAMPLES 3.9. Consider the real analytic map:

$$f\colon \mathbb{R}^3 \to \mathbb{R}^2$$
$$(x,y,z) \mapsto (xz+y^3, x)$$

One can check that:

$$V := f^{-1}(0) = \{(0,0,z),\ z \in \mathbb{R}\}$$

and V coincides with the critical locus of f. So it has an isolated critical value at the origin of $\mathbb{R}^2$.

Now we apply Proposition 3.8. So the points where the $X_{\mathcal{L}}$'s are not transverse to the corresponding spheres are the points where the matrix

$$\begin{pmatrix} -y^3 & 3xy^2 & x^2 \\ x & y & z \end{pmatrix}$$

has rank strictly less than 2. This is achieved only on V. So this map is d-regular.

Consider now the slightly different map

$$g\colon \mathbb{R}^3 \to \mathbb{R}^2$$
$$(x,y,z) \mapsto (x^2z+y^3, x)$$

this map still has isolated critical value. However, all along the line $\{(x,0,x), x \in \mathbb{R}\}$, the spaces $X_{\mathcal{L}}$ are not transverse to the corresponding spheres, so it is not d-regular.

REMARK 3.10.
- When $f: \mathbb{R}^{2n} \to \mathbb{R}^2$ is given by the real and imaginary part of a holomorphic map, then it has an isolated critical value. By Remark 2.11 every such map is d-regular.
- From the examples in Section 1 we see that polar weighted homogeneous polynomials and real weighted homogeneous maps with an isolated critical value, are d-regular maps (see [**10, 22**]).
- In [**27**] it is proved that when f and g are holomorphic maps $\mathbb{C}^2 \to \mathbb{C}$ such that the product $f\bar{g}$ has an isolated critical value at the origin, then the map $f\bar{g}$ is d-regular. The same statement holds when f, g are defined in $n > 2$ complex variables provided the meromorphic map f/g is semi-tame (see [**7**]).
- In [**28**] it is proved that every real analytic map $\mathbb{R}^{n+2} \to \mathbb{R}^2$ with an isolated critical point such that the family of hypersurfaces X_t (defined as above) satisfy that for any t the strata $(X_t \setminus \{0\}, \{0\})$ is (c)-regular with respect to the function distance to the origin, is also d-regular. Presumably the same statement holds for real analytic maps with an isolated critical value.
- The strongly non-degenerate mixed functions in [**23**] are all d-regular.

Notice that the projection map of a locally trivial (differentiable) fibration is *a fortiori* a submersion. Hence, if we expect ϕ to be a fibration it needs to be a submersion. The following statement is proved in [**12**, Proposition 3.2]. An equivalent result is proved in [**5**] for maps with an isolated critical point.

PROPOSITION 3.11. *The map f is d-regular if and only if there exists ε_0 such that for any $\varepsilon \leq \varepsilon_0$ the map ϕ induced on $\mathbb{S}_\varepsilon^{n-1}$ is a submersion.*

Hence, given a locally surjective real analytic map f with an isolated critical value and with the Thom a_f-property, d-regularity is necessary for the associated map $\phi = \frac{f}{\|f\|} : \mathbb{S}_\varepsilon^{n-1} \to \mathbb{S}^{p-1}$ to be a locally trivial fibration. We now explain why and how d-regularity is also a sufficient condition for ϕ to be the projection map of a locally trivial fibration, [**12**, Theorem 5.3]. The method we expose here has the advantage that we get for free that whenever the two types of fibrations exist, one on a Milnor tube, the other on a small sphere minus the link, these are equivalent. For this we use the following:

DEFINITION 3.12. Consider the map

$$\mathfrak{F} \colon U \setminus V \to \mathbb{R}^p$$
$$x \mapsto \|x\| \frac{f(x)}{\|f(x)\|} .$$

We will call it the *spherification map* associated to f.

This map has the peculiarity that its Milnor tubes are the spheres (minus the link), for it carries spheres in $\mathbb{R}^n$ into spheres in $\mathbb{R}^p$ (preserving the radius). Also, each of its fibers is contained in a single element of the canonical pencil of f. In [**12**, Proposition 3.2] we proved that the map f is d-regular if and only if the spherification map is a submersion in a sufficiently small neighborhood of the origin.

This will allow us to lift correctly some particular vector fields.

From now on we assume $f \colon \mathbb{R}^n \to \mathbb{R}^p$ has an isolated critical value, it is locally surjective, d-regular, and with the Thom a_f-property.

Let v be the radial vector field in $\mathbb{R}^p$, $v(y_1, \cdots, y_p) = (y_1, \cdots, y_p)$. Since f and $\mathfrak{F}$ have no critical point outside V, we can lift v to obtain two vector fields v_f and $v_{\mathfrak{F}}$ using respectively f and $\mathfrak{F}$.

The vector v_f is transverse to all sufficiently small tubes $N(\varepsilon, \delta)$ and tangent to all $X_{\mathcal{L}}$'s. Meanwhile the vector field $v_{\mathfrak{F}}$ is transverse to all spheres and tangent to each $X_{\mathcal{L}}$. One can easily prove that sufficiently near the origin these vector fields cannot point in opposite directions. So their sum $w = v_f + v_{\mathfrak{F}}$ is a never vanishing smooth vector field, which is tangent to the $X_{\mathcal{L}}$'s and transverse to the spheres and to $f^{-1}(\partial \mathbb{D}_\delta)$. The details of this proof are done in [**12**], Proposition 5.2. One has:

LEMMA 3.13. *Let an analytic map-germ $f : (\mathbb{R}^n, \underline{0}) \to (\mathbb{R}^p, 0)$ have an isolated critical value at 0. The map f is d-regular if and only if there exists a smooth vector field w on $\mathbb{B}_\varepsilon \setminus V$ which has the following properties:*

(i) It is radial, i.e., it is transverse to all spheres in $\mathbb{B}_\varepsilon$ centred at $\underline{0}$.
(ii) It is tangent to each $X_{\mathcal{L}} \setminus V$.
(iii) It is transverse to all the tubes $f^{-1}(\partial \mathbb{D}_\delta)$.

The flow generated by the vector field w allows us to deform the fibration in the tube into the desired one in the sphere. In fact, if as before we call $\overset{\circ}{T}(\varepsilon, \delta)$ to $\mathbb{S}_\varepsilon^{n-1} \cap f^{-1}(\overset{\circ}{\mathbb{D}}_\delta \setminus \{0\})$. The space $\overset{\circ}{T}(\varepsilon, \delta)$ should be thought of as being the "cap" of the tube $f^{-1}(\overset{\circ}{\mathbb{D}}_\delta \setminus\{0\}) \cap \mathbb{B}_\varepsilon$, since $\overset{\circ}{T}(\varepsilon,\delta)$ is the set where $f^{-1}(\overset{\circ}{\mathbb{D}}_\delta \setminus\{0\}) \cap \mathbb{B}_\varepsilon$ meets the boundary sphere $\mathbb{S}_e$.

Any point in $N(\varepsilon,\delta)$ can be moved following the flow generated by the vector field w until it reaches (in a unique way) a point in $\mathbb{S}_\varepsilon^{n-1}\setminus \overset{\circ}{T}(\varepsilon,\delta)$. We obtain then a fibration:

$$\phi_1 \colon \mathbb{S}_\varepsilon^{n-1}\setminus \overset{\circ}{T}(\varepsilon,\delta) \to \partial\mathbb{D}_\delta \cong \mathbb{S}^{p-1}.$$

Since the flow is tangent to the $X_\mathcal{L}$'s, any point moves along the same $X_\mathcal{L}$. So the map ϕ_1 coincides with $f/\|f\|$.

On the other hand we have already seen in Equation (32), that we have a fibration on the complement

$$\phi_T\colon\ \overset{\circ}{T}(\varepsilon,\delta) \to \mathbb{S}^{p-1}.$$

These two fibrations glue together giving the desired one

$$\phi\colon \mathbb{S}_\varepsilon^{n-1}\setminus K \to \mathbb{S}^{p-1}.$$

Let us summarize the previous discussion as follows. We use the same notation. We consider an analytic map-germ $f:(\mathbb{R}^n,\underline{0}) \to (\mathbb{R}^p,0)$ with an isolated critical value at 0.

- If f has the Thom a_f-property then given $\varepsilon > 0$ sufficiently small, there exists $\delta > 0$ such that all fibers $f^{-1}(t)$ with $\|t\| \le \delta$ intersect transversally the sphere $\mathbb{S}_\varepsilon$.
- If given $\varepsilon > 0$ sufficiently small, there exists $\delta > 0$ such that all fibers $f^{-1}(t)$ with $\|t\| \le \delta$ intersect transversally the sphere $\mathbb{S}_\varepsilon$, then one has Milnor-Lê fibrations:

$$f:\widehat{N}(\varepsilon,\delta) \to \mathbb{D}_\delta\setminus\{0\}\ , \qquad \text{where } \widehat{N}(\varepsilon,\delta) = f^{-1}(\mathbb{D}_\delta\setminus\{0\})\cap\mathbb{B}_\varepsilon\,,$$

and

$$f:N(\varepsilon,\delta) \to \partial\mathbb{D}_\delta \cong \mathbb{S}^{p-1}\ , \qquad \text{where } \widehat{N}(\varepsilon,\delta) = f^{-1}(\partial\mathbb{D}_\delta)\cap\mathbb{B}_\varepsilon\,.$$

- If f is d-regular, then there exists a smooth vector field w as in Lemma 3.13. The integral lines of this vector field are transverse to all spheres around $\underline{0}$, transverse to all Milnor tubes $f^{-1}(\partial\mathbb{D}_\eta)\cap\mathbb{B}_\varepsilon$, and tangent to each element $X_\mathcal{L}$ of the canonical pencil.
- If f is d-regular and it has the above fibration $f:\widehat{N}(\varepsilon,\delta)\to\mathbb{D}_\delta\setminus\{0\}$, then the vector field w allows us to identify its boundary $N(\varepsilon,\delta)$ with the complement of the cap $\overset{\circ}{T}(\varepsilon,\delta)$ in the sphere. One thus gets that f has a Milnor fibration

$$\phi = \frac{f}{\|f\|} : \mathbb{S}_\varepsilon^{n-1}\setminus V \to \mathbb{S}^{p-1},$$

and this fibration is equivalent to the one on the tube $N(\varepsilon,\delta)$.

We remark that the above construction can also be used conversely: Given f with isolated critical value and d-regular, we have a vector field w as in Lemma 3.13. If we know that f has a Milnor fibration $\mathbb{S}_\varepsilon^{n-1}\setminus V \overset{f/\|f\|}{\longrightarrow} \mathbb{S}^{p-1}$, and f has a Milnor-Lê fibration on the tube $\widehat{N}(\varepsilon,\delta)$, then we can use the integral lines of w to identify the fibers of ϕ on the sphere, with those of f on the tube. Thus we arrive to the following theorem, which is implicit in [**12**] and completes previous results by various authors in [**29, 3, 4, 27, 11, 10**], including the classical holomorphic case by Milnor, which is the paradigm of this whole discussion.

THEOREM 3.14. *Let an analytic map-germ* $f : (\mathbb{R}^n, \underline{0}) \to (\mathbb{R}^p, 0)$ *have an isolated critical value at* 0. *If the two fibrations exist (one on a Milnor tube, another on the sphere minus the link), then these fibrations are smoothly equivalent. That is, there exists a diffeomorphism between their corresponding total spaces, carrying fibers into fibers.*

References

[1] N. A'Campo. Le nombre de Lefschetz d'une monodromie. *Nederl. Akad. Wetensch. Proc. Ser. A* **76**, *Indag. Math.*, 35:113–118, 1973. MR0320364 (47:8903)

[2] N. A'Campo. Monodromy of real isolated singularities. *Topology*, 42(6):1229–1240, 2003. MR1981355 (2004c:14011)

[3] R. N. Araujo dos Santos. Uniform (m)-condition and strong Milnor fibrations. In *Singularities II*, volume 475 of *Contemp. Math.*, pages 189–198. Amer. Math. Soc., Providence, RI, 2008. MR2454367 (2010c:58046)

[4] R. N. Araujo dos Santos. Equivalence of real Milnor's fibrations for quasi homogeneous singularities. *Singularities II*, volume 475 of *Contemp. Math.*, pp. 189–198. Amer. Math. Soc., Providence, RI, 2008. MR2454367 (2010c:58046)

[5] R. N. Araujo dos Santos and M. Tibăr. Real map germs and higher open book structures. Geom. Dedicata, 147, 177-185 (2010). MR2660575 (2011g:58062)

[6] K. Bekka. C-régularité et trivialité topologique In *Singularity theory and its applications, Part I (Coventry, 1988/1989)*, Lecture Notes in Math. (1462), 42–62, 1991. MR1129023 (92h:58012)

[7] A. Bodin and A. Pichon. Meromorphic functions, bifurcation sets and fibred links. *Math. Res. Lett.*, 14(3):413–422, 2007. MR2318645 (2008f:32038)

[8] D. Burghelea and A. Verona. Local homological properties of analytic sets. *Manuscripta Math.*, 7:55–66, 1972. MR0310285 (46:9386)

[9] P. T. Church and K. Lamotke. Non-trivial polynomial isolated singularities. *Nederl. Akad. Wetensch. Proc. Ser. A* **78**=*Indag. Math.*, 37:149–154, 1975. MR0365592 (51:1844)

[10] J. L. Cisneros-Molina. Join theorem for polar weighted homogeneous singularities. In *Singularities II*, volume 475 of *Contemp. Math.*, pages 43–59. Amer. Math. Soc., Providence, RI, 2008. MR2454370 (2010e:32030)

[11] J. L. Cisneros-Molina, J. Seade, and J. Snoussi. Refinements of Milnor's fibration theorem for complex singularities. *Adv. Math.*, 222(3):937–970, 2009. MR2553374 (2010k:32042)

[12] J. L. Cisneros-Molina, J. Seade, and J. Snoussi. Milnor Fibrations and d-regularity for real analytic Singularities. International Journal of Mathematics, vol. 21, No. 4, 419-434 (2010). MR2647448 (2011e:58063)

[13] J. L. Cisneros-Molina, J. Seade, and J. Snoussi. Milnor Fibrations for Real and Complex Singularities In *Topology of Algebraic Varieties and Singularities*, volume 538 of *Contemp. Math.*, pages 345–362. Amer. Math. Soc., Providence, RI, 2011. MR2768672

[14] A. H. Durfee. Neighborhoods of algebraic sets. *Trans. Amer. Math. Soc.*, 276(2):517–530, 1983. MR688959 (84m:32013)

[15] M. Hirasawa and L. Rudolph. Constructions of Morse maps for knots and links, and upper bounds on the Morse-Novikov number. Preprint 2003, math.GT/0311134. To appear in Journal of Knot Theory and its ramifications.

[16] H. Hironaka. Stratification and flatness. In *Real and complex singularities (Proc. Ninth Nordic Summer School/NAVF Sympos. Math., Oslo, 1976)*, pages 199–265. Sijthoff and Noordhoff, Alphen aan den Rijn, 1977. MR0499286 (58:17187)

[17] A. Jacquemard. Fibrations de Milnor pour des applications réelles. *Boll. Un. Mat. Ital. B (7)*, 3(3):591–600, 1989. MR1010525 (91a:55019)

[18] D. T. Lê. Some remarks on relative monodromy. In *Real and complex singularities (Proc. Ninth Nordic Summer School/NAVF Sympos. Math., Oslo, 1976)*, pages 397–403. Sijthoff and Noordhoff, Alphen aan den Rijn, 1977. MR0476739 (57:16296)

[19] E. Looijenga. A note on polynomial isolated singularities. *Nederl. Akad. Wetensch. Proc. Ser. A* **74**=*Indag. Math.*, 33:418–421, 1971. MR0303557 (46:2694)

[20] J. Milnor. On isolated singularities of hypersurfaces. Preprint, June 1966.

[21] J. Milnor. *Singular points of complex hypersurfaces.* Annals of Mathematics Studies, No. 61. Princeton University Press, Princeton, N.J., 1968. MR0239612 (39:969)

[22] M. Oka. Topology of polar weighted homogeneous hypersurfaces. *Kodai Math. J.*, 31(2):163–182, 2008. MR2435890 (2009g:32061)

[23] M. Oka. Non-degenerate mixed functions. *Kodai Math J.*, 33(1):1–62, 2010. MR2732230 (2012a:32032)

[24] M. Oka. On Mixed Brieskorn Variety. Preprint, arXiv:0909.4605, 2009.

[25] B. Perron. Le nœud "huit" est algébrique réel. *Invent. Math.*, 65(3):441–451, 1982. MR643562 (84d:57005)

[26] A. Pichon and J. Seade. Real singularities and open-book decompositions of the 3-sphere. *Ann. Fac. Sci. Toulouse Math. (6)*, 12(2):245–265, 2003. MR2123256 (2005m:57031)

[27] A. Pichon and J. Seade. Fibred multilinks and singularities $f\bar{g}$. *Math. Ann.*, 342(3):487–514, 2008. MR2430986 (2009g:32062)

[28] M. A. S. Ruas and R. N. Araujo dos Santos. Real Milnor fibrations and (c)-regularity. *Manuscripta Math.*, 117(2):207–218, 2005. MR2150482 (2006g:32047)

[29] M. A. S. Ruas, J. Seade, and A. Verjovsky. On real singularities with a Milnor fibration. In *Trends in singularities*, Trends Math., pages 191–213. Birkhäuser, Basel, 2002. MR1900787 (2003c:32031)

[30] L. Rudolph. Isolated critical points of mappings from $\mathbf{R}^4$ to $\mathbf{R}^2$ and a natural splitting of the Milnor number of a classical fibered link. I. Basic theory; examples. *Comment. Math. Helv.*, 62(4):630–645, 1987. MR920062 (88k:57009)

[31] J. Seade. Open book decompositions associated to holomorphic vector fields. *Bol. Soc. Mat. Mexicana (3)*, 3(2):323–335, 1997. MR1679312 (2000c:32085)

[32] J. Seade. *On the topology of isolated singularities in analytic spaces*, volume 241 of *Progress in Mathematics.* Birkhäuser Verlag, Basel, 2006. MR2186327 (2007a:32031)

[33] J.-L. Verdier. Stratifications de Whitney et théorème de Bertini-Sard. *Invent. Math.*, 36:295–312, 1976. MR0481096 (58:1242)

Instituto de Matemáticas, Unidad Cuernavaca, Universidad Nacional Autónoma de México, Avenida Universidad s/n, Colonia Lomas de Chamilpa, Cuernavaca, Morelos, México.

E-mail address: jlcm@matcuer.unam.mx

Instituto de Matemáticas, Unidad Cuernavaca, Universidad Nacional Autónoma de México, Avenida Universidad s/n, Colonia Lomas de Chamilpa, Cuernavaca, Morelos, México.

E-mail address: jseade@matcuer.unam.mx

Instituto de Matemáticas, Unidad Cuernavaca, Universidad Nacional Autónoma de México, Avenida Universidad s/n, Colonia Lomas de Chamilpa, Cuernavaca, Morelos, México

E-mail address: jsnoussi@matcuer.unam.mx

Contemporary Mathematics
Volume **569**, 2012
http://dx.doi.org/10.1090/conm/569/11248

Bi-Lipschitz $\mathcal{G}$-triviality and Newton polyhedra, $\mathcal{G} = \mathcal{R}$, $\mathcal{C}$, $\mathcal{K}$, $\mathcal{R}_V$, $\mathcal{C}_V$, $\mathcal{K}_V$

J. C. F. Costa, M. J. Saia, and C. H. Soares Júnior

Abstract. In this work we provide estimates for the bi-Lipschitz $\mathcal{G}$-triviality, $\mathcal{G} = \mathcal{C}$ or $\mathcal{K}$, for a family of map germs satisfying a Lojasiewicz condition. We work with two cases: the class of weighted homogeneous map germs and the class of non-degenerate map germs with respect to some Newton polyhedron. We also consider the bi-Lipschitz triviality for families of map germs defined on an analytic variety V. We give estimates for the bi-Lipschitz $\mathcal{G}_V$-triviality where $\mathcal{G} = \mathcal{R}, \mathcal{C}$ or $\mathcal{K}$ in the weighted homogeneous case. Here we assume that the map germ and the analytic variety are both weighted homogeneous with respect to the same weights. The method applied in this paper is based in the construction of controlled vector fields in the presence of a suitable Lojasiewicz condition. In the last section of this work we compare our results with other results related to this work showing tables with all estimates that we know, including ours.

1. Introduction

A basic problem in Singularity Theory is the local classification of mappings up to diffeomorphisms. However, this problem is too rigid. Therefore, it seems natural to investigate classification of mappings by morphisms weaker than diffeomorphisms. In this paper we are interested in the study of mappings up to bi-Lipschitz maps.

A mapping $\phi : U \subset \mathbb{R}^\ell \to \mathbb{R}^s$ is called *Lipschitz* if there exists a constant $c > 0$ such that

$$\| \phi(x) - \phi(y) \| \leq c \| x - y \|, \quad \forall x, y \in U.$$

When $\ell = s$ and ϕ has a Lipschitz inverse, we say that ϕ is *bi-Lipschitz.*

Equivalence relations defined in terms of bi-Lipschitz maps are important tools in the study of equisingularity of mappings and sets from the metric point of view. Metric classification refers to classification up to bi-Lipschitz maps.

We can define Lipschitz versions of the equivalence relations involving the classical Mather's groups $\mathcal{R}, \mathcal{C}$ and $\mathcal{K}$ as follows:

i) Two map germs $f, g : (\mathbb{R}^n, 0) \to (\mathbb{R}^p, 0)$ are *bi-Lipschitz $\mathcal{R}$-equivalent* if there exists a germ of a bi-Lipschitz homeomorphism $h : (\mathbb{R}^n, 0) \to (\mathbb{R}^n, 0)$ such that $g = f \circ h^{-1}$.

2010 *Mathematics Subject Classification.* Primary 58K40,58K60.

Key words and phrases. bi-Lipschitz determinacy, Newton filtration, controlled vector fields.

This work is partially supported by CAPES, CNPq, FAPESP and FUNCAP.

ii) Two map germs $f, g : (\mathbb{R}^n, 0) \to (\mathbb{R}^p, 0)$ are *bi-Lipschitz $\mathcal{K}$-equivalent* if there exists a pair of bi-Lipschitz homeomorphisms (h, H), with $h : (\mathbb{R}^n, 0) \to (\mathbb{R}^n, 0)$, $H = (H_1, H_2) : (\mathbb{R}^n \times \mathbb{R}^p, 0) \to (\mathbb{R}^n \times \mathbb{R}^p, 0)$ satisfying the conditions: $H_1 = h$, $H(\mathbb{R}^n \times \{0\}) = \mathbb{R}^n \times \{0\}$ and $H \circ (Id, f) = (Id, g) \circ h$, where Id denotes the germ of the identity in $\mathbb{R}^n$. When $h = Id$ we call the bi-Lipschitz $\mathcal{K}$-equivalence by *bi-Lipschitz $\mathcal{C}$-equivalence.*

In this article we consider the bi-Lipschitz $\mathcal{G}$-triviality along of a family of map germs f_t with $f_0 = f$ satisfying a Lojasiewicz condition.

The bi-Lipschitz $\mathcal{R}$-triviality of families of function germs was studied by Fernandes and Ruas in [**10**] for the weighted homogeneous case and by Fernandes and Soares Jr. in [**11**] for the Newton non-degenerate case. However the study of bi-Lipschitz triviality with respect to the groups $\mathcal{C}$ and $\mathcal{K}$ has a gap in the literature, which is the main subject of this article.

With respect to the class of weighted homogeneous map germs, our main results are Theorems 2.3 and 2.6 in which we obtain sufficient conditions for the bi-Lipschitz $\mathcal{K}$ and bi-Lipschitz $\mathcal{C}$-triviality of families, respectively. These results are given in terms of the weights and degrees of weighted homogeneity. For the class of non-degenerate map germs with respect to some Newton polyhedron, the main results are Theorems 2.14 and 2.15 in which we also obtain sufficient conditions for triviality of families, similar to the weighted homogeneous case. We also consider the bi-Lipschitz $\mathcal{G}_V$-triviality for families of map germs defined on an analytic variety V, where $\mathcal{G} = \mathcal{R}, \mathcal{C}$ or $\mathcal{K}$. In this case our main results are Theorems 4.1, 4.3 and 4.4, respectively.

It is worth mentioning here the difference between the classification questions considered in this paper. For example, the bi-Lipschitz $\mathcal{R}$-equivalence admits moduli (cf. [**12**]) while the bi-Lipschitz $\mathcal{K}$-equivalence does not have moduli, this is a recent result due to Valette and Ruas [**18**]. The results presented here and in the work of Ruas and Valette on the finiteness theorems can be useful for the the study of Lipschitz classification of map germs.

The method applied to obtain our results is an application of a Thom-Levine type theorem with the construction of controlled vector fields in the presence of a suitable Lojasiewicz condition. Consider a deformation $f_t(x)$ of f, then the bi-Lipschitz $\mathcal{G}$-triviality of f_t is proven by solving localized equations of type $-\rho\frac{\partial f_t}{\partial t} = \xi(f)$ where ρ is a control function and ξ is a germ of smooth vector field in the $\mathcal{G}$-tangent space of f. If the deformation f_t satisfies an appropriate filtration condition, we obtain the bi-Lipschitz property of the vector field $\rho^{-1}\xi$.

This method was used by several authors to obtain estimates for the C^ℓ-$\mathcal{G}$-triviality of map germs, $0 \leq \ell < \infty$. For instance, the works of Damon [**9, 8**] and Kuo [**13**], among others, treated the C^0-$\mathcal{G}$-triviality of map germs.

More recently Abderrahmane [**1**] studies the C^0-$\mathcal{R}$-triviality for function germs which satisfy a Newton non-degeneracy condition. Estimates for the C^ℓ-$\mathcal{R}$-triviality in families of real functions germs of class at least $C^{\ell+1}$ appear in the work of Bromberg and Lopes de Medrano [**3**], for the weighted homogeneous case. The C^ℓ-$\mathcal{G}$-triviality, with $0 \leq \ell < \infty$ and $\mathcal{G} = \mathcal{R}$, $\mathcal{C}$ or $\mathcal{K}$ appear in [**15**] for homogeneous map germs, in [**16**] for weighted homogeneous map germs and in [**19**] for map germs which satisfy a Newton non-degeneracy condition.

Concerning the $\mathcal{G}_V$-triviality for families of map germs defined on an analytic variety V, we see estimates for the C^ℓ-$\mathcal{G}_V$-triviality with $\ell > 1$ and $\mathcal{G} = \mathcal{R}$, $\mathcal{C}$ or $\mathcal{K}$

in [**14**], but again the bi-Lipschitz case is not treated. Then in this paper we also consider this case. We give estimates for the bi-Lipschitz $\mathcal{G}_V$-triviality, $\mathcal{G} = \mathcal{R}, \mathcal{C}$ or $\mathcal{K}$, in the weighted homogenous case, that is, when the map germ and the analytic variety are both weighted homogeneous with respect to the same set of fixed weights.

In the last section we compare our results with others which appear in the literature as a consequence of the method used here. All this results are shown in tables which allows us to compare them easily.

Notice that the bi-Lipschitz triviality with respect to the $\mathcal{A}$ group is not treated in this paper. In [**6**] we give estimates for the bi-Lipschitz $\mathcal{A}$-triviality of family of map germs. The main difference is that here we show an explicit construction of the controlled vector fields while in [**6**] the estimates are given in terms of the vector fields that satisfy a suitable filtration condition.

2. Bi-Lipschitz $\mathcal{G}$-triviality, $\mathcal{G} = \mathcal{R}, \mathcal{C}$ or $\mathcal{K}$

We denote by $\mathcal{E}_n$ the local ring of real analytic function germs $f : (\mathbb{R}^n, 0) \to \mathbb{R}$ and by m_n its maximal ideal. The $\mathcal{E}_n$-module of analytic map germs $f : (\mathbb{R}^n, 0) \to (\mathbb{R}^p, 0)$ will be denoted by $\mathcal{E}_{n,p}$. We consider along this article $n \geq p$.

For any analytic map germ $f : (\mathbb{R}^n, 0) \to (\mathbb{R}^p, 0)$, we denote by $I_{\mathcal{R}} f$ the ideal of $\mathcal{E}_n$ generated by the $p \times p$ minors of the jacobian matrix of f, by $I_{\mathcal{C}} f$ the ideal generated by the coordinate functions of f and by $I_{\mathcal{K}} f$ the ideal $I_{\mathcal{R}} f + I_{\mathcal{C}} f$.

Let $N_{\mathcal{C}}\, f(x) = |f(x)|^2$, $N_{\mathcal{R}}\, f(x) = \sum_j M_j^2$, where the M_j are the generators of $I_{\mathcal{R}} f$ and $N_{\mathcal{K}}\, f = N_{\mathcal{R}}\, f + N_{\mathcal{C}}\, f$. For $\mathcal{G} = \mathcal{R}, \mathcal{C}$ or $\mathcal{K}$ we say that $N_{\mathcal{G}}\, f$ satisfies a Lojasiewicz condition if there exist constants $c > 0$ and $\alpha > 0$ with $N_{\mathcal{G}}\, f(x) \geq c|x|^{\alpha}$.

2.1. The Weighted Homogeneous Case.

DEFINITION 2.1. *A map germ* $f : (\mathbb{R}^n, 0) \to (\mathbb{R}^p, 0)$ *in* $\mathcal{E}_{n,p}$ *is* weighted homogeneous of type $(r_1, \ldots, r_n; d_1, \ldots, d_p)$ *with* r_i, $d_j \in \mathbb{Q}^+$, *if for all* $\lambda \in \mathbb{R}_+$*:*

$$f(\lambda^{r_1} x_1, \ldots, \lambda^{r_n} x_n) = (\lambda^{d_1} f_1(x), \ldots, \lambda^{d_p} f_p(x)).$$

In this subsection we fix an n-tuple of weights $\overline{r} = (r_1, \ldots, r_n)$ in such a way that we can define a weighted filtration in the ring $\mathcal{E}_n$.

DEFINITION 2.2. *Given* $\overline{r} = (r_1, \ldots, r_n)$*, for any monomial* $x^{\alpha} = x_1^{\alpha_1} x_2^{\alpha_2} \ldots x_n^{\alpha_n}$*, define* $\mathrm{fil}_{\overline{r}}(x^{\alpha}) = \sum_{i=1}^n \alpha_i r_i$*. We define the* $\overline{r}$*-filtration in* $\mathcal{E}_n$ *via the function defined for any* f *in* $\mathcal{E}_n$ *by*

$$\mathrm{fil}_{\overline{r}}(f) = \inf_{\alpha} \left\{ \mathrm{fil}_{\overline{r}}(x^{\alpha}) \mid \left(\frac{\partial^{\alpha} f}{\partial x^{\alpha}} \right) (0) \neq 0 \right\}.$$

This definition can be extended to $\mathcal{E}_{n+s}$, the ring of s-parameter families of germs in n-variables, by defining $\mathrm{fil}_{\overline{r}}(x^{\alpha} t^{\beta}) = \mathrm{fil}_{\overline{r}}(x^{\alpha})$.

Let $(r_1, \ldots, r_n; 2\ell)$ be fixed. The standard control function of type $(r_1, \ldots, r_n; 2\ell)$ is

$$\rho_{\ell}(x) = x_1^{2\alpha_1} + \cdots + x_n^{2\alpha_n},$$

where the $\alpha_i = \frac{\ell}{r_i}$ are chosen such that the function ρ_{ℓ} is weighted homogeneous of type $(r_1, \ldots, r_n; 2\ell)$.

For any map germ $g = (g_1, \ldots, g_p)$ in $\mathcal{E}_{n,p}$, we introduce $\mathrm{fil}_{\overline{r}}(g) = (d_1, \ldots, d_p)$, where $d_i = \mathrm{fil}_{\overline{r}}(g_i)$, for each $i = 1, \ldots, p$. As we are working with a fixed set of weights $\overline{r} = (r_1, \ldots, r_n)$, from now on we shall omit the subindex $\mathrm{fil}_{\overline{r}}$ and call $\mathrm{fil}_{\overline{r}}(f) = \mathrm{fil}(f)$.

Analogously to the previous definitions of the functions $N_{\mathcal{G}}\, f$, for any polynomial map germ $f : (\mathbb{R}^n, 0) \to (\mathbb{R}^p, 0)$ we define the weighted homogeneous versions of them. That is, $N^{\star}_{\mathcal{R}}\, f := \sum_I M_I^{2k_I}$, where M_I denotes a $p\times p$ minor of the jacobian matrix of f with $I = \{i_1, \ldots, i_p\}$, $1 \leq i_1 \leq \cdots \leq i_p \leq n$, $k_I = \dfrac{k}{s_I}$ with $s_I = \mathrm{fil}(M_I)$ and $k = lcm\,\{s_I\}$. We call $N^{\star}_{\mathcal{C}}\, f := \sum_{i=1}^{p} (f_i)^{2\beta_i}$, where $\beta_i = \frac{lcm\,\{\mathrm{fil}\,(f_j),\, j=1,\ldots,p\}}{\mathrm{fil}\,(f_i)}$.

To define the weighted homogeneous version of the function $N_{\mathcal{K}}\, f$, let $a, b \in \mathbb{N}$ be prime numbers with $gcd(a, b) = 1$ such that $\mathrm{fil}\,((N^{\star}_{\mathcal{R}}\, f)^a) = \mathrm{fil}\,((N^{\star}_{\mathcal{C}}\, f)^b)$.

We call $N^{\star}_{\mathcal{K}}\, f := (N^{\star}_{\mathcal{R}}\, f)^a + (N^{\star}_{\mathcal{C}}\, f)^b$.

Now, for any family of polynomial map germs $f_t : (\mathbb{R}^n, 0) \to (\mathbb{R}^p, 0)$, $t \in [0, 1]$, we set $N^{\star}_{\mathcal{R}}\, f_t := \sum_I M_{t_I}^{2k_I}$, $N^{\star}_{\mathcal{C}}\, f_t := \sum_{i=1}^{p} (f_{t_i})^{2\beta_i}$ and $N^{\star}_{\mathcal{K}}\, f_t := (N^{\star}_{\mathcal{R}}\, f_t)^a + (N^{\star}_{\mathcal{C}}\, f_t)^b$, where a, b, k_I and β_i are as before.

Then we have the following theorem:

THEOREM 2.3. *Let $f : (\mathbb{R}^n, 0) \to (\mathbb{R}^p, 0)$ be a weighted homogeneous polynomial map germ of type $(r_1, \ldots, r_n; d_1, \ldots, d_p)$ with $r_1 \leq \cdots \leq r_n$, $d_1 \leq \cdots \leq d_p$, satisfying a Lojasiewicz condition $N^{\star}_{\mathcal{K}}\, f(x) \geq c|x|^{\alpha}$, for constants c and α. Then, small deformations of f defined by $f_t = f + t\Theta$, $\Theta = (\Theta_1, \ldots, \Theta_p)$, with* $\mathrm{fil}\,(\Theta_i) \geq d_p + r_n - r_1$ *are bi-Lipschitz $\mathcal{K}$-trivial.*

PROOF. The idea of this proof is to construct two Lipschitz vector fields ξ and $\tilde{\eta}$ associated to the functions $N^{\star}_{\mathcal{R}}\, f_t$ and $N^{\star}_{\mathcal{C}}\, f_t$, respectively, to obtain the bi-Lipschitz $\mathcal{K}$-triviality of the family f_t.

Let $B = lcm\,\{\mathrm{fil}\,(f_j),\, j = 1, \ldots, p\}$ and $k = lcm\,\{s_I\}$, then

$$\mathrm{fil}\,(N^{\star}_{\mathcal{K}}\, f_t) = \mathrm{fil}\,(N^{\star}_{\mathcal{R}}\, f_t)^a = \mathrm{fil}\,(N^{\star}_{\mathcal{C}}\, f_t)^b = 2ka = 2Bb \text{ and}$$

$$\begin{aligned}(N^{\star}_{\mathcal{K}}\, f_t)\frac{\partial f_t}{\partial t} &= (N^{\star}_{\mathcal{R}}\, f_t)^a\frac{\partial f_t}{\partial t} + (N^{\star}_{\mathcal{C}}\, f_t)^b\frac{\partial f_t}{\partial t} = \\ &= (N^{\star}_{\mathcal{R}}\, f_t)^{a-1}\, N^{\star}_{\mathcal{R}}\, f_t\,\frac{\partial f_t}{\partial t} + (N^{\star}_{\mathcal{C}}\, f_t)^{b-1}\, N^{\star}_{\mathcal{C}}\, f_t\,\frac{\partial f_t}{\partial t} = \\ &= (N^{\star}_{\mathcal{R}}\, f_t)^{a-1}\, \mathrm{d}f_t(W_{\mathcal{R}}) + (N^{\star}_{\mathcal{C}}\, f_t)^{b-1} \textstyle\sum_{i=1}^{p} L_i\, f_{ti},\end{aligned}$$

here $\mathrm{W}_{\mathcal{R}} = \sum_I M_I^{2k_I-1} W_I$, $W_I = \sum_{i=1}^{n} w_i \frac{\partial}{\partial x_i}$ and $\left\{\begin{array}{lcl} w_i & = & 0, \quad i \notin I \\ w_{im} & = & \sum_{j=1}^{p} N_{ji_m}(\frac{\partial f_t}{\partial t})_j,\, i_m \in I, \end{array}\right.$

where N_{ji_m} is the $(p-1)\times(p-1)$ minor cofactor of $\dfrac{\partial f_j}{\partial x_{i_m}}$ in the differential $\mathrm{d}f$ of f and $L_i = f_{ti}^{2\beta_i - 1}(\frac{\partial f_t}{\partial t})$, with $\beta_i = \frac{B}{fil\,(f_i)}$, for $j = 1, \ldots, p$ and $i = 1, \ldots, p$.

$$\text{Then} \quad (N^{\star}_{\mathcal{K}}\, f_t)\frac{\partial f_t}{\partial t} = \mathrm{d}f_t((N^{\star}_{\mathcal{R}}\, f_t)^{a-1}\, W_{\mathcal{R}}) + \sum_{i=1}^{p}(N^{\star}_{\mathcal{C}}\, f_t)^{b-1} f_{ti}^{2\beta_i - 1}\left(\frac{\partial f_t}{\partial t}\right) f_{ti}.$$

We divide the above expression by $N^{\star}_{\mathcal{K}}\, f_t$ to obtain:

$$\frac{\partial f_t}{\partial t} = \mathrm{d}f_t\left(\frac{(N^{\star}_{\mathcal{R}}\, f_t)^{a-1}\, W_{\mathcal{R}}}{N^{\star}_{\mathcal{K}}\, f_t}\right) + \sum_{i=1}^{p} \frac{(N^{\star}_{\mathcal{C}}\, f_t)^{b-1} f_{ti}^{2\beta_i - 1}(\frac{\partial f_t}{\partial t}) f_{ti}}{N^{\star}_{\mathcal{K}}\, f_t}.$$

We define the vector field with respect to the group $\mathcal{R}$ by

$$\xi = \frac{(N^{\star}_{\mathcal{R}}\, f_t)^{a-1}\, W_{\mathcal{R}}}{N^{\star}_{\mathcal{K}}\, f_t}.$$

To define the vector field $\eta(x, y, t)$ with respect to the group $\mathcal{C}$ we call

$$\eta_j = \sum_{i=1}^{p} \frac{(N_{\mathcal{C}}^{\star} f_t)^{b-1} f_{ti}^{2\beta_i - 1} \Theta_j y_i}{N_{\mathcal{K}}^{\star} f_t} \tag{2.1}$$

and define $\eta(x, y, t) = (\eta_1, \ldots, \eta_p)(x, y, t)$.

First, we will show that the vector field ξ is Lipschitz. In fact observe that

$$\text{fil}\,((N_{\mathcal{R}}^{\star} f_t)^{a-1} W_{\mathcal{R}}) = \text{fil}\,(N_{\mathcal{R}}^{\star} f_t)^{a-1} + \text{fil}\,(W_{\mathcal{R}})$$

$$\geq (a-1)\,\text{fil}\,((N_{\mathcal{R}}^{\star} f_t)) + \min\{\text{fil}\,(M_I^{2k_I - 1}) + \text{fil}\,(W_I)\}$$

$$\geq 2ka - 2k + \min\{(2\tfrac{k}{\text{fil}\,(M_I)} - 1)\,\text{fil}\,(M_I) + \text{fil}\,(N_{ji_m}) + \text{fil}\,(\Theta_j)\}$$

$$\geq 2ka - 2k + \min\{2k - \text{fil}\,(\tfrac{\partial f_j}{\partial x_{i_m}}) + \text{fil}\,(\Theta_j)\}$$

$$\geq 2ka - d_p + r_1 + \text{fil}\,(\Theta_j)$$

$$\geq 2ka + r_n.$$

As by hypothesis $\text{fil}\,(N_{\mathcal{K}}^{\star} f)$ satisfies a Lojasiewicz condition and fil $(\Theta_i) \geq d_p + r_n - r_1$, it follows from Lemma 3 of [**16**] that there exist constants $0 < c_1 < c_2$ such that

$$c_2 \rho_k(x) \leq N_{\mathcal{R}}^{\star} f_t \leq c_1 \rho_k(x).$$

Then, as the filtration of the numerator of $\xi = \dfrac{(N_{\mathcal{R}}^{\star} f_t)^{a-1} W_{\mathcal{R}}}{N_{\mathcal{K}}^{\star} f_t}$ is greater than or equal to fil $(N_{\mathcal{K}}^{\star} f_t) + r_n$, it follows from Lemma 3.2. of [**10**] that the vector field ξ is Lipschitz.

On the other hand, considering the vector field η we have:

$$\text{fil}\,((N_{\mathcal{C}}^{\star} f_t)^{b-1} f_{ti}^{2\beta_i - 1}(\tfrac{\partial f_t}{\partial t})) \geq (b-1)2B + (2\beta_i - 1)\,\text{fil}\,(f_{ti}) + \text{fil}\,(\Theta_j)$$

$$\geq 2Bb - 2B + 2\tfrac{B}{\text{fil}\,(f_{ti})}\text{fil}\,(f_{ti}) - \text{fil}\,(f_{ti}) + \text{fil}\,(\Theta_j)$$

$$\geq 2Bb - d_p + (d_p + r_n - r_1) = 2Bb + r_n - r_1.$$

However we can not apply Lemma 3.2. of [**10**] to the vector field η. Thus we do not know if η is Lipschitz or not.

But we can modify η in order to obtain a Lipschitz vector field $\tilde{\eta}$ using a convenient bump function. The strategy used here is based on [**16**]. The construction of the bump function is recovered here in order to clarify the definition of the vector field $\tilde{\eta}$.

REMARK 2.4. In Theorem 2.3 we have that $d_i \geq d_1$ for all $i = 1, \ldots, p$. Then there exists a constant c such that $|f_{ti}(x)|^2 \leq c\rho_{d_1}(x)$ for all $i = 1, \ldots, p$, where

$$\rho_d(x) = x_1^{2\alpha_1} + \cdots + x_n^{2\alpha_n}$$

is the standard control function of type $(r_1, \ldots, r_n; 2d)$.

LEMMA 2.5. *([**16**]) Let c be the constant from Remark 2.4. We define neighborhoods V and U of the region $|y| < c\rho_{d_1}^{\frac{1}{2}}(x)$ in $\mathbb{R}^n \times \mathbb{R}^p \times \mathbb{R} - \{0,0,t\}$, as follows*

$$V = \{(x,y,t) \mid |y| \le c_1 \rho_{d_1}^{\frac{1}{2}}(x), \text{ with } c_1 > c\}$$

and U is chosen in such a way that $U \subset \overline{U} \subset V$. Then, there exists a conic bump function $p : \mathbb{R}^n \times \mathbb{R}^p \times \mathbb{R} \to \mathbb{R}$ such that $p_{|\mathbb{R}^n \times \mathbb{R}^p \times \mathbb{R} - \{0,0,t\}}$ is smooth and

$$\begin{array}{ll} p(x,y,t) = 1, & (x,y,t) \in \overline{U} \\ p(x,y,t) = 0, & (x,y,t) \in V^c \\ 0 \le p(x,y,t) \le 1, & (x,y,t) \in V - \overline{U} \\ p(0,0,t) = 0, & \forall t. \end{array}$$

To finish the proof of Theorem 2.3 we use the bump function p defined in Lemma 2.5 to define the vector field

$$\tilde{\eta}(x,y,t) = p(x,y,t)\eta(x,y,t).$$

As $p = 1$ for all $(x,y,t) \in \overline{U}$, $\tilde{\eta}$ coincides with η in a conic neighborhood of the graph of f_t, then the equation

$$\frac{\partial f_t}{\partial t} = \mathrm{d}f_t(\xi) + (p\eta) \circ f_t$$

also holds in this neighborhood.

Moreover, we can redefine η so that $\tilde{\eta}$ is zero outside of this conic neighborhood of the graph of f_t. With the bump function, using the terms that appear in the expression (2.1), we can write

$$\tilde{\eta}_{ji} = \frac{(N_{\mathcal{C}}^{\star} f_t)^{b-1} f_{t_i}^{2\beta_j - 1} \Theta_i \, p \, y_i}{N_{\mathcal{K}}^{\star} f_t}. \tag{2.2}$$

Then from the composition $\eta \circ f_t$ we obtain that the filtration of the numerator of each $\tilde{\eta}_{ij}(f_{t_j})$ is greater than or equal to $fil\ (N_{\mathcal{K}}^{\star} f_t) + r_n$. Therefore, we apply Lemma 3.2 of [**10**] to conclude that $\tilde{\eta}_{ji}$ is Lipschitz. Hence $\tilde{\eta}$ is a Lipschitz vector field in that conic neighborhood of the graph of f_t. As $p = 0$ in V^c, $\tilde{\eta} = p\eta$ is zero out of another conic neighborhood of the graph of f_t. Then it follows that $\tilde{\eta}$ is well defined in all its domain and $\tilde{\eta}$ is Lipschitz.

By integrating the vector fields ξ and $\tilde{\eta}$, we obtain the required bi-Lipschitz $\mathcal{K}$-triviality and the result follows. □

We also have an estimate for the bi-Lipschitz $\mathcal{C}$-triviality given in the following theorem:

THEOREM 2.6. *Let $f : (\mathbb{R}^n, 0) \to (\mathbb{R}^p, 0)$ be a weighted homogeneous polynomial map germ of type $(r_1, \ldots, r_n; d_1, \ldots, d_p)$ with $r_1 \le \cdots \le r_n$, $d_1 \le \cdots \le d_p$, satisfying a Lojasiewicz condition $N_{\mathcal{C}}^{\star} f(x) \ge c|x|^{\alpha}$, for constants c and α. Then, small deformations of f defined by $f_t = f + t\Theta$ with $\Theta = (\Theta_1, \ldots, \Theta_p)$ and fil $(\Theta_i) \ge d_p + r_n - r_1$ are bi-Lipschitz $\mathcal{C}$-trivial.*

PROOF. The proof of this result is analogous to the proof of the Theorem 2.3. We show here the main difference, which appear in the Thom-Levine type equation.

As we are considering the group $\mathcal{C}$, we work only with the function $N_{\mathcal{C}}^{\star} f$ to write the following equation:

$$(N_{\mathcal{C}}^{\star} f_t)\frac{\partial f_t}{\partial t} = \sum_{i=1}^{p} f_{ti}^{2\beta_i - 1}\left(\frac{\partial f_t}{\partial t}\right) f_{ti}. \tag{2.3}$$

Here we need to consider the vector field η which we obtain dividing the equation (2.3) by $N_{\mathcal{C}}^{\star} f_t$. That is, as $\frac{\partial f_t}{\partial t} = \sum_{i=1}^{p} \frac{f_{ti}^{2\beta_i-1}(\frac{\partial f_t}{\partial t}) f_{ti}}{N_{\mathcal{C}}^{\star} f_t}$ we define the vector field $\eta(x, y, t) = (\eta_1, \dots, \eta_p)(x, y, t)$ where each η_j is

$$\eta_j = \sum_{i=1}^{p} \frac{f_{ti}^{2\beta_i - 1}\Theta_j y_i}{N_{\mathcal{C}}^{\star} f_t}, \qquad \text{for } j = 1, \dots, p.$$

Now we use the bump function to construct the vector field $\tilde{\eta}(x, y, t) = p(x, y, t)\eta(x, y, t)$ and following the same steps as done in the proof of Theorem 2.3 the result follows. □

2.2. The Newton Non Degenerate Case.

In this subsection we consider a Newton filtration for $\mathcal{E}_n$ and provide estimates for the bi-Lipschitz $\mathcal{G}$-triviality, $\mathcal{G} = \mathcal{C}$ or $\mathcal{K}$ of a family of analytic map germs satisfying some non-degeneracy condition with respect to this filtration.

To define the Newton filtration we follow [**1**]. We construct a Newton polyhedron from a fixed $n \times m$ matrix $A = (a_i^j)$, with $i = 1, ..., n$, $j = 1, ..., m$, $a^j = (a_1^j, \dots, a_n^j) \in \mathbb{Q}_+^n$ and $m \geq n$, such that the first n columns of A are $(0, ..., 0, a_j^j, 0, ..., 0)$ with $a_j^j > 0$, for all $j = 1, ..., n$.

DEFINITION 2.7. *The Newton polyhedron $\Gamma_+(A)$ is the convex hull in $\mathbb{R}_+^n$ of $Supp(A) + \mathbb{R}_+^n$, where $Supp(A) = \{a^j,\ j = 1, ..., m\}$. The Newton diagram of A, denoted $\Gamma(A)$, is the union of the compact faces of $\Gamma_+(A)$.*

It is usual to consider Newton polyhedra associated to a function or to a map germ f. Here we are using this more general definition because we sometimes can use Newton polyhedra which are different from the Newton polyhedron of f.

Now we define the control functions associated to a Newton polyhedron $\Gamma_+(A)$.

For any vector a^j of the matrix A and $k \in \mathbb{R}_+$, we denote $ka^j = (ka_1^j, ka_2^j, \dots, ka_n^j)$. We fix the smallest integer p such that pa_i^j is integer for all i, j. For any non-negative rational number d we define the standard control function $\rho^d : \mathbb{R}^n \to \mathbb{R}$ as:

$$\rho^d(x) = \left(\sum_{j=1}^{m} x_1^{2pa_1^j} x_2^{2pa_2^j} \cdots x_n^{2pa_n^j}\right)^{\frac{d}{2p}}. \tag{2.4}$$

We remark that the function $\rho^{2p}(x) = \sum_{j=1}^{m} x_1^{2pa_1^j} x_2^{2pa_2^j} \cdots x_n^{2pa_n^j}$ is a polynomial.

We denote by $\Gamma_+(\rho^d)$ the Newton polyhedron of the matrix $dA = (da_i^j)$, i.e, $\Gamma_+(\rho^d) = \Gamma_+(dA)$.

EXAMPLE 2.8. For $n = 2$, any matrix A is written as

$$A = \begin{pmatrix} a_1^1 & 0 & a_1^3 & \cdots & a_1^m \\ 0 & a_2^2 & a_2^3 & \cdots & a_2^m \end{pmatrix}.$$

For instance, fix $A = \begin{pmatrix} \frac{1}{b} & 0 & \frac{b-1}{(b+1)b} \\ 0 & \frac{1}{b+1} & \frac{1}{(b+1)b} \end{pmatrix}$ with b a positive integer. Then we obtain $p = b(b+1)$ and the control function ρ of $\Gamma_+(A)$ is given as $\rho(x,y) = \left(x^{2b+2} + y^{2b} + x^{2b-2}y^2\right)^{\frac{1}{2b(b+1)}}$.

The corresponding Newton polyhedron $\Gamma_+(\rho^{2b(b+1)})$ has two faces with vertices $\{(2b+2, 0), (2b-2, 1), (0, 2b)\}$.

We shall consider this matrix later in Example 3.1, in order to obtain estimates for the bi-Lipschitz $\mathcal{K}$-triviality of the germ $f(x,y) = (xy, x^{2b+2} - y^{2b} + x^{2b-2}y^2)$. In this case the matrix A is associated to the second coordinate of f.

Notice that if A is a square $n \times n$ diagonal matrix, the corresponding Newton polyhedron has only one face and in particular we have a weighted homogeneous filtration.

The main condition to guarantee the bi-Lipshitz triviality in this case is given in terms of a non-degeneracy condition, called A-isolatedness, which we describe here.

Let $g : (\mathbb{R}^n, 0) \to (\mathbb{R}, 0)$ be an analytic function germ written as $g(x) = \sum a_\gamma x^\gamma$, where x^γ denotes the monomial $x^\gamma = x_1^{\gamma_1} \ldots x_n^{\gamma_n}$ and γ denotes the multi-index $\gamma = (\gamma_1, \ldots, \gamma_n)$.

For any Newton Polyhedron $\Gamma_+(A)$, if Δ is a subset of $\Gamma_+(A)$, we call g_Δ the germ $g_\Delta(x) = \sum_{\gamma \in \Delta} a_\gamma x^\gamma$. For any fixed germ $f(x) = \sum a_\gamma x^\gamma$, call d the biggest integer such that $\gamma = (\gamma_1, \ldots, \gamma_n) \in \Gamma_+(\rho^d)$ for all $a_\gamma \neq 0$.

DEFINITION 2.9 ([**1**], p. 525). *The origin is an A-isolated point of a germ f if for each compact face Δ of $\Gamma(\rho^d)$, the equation $f_\Delta(x) = 0$ does not have solution in $(\mathbb{R} - \{0\})^n$.*

For a Newton polyhedron $\Gamma_+(A)$ and for each $w = (w_1, \ldots, w_n) \in \mathbb{R}_+^n$ define:

DEFINITION 2.10. (a) $\ell(w) = \min\{\langle w, k\rangle : k \in \Gamma_+(A)\}$, $\langle w, k\rangle = \sum_{i=1}^n w_i k_i$.
(b) $\Delta(w) = \{k \in \Gamma_+(A) : \langle w, k\rangle = \ell(w)\}$.
(c) *Two vectors $a, a' \in (\mathbb{R}_+^\star)^n$ are equivalent if $\Delta(a) = \Delta(a')$.*

Fix a Newton polyhedron $\Gamma_+(A)$. For each $(n-1)$-dimensional compact face Δ_k of $\Gamma(A)$ we denote by $v^k = (v_1^k, ..., v_n^k)$, the vector in $\mathbb{Z}_+^n - \{0\}$ with minimum length which is associated to Δ_k i.e, $\Delta_k = \Delta(v^k)$. This vector v^k is called primitive integer vector associated to Δ_k.

DEFINITION 2.11. *For any monomial x^γ set* $\mathrm{fil}(\gamma) = \min\limits_k \left\{ \frac{M}{\ell(v^k)} \langle \gamma, v^k\rangle \right\}$.

For any analytic real germ $f(x) = \sum a_\gamma x^\gamma$ define $\mathrm{fil}(f) = \inf\{\mathrm{fil}(\gamma) \mid a_\gamma \neq 0\}$, *where $M = lcm\{\ell(v^k)\}$, $R = \max\limits_k \max\limits_i \left\{ \frac{M}{\ell(v^k)} v_i^k \right\}$ and $r = \min\limits_k \min\limits_i \left\{ \frac{M}{\ell(v^k)} v_i^k \right\}$.*

DEFINITION 2.12. *For a fixed matrix A, a germ f is called A-*homogeneous *of degree d if* $f(x) = \sum_{\nu \in \Gamma(\rho^d)} c_\nu x^\nu$.

Observe that if A is the identity matrix, the definition above recovers the usual definition of a homogeneous germ of degree d.

DEFINITION 2.13. *An analytic map germ $f = (f_1, \dots, f_p) : (\mathbb{R}^n, 0) \to (\mathbb{R}^p, 0)$ is A-homogeneous of degree $d = (d_1, \dots, d_p)$ if each f_i is A-homogeneous of degree d_i.*

For any polynomial map germ $f : (\mathbb{R}^n, 0) \to (\mathbb{R}^p, 0)$ we define, for some fixed Newton filtration, the appropriate control functions

$$N^\star_{\mathcal{R}} f = \sum_I M_I^{2k_I}, \quad N^\star_{\mathcal{C}} f = \sum_{i=1}^p (f_i)^{2\beta_i} \text{ and } N^\star_{\mathcal{K}} f = (N^\star_{\mathcal{R}} f)^a + (N^\star_{\mathcal{C}} f)^b,$$

with $k_I = \frac{k}{s_I}$, $s_I = \text{fil}\,(M_I)$, $k = lcm\,\{s_I\}$ and $a, b \in \mathbb{N}$ being prime numbers with $gcd(a, b) = 1$ such that $\text{fil}\,(N^\star_{\mathcal{R}} f)^a = \text{fil}\,(N^\star_{\mathcal{C}} f)^b$.

Then for any family of polynomial map germs $f_t : (\mathbb{R}^n, 0) \to (\mathbb{R}^p, 0)$, $t \in [0, 1]$, with $f_o(x) = f(x)$, we also define

$$N^\star_{\mathcal{R}} f_t = \sum_I M_{t_I}^{2k_I}, \quad N^\star_{\mathcal{C}} f_t = \sum_{i=1}^p (f_{t_i})^{2\beta_i} \text{ and } N^\star_{\mathcal{K}} f_t = (N^\star_{\mathcal{R}} f_t)^a + (N^\star_{\mathcal{C}} f_t)^b,$$

where a, b, k_I, β_i are as above.

Notice that we are using the notations used in the weighted homogeneous case but this does not cause any confusion.

In the next theorems we give the estimates for the bi-Lipschitz $\mathcal{G}$-triviality, $\mathcal{G} = \mathcal{C}$ or $\mathcal{K}$, in this context.

THEOREM 2.14. *Let $f : (\mathbb{R}^n, 0) \to (\mathbb{R}^p, 0)$ be a polynomial map germ. Suppose that $N^\star_{\mathcal{C}} f$ is A-isolated for some matrix A. If $f_t = f + t\Theta$ is a deformation of f, $\Theta = (\Theta_1, \dots, \Theta_p)$, with $fil(\Theta_i) \geq fil(f_p) + R - r$, then f_t is bi-Lipschitz $\mathcal{C}$-trivial.*

THEOREM 2.15. *Let $f : (\mathbb{R}^n, 0) \to (\mathbb{R}^p, 0)$ be a polynomial map germ. Suppose that $N^\star_{\mathcal{K}} f$ is A-isolated for some matrix A. If $f_t = f + t\Theta$ is a deformation of f, $\Theta = (\Theta_1, \dots, \Theta_p)$, with $fil(\Theta_i) \geq fil(f_p) + R - r$, then f_t is bi-Lipschitz $\mathcal{K}$-trivial.*

The numbers R and r in Theorems 2.14 and 2.15 are given in Definition 2.11.

The proofs of these theorems almost repeat the proofs in the weighted homogeneous case. The main difference is that here we are working with the Newton filtration and the controls are associated to this filtration. Therefore, we omit the new proofs.

REMARK 2.16. In [**19**] there are shown estimates for the C^ℓ-$\mathcal{G}$-triviality, with $\mathcal{G} = \mathcal{C}$ or $\mathcal{K}$ and $0 \leq \ell < \infty$ in the A-isolated case. However these estimates can be improved using the bump function described here. In the last section, we shall include these new estimates for all C^ℓ-$\mathcal{G}$-trivialities, with $\mathcal{G} = \mathcal{C}$ or $\mathcal{K}$ and $0 \leq \ell < \infty$, in tables of all known estimates.

3. Examples

EXAMPLE 3.1. Let $f : (\mathbb{R}^2, 0) \to (\mathbb{R}^2, 0)$; $f(x, y) = (xy, x^{2b+2} - y^{2b} + x^{2b-2}y^2)$, with $b > 2$ and fix the matrix $A = \begin{pmatrix} \frac{1}{b} & 0 & \frac{b-1}{(b+1)b} \\ 0 & \frac{1}{b+1} & \frac{1}{(b+1)b} \end{pmatrix}$.

Here $lcm\{\ell(v_1), \ell(v_2)\} = 2b(b+1)$, $\text{fil}(xy) = 2b+2$, $\text{fil}(y^{2b}) = 2b(b+1)$, $\text{fil}(x^{2b-2}y^2) = 2b(b+1)$ and $\text{fil}(x^{2b+2}) = 2b(b+1)$, then we obtain $R = 2b$ and $r = b$.

The jacobian matrix of f has only one 2×2 minor,

$$M(x, y) = -2((b+1)x^{2b+2} + by^{2b} + bx^{2b-2}y^2)$$

which is A-homogeneous of degree $2b(b+1)$. Notice that the second coordinate function of f is also A-homogeneous of the same degree $2b(b+1)$.

We have in this case,

$$N^{\star}_{\mathcal{R}}f(x, y) = |M(x, y)|^2, \quad N^{\star}_{\mathcal{C}}f(x, y) = xy^{2b(b+1)} + (x^{2b+2} - y^{2b} + x^{2b-2}y^2)^{2b+2}$$

and

$$N^{\star}_{\mathcal{K}}f(x, y) = |M(x, y)|^{2b+2} + xy^{2b(b+1)} + (x^{2b+2} - y^{2b} + x^{2b-2}y^2)^{2b+2},$$

which is A-isolated.

From Theorem 2.15 we obtain that a family $f_t = f + t\Theta$ with $\Theta = (\Theta_1, \Theta_2)$ is bi-Lipschitz $\mathcal{K}$-trivial if $\text{fil}(\Theta_i) \geq 2b(b+1) + b$.

It is interesting to compare this estimate with the C^1-$\mathcal{K}$ and C^0-$\mathcal{K}$ estimates. From Theorem 3.16 of [**19**] this family f_t is C^1-$\mathcal{K}$-trivial if $fil(\Theta_i) \geq 2b(b+1)+2b+1$ and it is C^0-$\mathcal{K}$-trivial if $\text{fil}(\Theta_i) \geq 2b(b+1) + 1$.

EXAMPLE 3.2. Let $f : (\mathbb{R}^2, 0) \to (\mathbb{R}^2, 0)$ be given by $f(x, y) = (xy + x^2y^2, x^{2(c+1)} + xy - y^{2c})$ with $c \geq 3$.

Fix $A = \begin{pmatrix} 2(c+1) & 0 & 1 \\ 0 & 2c & 1 \end{pmatrix}$. Then $lcm\,\{\ell(v^1), \ell(v^2)\} = 2c(c+1)$, $R = 2c^2+c$, $r = c$ and $\text{fil}(f_1, f_2) = (2c(c+1), 2c(c+1))$.

Since $\text{fil}(f_1) = \text{fil}(f_2)$, we have that the origin is A-isolated for $N^{\star}_{\mathcal{C}}f$. Applying Theorem 2.14 we have that the family

$$f_t(x, y) = (xy + x^2y^2 + t\Theta_1(x, y), x^{2(c+1)} + xy - y^{2c} + t\Theta_2(x, y))$$

is bi-Lipschitz $\mathcal{C}$-trivial if $fil(\Theta_i)(x, y) \geq 4c^2 + 2c$.

EXAMPLE 3.3. Let $f : (\mathbb{R}^3, 0) \to (\mathbb{R}^2, 0)$, $f(x, y, z) = (3x^6 + 2y^6 + xz^4, x^6 + y^6 + yz^3)$.

Then $df = \begin{bmatrix} 18x^5 + z^4 & 12y^5 & 4xz^3 \\ 6x^5 & 6y^5 + z^3 & 3yz^2 \end{bmatrix}$ and its minors are $M_{12} = 36x^5y^5 + 18x^5z^3 + 6y^5z^4 + z^7$, $M_{13} = 54x^5yz^2 + 3yz^6 - 24x^6z^3$ and $M_{23} = 36y^6z^2 - 24xy^5z^3 - 4xz^6$.

Here the origin is not A-isolated for $N^{\star}_{\mathcal{C}}f$, for any matrix A. Hence the Theorem 2.14 does not apply to this example. However we obtain an estimate for the $\mathcal{K}$

group since the origin is A-isolated for $N^{\star}_{\mathcal{K}} f$ with to respect to the matrix $A = \begin{pmatrix} 6 & 0 & 0 & 1 & 0 \\ 0 & 6 & 0 & 0 & 1 \\ 0 & 0 & 7 & 4 & 3 \end{pmatrix}$. Then we obtain $R = 288$, $r = 84$ and $\mathrm{fil}(f_1) = \mathrm{fil}(f_2) = 504$.

Therefore we conclude that a family

$$f_t(x,y,z) = (3x^6 + 2y^6 + xz^4 + t\Theta_1(x,y,z), x^6 + y^6 + yz^3 + ty^5z^3t\Theta_2(x,y,z))$$

is bi-Lipschitz $\mathcal{K}$-trivial if $fil(\Theta_i) \geq 504 + 288 - 84 = 708$. For instance, if we consider the monomial y^4z^3, we obtain that the family

$$f_t(x,y,z) = (3x^6 + 2y^6 + xz^4, x^6 + y^6 + yz^3 + ty^4z^3)$$

is bi-Lipschitz $\mathcal{K}$-trivial.

4. Bi-Lipschitz $\mathcal{G}_V$-triviality, $\mathcal{G} = \mathcal{R}$, $\mathcal{C}$ or $\mathcal{K}$

In this section we give estimates for the bi-Lipschitz $\mathcal{G}_V$-triviality, where $\mathcal{G} = \mathcal{R}$, $\mathcal{C}$ or $\mathcal{K}$ of map germs defined on a germ of real analytic variety $(V, 0)$.

However, there is a different approach for each group investigated. If we work with the group $\mathcal{R}_V$, we consider a real analytic germ of variety $(V, 0)$ in $(\mathbb{R}^n, 0)$ defined by $V = \{x \in \mathbb{R}^n : g_1(x) = \cdots = g_s(x) = 0\}$ where each $g_i(x)$ is a germ of real analytic function for $i = 1, \ldots, s$.

Two germs $f_1, f_2 : (\mathbb{R}^n, 0) \to (\mathbb{R}, 0)$ are *bi-Lipschitz $\mathcal{R}_V$-equivalent* if and only if there exists a germ of a bi-Lipschitz homeomorphism $h : (\mathbb{R}^n, 0) \to (\mathbb{R}^n, 0)$ with $h(V) = V$ and $f_2 = f_1 \circ h^{-1}$.

When our interest are the groups $\mathcal{C}_V$ and $\mathcal{K}_V$, we consider $(V, 0)$ as a germ of a real subvariety of $(\mathbb{R}^p, 0)$ defined by a finitely generated ideal I of $\mathcal{E}_p$.

The group $\mathcal{K}_V$ is the subgroup of $\mathcal{K}$ consisting of elements $(h, H) \in \mathcal{K}$ such that $H(\mathbb{R}^n \times V) = \mathbb{R}^n \times V$. It is a geometric subgroup of $\mathcal{K}$ in the sense of [**17**]. In particular, if $V = \{0\}$ then this is just the contact group. Two map germs $f_1, f_2 : (\mathbb{R}^n, 0) \to (\mathbb{R}^p, 0)$ are *bi-Lipschitz $\mathcal{K}_V$-equivalent* iff there exists an element $(h, H) \in \mathcal{K}_V$ such that $(h, H) \cdot f_1 = f_2$, where the action is that of contact equivalence. When $h = Id$, we call the bi-Lipschitz $\mathcal{K}_V$-equivalence by *bi-Lipschitz $\mathcal{C}_V$-equivalence.*

4.1. Bi-Lipschitz $\mathcal{R}_V$-triviality.

Consider a germ of a real analytic variety $(V, 0) \subset (\mathbb{R}^n, 0)$, defined by $V = \{x : g_1(x) = \cdots = g_s(x) = 0\}$ where each $g_i(x)$ is a real analytic function germ, for $i = 1, \ldots, s$.

As usual, we denote by θ_n the set of germs of vector fields in $(\mathbb{R}^n, 0)$ and we denote the submodule of germs of vector fields tangent to V by $\theta_V = \{\psi \in \theta_n : \psi(I(V)) \subset I(V)\}$, where $I(V)$ is the ideal in $\mathcal{E}_n$ consisting of germs of real analytic functions vanishing on V (see [4] for more details).

We denote by θ_V^0 the tangent space of $\mathcal{R}_V \cdot f$. From [**8**] or [**17**]

$$\theta_V^0 = m_n\{\zeta_j\}_{1 \leq j \leq p} + I(V)\mathcal{E}_n \left\{ \frac{\partial}{\partial x_i} \right\},$$

where $\{\zeta_j\}_{1 \leq j \leq p}$ together with $I(V) \cdot \left\{ \frac{\partial}{\partial x_i} \right\}$ generate the $\mathcal{E}_n$-module of vector fields tangent to V.

Let $(V,0)$ be a weighted homogeneous analytic subvariety of $(\mathbb{R}^n,0)$. A set of generators $\{\gamma_1,\ldots,\gamma_m\}$ of θ_V is called weighted homogeneous of type $(r_1,\ldots,r_n; S_1,\ldots,S_m)$ if $\gamma_i = \sum_{j=1}^n \gamma_{ij}\frac{\partial}{\partial x_i}$ where the γ_{ij}, $i=1,\ldots,m$, are weighted homogeneous polynomials of type $(r_1,\ldots,r_n; S_i+r_j)$ whenever $\gamma_{ij}\neq 0$.

When V is a weighted homogeneous variety, we can always choose weighted homogeneous generators for θ_V. We say that a function germ f is *consistent* with a weighted homogeneous variety V if f is weighted homogeneous of the same type as V, or in other words, if V is defined by weighted homogeneous polynomials g_i of type $(r_1,\ldots,r_n;\alpha_i)$ for some α_i. Then f is also weighted homogenous of type $(r_1,\ldots,r_n;d)$ for some d.

Let $f:(\mathbb{R}^n,0)\to(\mathbb{R},0)$ be a weighted homogeneous function germ of type $(r_1,\ldots,r_n;d)$ which is consistent with V. We set

$$N_{\mathcal{R}_V}f = (df(\gamma_j))^2 = \sum_{j=1}^{m}\left(\sum_{i=1}^{n}\frac{\partial f}{\partial x_i}\gamma_{ji}\right)^2.$$

Then we have the following:

THEOREM 4.1. *Under the above conditions, suppose that $N_{\mathcal{R}_V}f$ satisfies a Lojasiewicz condition $N_{\mathcal{R}_V}f(x)\geq c|x|^\alpha$ for some constants c and α. Then deformations of f defined by $f_t(x)=f(x)+t\Theta(x)$ with fil$(\Theta)\geq d+r_n-r_1$ are bi-Lipschitz $\mathcal{R}_V$-trivial for all $t\in[0,1]$.*

The proof of Theorem 4.1 is analogous to the proof of Theorem 3.1. given by H. Liu and D. Zhang in [**14**], where they give estimates for the C^ℓ-$\mathcal{R}_V$-determinacy, with $\ell\geq 2$, for weighted homogeneous function germs on a class of weighted homogeneous real analytic varieties. This proof is done using the same method of constructing of controlled vector fields.

REMARK 4.2. An estimate for the C^0-$\mathcal{R}_V$-triviality was first given by Damon in [**8**], see also [**17**]. We do not know by now any estimate for the C^1-$\mathcal{R}_V$-triviality, however it is also possible to show that the estimate given in Theorem 4.1 coincides with the estimate for the C^1-$\mathcal{R}_V$-triviality. This case is the unique situation where we find equal estimates for the C^1-triviality and the bi-Lipschitz triviality. Our method does not allow us to improve this estimate or show if there are germs which are bi-Lipschitz $\mathcal{R}_V$-trivial and not C^1-$\mathcal{R}_V$-trivial.

4.2. Bi-Lipschitz $\mathcal{G}_V$-triviality, $\mathcal{G}=\mathcal{K}$ or $\mathcal{C}$.

Let $(V,0)$ be a germ of an analytic subvariety of $(\mathbb{R}^p,0)$ defined by a weighted homogeneous polynomial g of type $(w_1,\ldots,w_p;L)$. If θ_V is a free module over $\mathcal{E}_y$, then it has a basis consisting of p weighted homogeneous elements. Let $\zeta_1,\ldots,\zeta_p$ be such basis of type $(w_1,\ldots,w_p;d_1,\ldots,d_p)$, with $d_1\leq d_2\leq\cdots\leq d_p$, for θ_V, where

$$\zeta_i=\sum_{j=1}^{p}\zeta_{ij}(y)\frac{\partial}{\partial y_j}.$$

Let $f:(\mathbb{R}^n,0)\to(\mathbb{R}^p,0)$ be a weighted homogeneous map germ of type $(r_1,\ldots,r_n;w_1,\ldots,w_p)$ with $r_1\leq\cdots\leq r_n$ and $w_1\leq\cdots\leq w_p$.

Here we follow [**14**] and set $N_{\mathcal{C}_V}f=\sum_{i=1}^{p}\left(\sum_{j=1}^{p}(\gamma_{ij}\circ f)^2\right)$, where the ζ_{ij} are weighted homogeneous polynomials of type $(w_1,\ldots,w_p;d_i+w_j)$. Set $N_{\mathcal{K}_V}f = N_{\mathcal{R}}f+N_{\mathcal{C}_V}f$.

THEOREM 4.3. *Under the above conditions, suppose that $N_{\mathcal{K}_V} f$ satisfies a Lojasiewicz condition $N_{\mathcal{K}_V} f(x) \geq c|x|^\alpha$ for some constants c and α. Then deformations of f defined by $f_t(x) = f(x) + t\Theta(x)$, $\Theta = (\Theta_1, \ldots, \Theta_p)$, with fil$(\Theta_i) \geq d_i + w_p + r_n - r_1$ are bi-Lipshitz $\mathcal{K}_V$-trivial for all $t \in [0,1]$.*

The proof of Theorem 4.3 is analogous to the proof of Theorem 4.4. given by H. Liu and D. Zhang in [**14**], where there are given estimates for the C^ℓ-$\mathcal{K}_V$-determinacy, with $\ell \geq 2$. We observe that here we are applying the bump function to obtain this estimate.

It is interesting to remark here that this estimate depends of the weighted degrees of the generators $\zeta_1, \ldots, \zeta_p$ of θ_V, while in the estimate given for the $\mathcal{R}_V$-triviality this does not occurs.

We also have an estimate for the $\mathcal{C}_V$-triviality considering the notation used in Theorem 4.3.

THEOREM 4.4. *Suppose that $N_{\mathcal{C}_V} f$ satisfies a Lojasiewicz condition $N_{\mathcal{C}_V} f(x) \geq c|x|^\alpha$ for some constants c and α. Then deformations $f_t(x) = f(x) + t\Theta(x)$, $\Theta = (\Theta_1, \ldots, \Theta_p)$, with fil$(\Theta_i) \geq d_i + w_p + r_n - r_1$ are bi-Lipschitz $\mathcal{C}_V$-trivial for all $t \in [0,1]$.*

The proof of Theorem 4.4 is analogous to the proof of Theorem 4.3. The difference is that here we consider only a vector field associated to the $\mathcal{C}_V$ group, as in the proof of Theorem 2.6. Here we are also using the bump function to obtain the estimate.

5. Tables of estimates

In the tables below we show all estimates which can be obtained using the method of constructing controlled vector fields in the presence of a suitable Lojasiewicz condition. In all tables, we consider the triviality of a family f_t, where $f_t = f + t\Theta$, $\Theta = (\Theta_1, \ldots, \Theta_p)$, is a deformation of f.

5.1. The weighted homogeneous case.

Let f be a weighted homogeneous map germ of type $(r_1, \ldots, r_n; d_1, \ldots, d_p)$ with $r_1 \leq \cdots \leq r_n$ and $d_1 \leq d_2 \leq \cdots \leq d_p$.

Group $\mathcal{R}$	**Estimate**	**Reference**
C^ℓ-$\mathcal{R}$-trivial, $\ell \geq 2$	$\text{fil}(\Theta_i) \geq d_i - r_1 + \ell r_n + 1$	[**16**]
C^1-$\mathcal{R}$-trivial	$\text{fil}(\Theta_i) \geq d_i - r_1 + r_n + 1$	[**16**]
bi-Lipschitz $\mathcal{R}$-trivial	$\text{fil}(\Theta_i) \geq d_i - r_1 + r_n$	[**10**]
C^0-$\mathcal{R}$-trivial	$\text{fil}(\Theta_i) \geq d_i$	[**16**]

Groups $\mathcal{G} = \mathcal{C}$ or $\mathcal{K}$	**Estimate**	**Reference**
C^ℓ-$\mathcal{G}$-trivial, $\ell \geq 2$	$\text{fil}(\Theta_i) \geq d_p + \ell r_n$	[**16**]
C^1-$\mathcal{G}$-trivial	$\text{fil}(\Theta_i) \geq d_p + r_n$	[**16**]
bi-Lipschitz $\mathcal{G}$-trivial	$\text{fil}(\Theta_i) \geq d_p - r_1 + r_n$	$--$
C^0-$\mathcal{G}$-trivial	$\text{fil}(\Theta_i) \geq d_p$	[**16**]

In this table and the others in the sequel, the references $--$ mean that the results are given in this article.

5.2. The Newton non-degenerate case.

Let $f : (\mathbb{R}^n, 0) \to (\mathbb{R}^p, 0)$ be an A-homogeneous map germ of degree $d = (d_1, \ldots, d_p)$ with $d_1 \leq d_2 \leq \cdots \leq d_p$, $R = \max_k \max_i \left\{ \frac{M}{\ell(v^k)} v_i^k \right\}$, $r = \min_k \min_i \left\{ \frac{M}{\ell(v^k)} v_i^k \right\}$ and $M = lcm\{\ell(v^k)\}$ for a fixed Newton polyhedron $\Gamma_+(A)$.

Group $\mathcal{R}$	Estimate	Reference
C^ℓ-$\mathcal{R}$-trivial, $\ell \geq 2$	$\text{fil}(\Theta_i) \geq d_i - r + \ell R + 1$	[19]
C^1-$\mathcal{R}$-trivial	$\text{fil}(\Theta_i) \geq d_i - r + R + 1$	[19]
bi-Lipschitz $\mathcal{R}$-trivial	$\text{fil}(\Theta_i) \geq d_i - r + R$	[11]
C^0-$\mathcal{R}$-trivial	$\text{fil}(\Theta_i) \geq d_i$	[19]

Groups $\mathcal{G} = \mathcal{C}$ or $\mathcal{K}$	Estimate	Reference
C^ℓ-$\mathcal{G}$-trivial, $\ell \geq 2$	$\text{fil}(\Theta_i) \geq d_p + \ell R$	--
C^1-$\mathcal{G}$-trivial	$\text{fil}(\Theta_i) \geq d_p + R$	--
bi-Lipschitz $\mathcal{G}$ trivial	$\text{fil}(\Theta_i) \geq d_p - r + R$	--
C^0-$\mathcal{G}$-trivial	$\text{fil}(\Theta_i) \geq d_p$	--

In [19] the authors give also estimates for the C^ℓ-$\mathcal{G}$-triviality with $\mathcal{G} = \mathcal{C}$ or $\mathcal{K}$ and $0 \leq \ell < \infty$. However these estimates were obtained without the bump function. Hence those estimates are different from these shown in the table above by adding 1.

5.3. The $\mathcal{G}_V$-triviality.

Let f be a weighted homogeneous function germ of type $(r_1, \ldots, r_n; d)$ and let V be an analytic variety in $\mathbb{R}^n$ defined by weighted homogeneous polynomials g_i of type $(r_1, \ldots, r_n; D_i)$ for some D_i with $r_1 \leq \cdots \leq r_n$.

Group $\mathcal{R}_V$	Estimate	Reference
C^ℓ-$\mathcal{R}_V$-trivial, $\ell \geq 2$	$\text{fil}(\Theta) \geq d - r_1 + \ell r_n$	[14]
C^1-$\mathcal{R}_V$-trivial	$\text{fil}(\Theta) \geq d - r_1 + r_n$	--
bi-Lipschitz $\mathcal{R}_V$-trivial	$\text{fil}(\Theta) \geq d - r_1 + r_n$	--
C^0-$\mathcal{R}_V$-trivial	$\text{fil}(\Theta) \geq d$	[8]

For the groups $\mathcal{C}_V$ and $\mathcal{K}_V$, let V be a germ at 0 of an analytic subvariety of $(\mathbb{R}^p, 0)$ defined by a weighted homogeneous polynomial g of type $(w_1, \ldots, w_p; L)$. Let $f : (\mathbb{R}^n, 0) \to (\mathbb{R}^p, 0)$ be a weighted homogeneous map germ of type $(r_1, \ldots, r_n; w_1, \ldots, w_p)$ with $r_1 \leq \cdots \leq r_n$ and $w_1 \leq \cdots \leq w_p$. In this case, we also suppose that θ_V is a free $\mathcal{E}_p$ module and $\psi_1, \ldots \psi_p$ is a basis for θ_V formed by weighted homogeneous germs of type $(w_1, \ldots, w_p; d_1, \ldots, d_p)$ with $d_1 \leq d_2 \leq \cdots \leq d_p$.

Groups $\mathcal{G}_V = \mathcal{C}_V$ or $\mathcal{K}_V$	Estimate	Reference
C^ℓ-$\mathcal{G}_V$-trivial, $\ell \geq 2$	$\text{fil}(\Theta_i) \geq d_i + w_p + \ell r_n + 1$	[14]
C^1-$\mathcal{G}_V$-trivial	$\text{fil}(\Theta_i) \geq d_i + w_p + r_n + 1$	--
bi-Lipschitz $\mathcal{G}_V$-trivial	$\text{fil}(\Theta_i) \geq d_i + w_p + r_n - r_1$	--
C^0-$\mathcal{G}_V$-trivial	$\text{fil}(\Theta_i) \geq d_i + w_p$	--

Acknowledgements. The authors would like to thank the referee for her/his helpful comments. This paper was revised during the posdoctoral stage of the first named author at Universitat de València, Spain.

References

[1] O.M. Abderrahmane, *Polyèdre de Newton et Trivialité en Famille*, J. Math. Soc. Japan, 54, 3 (2002) 513–550. MR1900955 (2003d:58064)
[2] L. Birbrair, J.C.F. Costa, A. Fernandes, M.A.S. Ruas, *$\mathcal{K}$-bi-Lipschitz equivalence of real function-germs*, Proc. Amer. Math. Soc. 135 (2007) 1089–1095. MR2262910 (2007m:58048)
[3] S. Bromberg, L. Medrano, *C^r-sufficiency of quasihomogeneous functions*, Real and Complex Singularities, Pitman Res. Notes in Math. Series 333 (1995) 179–189.
[4] J.W. Bruce, M. Roberts, *Critical points of functions on analytic varities*, Topology 27 No.1 (1988) 57–90. MR935528 (89c:32019)
[5] M.A. Buchner, W. Kucharz, *Topological triviality of family of zero-sets*, Proc. Amer. Math. Soc. 102, 3 (1988) 699–705. MR929006 (89b:58029)
[6] J.C.F. Costa, M.J. Saia, C.H. Soares Jr. *Bi-Lipschitz $\mathcal{A}$-triviality of map germs and Newton filtrations*, Topology and its Applications 159 (2012) 430–436.
[7] J. Damon, *On the freeness of equisingular deformations of plane curve singularities*, Topology and its applications 118 (2002) 31–43. MR1877714 (2003a:32046)
[8] J. Damon, *Topological triviality and versality for subgroups of $\mathcal{A}$ and $\mathcal{K}$*, Nonlinearity 5 (1992) 373–412. MR1158379 (93f:58020)
[9] J. Damon, *Topological invariants of μ-constant deformations of complete intersection singularities*, Quart. J. Math. Oxford 2, 40 (1989) 139–159. MR997645 (90j:32012)
[10] A. Fernandes, M.A.S. Ruas, *Bi-Lipschitz determinacy of quasi-homogeneous germs*, Glasgow Math. J. 46, 1 (2004) 77–82. MR2034833 (2005f:58076)
[11] A. Fernandes, C.H. Soares Jr, *On the bi-Lipschitz triviality of families of real maps*, Real and Complex Singularities, Contemp. Math. 354, A.M.S., Providence, RI (2004) 95–103. MR2087806 (2005g:58077)
[12] J.-P. Henry, A. Parusinski, *Existence of moduli for bi-Lipschitz equivalence of analytic function-germs*, Compositio Math. 136 (2003), no. 2, 217-235. MR1967391 (2004d:32037)
[13] T.C. Kuo, *On C^0-sufficience of jets of potencial functions*, Topology 8 (1969) 167–171. MR0238338 (38:6614)
[14] H. Liu, D. Zhang, *C^ℓ-$\mathcal{G}_V$-determinacy of weighted homogeneous function germs on weighted homogeneous analytic varieties*, Hokkaido Maht. J. 37 (2008) 309–329. MR2415903 (2009c:58052)
[15] M.A.S. Ruas, *On the degree of C^ℓ-determinacy*, Math. Scand. 59 (1986) 59–70. MR873489 (88b:58019)
[16] M.A.S. Ruas, M.J. Saia, *C^ℓ-determinacy of weighted homogeneous germs*, Hokkaido Math. J. 26 (1997) 89–99. MR1432539 (97m:58019)
[17] M.A.S. Ruas, J.N. Tomazella, *Topological triviality of families of functions on analytic varieties*, Nagoya Math. J. 175 (2004) 39–50. MR2085309 (2005h:32068)
[18] M.A.S. Ruas, G. Valette, *C^0 and bi-Lipschitz $\mathcal{K}$-equivalence of mappings*, Math. Z. 269, no. 1–2 (2011) 293–308.
[19] M.J. Saia, C.H. Soares Jr, *C^l-$\mathcal{G}$-triviality of map germs and Newton polyhedra, $\mathcal{G} = \mathcal{R}, \mathcal{C}$ or $\mathcal{K}$*, Hokkaido Math. J. 27, 2 (2008) 331–348.

Departamento de Matemática, IBILCE-UNESP, São José do Rio Preto-SP, Brazil
E-mail address: `jcosta@ibilce.unesp.br`

Departamento de Matemática, ICMC-USP, São Carlos, SP, Brazil
E-mail address: `mjsaia@icmc.usp.br`

Departamento de Matemática, CCN-UFPI, Teresina, Piaui, Brazil
E-mail address: `humberto@ufpi.edu.br`

Contemporary Mathematics
Volume **569**, 2012
http://dx.doi.org/10.1090/conm/569/11243

Symplectic S_μ Singularities

Wojciech Domitrz and Żaneta Trębska

ABSTRACT. We study the local symplectic algebra of the 1-dimensional isolated complete intersection singularity of type S_μ. We use the method of algebraic restrictions to classify symplectic S_μ singularities. We distinguish these symplectic singularities by discrete symplectic invariants. We also give their geometric description.

1. Introduction

In this paper we study the symplectic classification of the 1-dimensional complete intersection singularity of type S_μ in the symplectic space $(\mathbb{R}^{2n}, \omega)$. We recall that ω is a symplectic form if ω is a smooth nondegenerate closed 2-form, and $\Phi : \mathbb{R}^{2n} \to \mathbb{R}^{2n}$ is a symplectomorphism if Φ is a diffeomorphism and $\Phi^*\omega = \omega$.

DEFINITION 1.1. Let N_1, N_2 be germs of subsets of symplectic space $(\mathbb{R}^{2n}, \omega)$. N_1, N_2 are **symplectically equivalent** if there exists a symplectomorphism-germ $\Phi : (\mathbb{R}^{2n}, \omega) \to (\mathbb{R}^{2n}, \omega)$ such that $\Phi(N_1) = N_2$.

The problem of symplectic classification of singular curves was introduced by V. I. Arnold in [**A1**]. Arnold proved that the A_{2k} singularity of a planar curve (the orbit with respect to the standard $\mathcal{A}$-equivalence of parameterized curves) split into exactly $2k+1$ symplectic singularities (orbits with respect to the symplectic equivalence of parameterized curves). He distinguished different symplectic singularities by different orders of tangency of the parameterized curve to the *nearest* smooth Lagrangian submanifold. Arnold posed a problem of expressing these new symplectic invariants in terms of the local algebra's interaction with the symplectic structure and he proposed to call this interaction the **local symplectic algebra**.

In [**IJ1**] G. Ishikawa and S. Janeczko classified symplectic singularities of curves in the 2-dimensional symplectic space. All simple curves in this classification are quasi-homogeneous. A symplectic form on a 2-dimensional manifold is a special case of a volume form on a smooth manifold. The generalization of results in [**IJ1**] to volume-preserving classification of singular varieties and maps in arbitrary dimensions was obtained in [**DR**]. The orbit of the action of all diffeomorphism-germs agrees with the volume-preserving orbit or splits into two volume-preserving

2010 *Mathematics Subject Classification.* Primary 53D05. Secondary 14H20, 58K50, 58A10.

Key words and phrases. Symplectic manifold, curves, local symplectic algebra, algebraic restrictions, relative Darboux theorem, singularities.

The work of W. Domitrz was supported by Polish MNiSW grant no. N N201 397237.

orbits (in the case $\mathbb{K} = \mathbb{R}$) for germs which satisfy a special weak form of quasi-homogeneity e.g. the weak quasi-homogeneity of varieties is a quasi-homogeneity with non-negative weights $\lambda_i \geq 0$ and $\sum_i \lambda_i > 0$.

A symplectic singularity is stably simple if it is simple and remains simple if the ambient symplectic space is symplectically embedded (i.e. as a symplectic submanifold) into a larger symplectic space. In [**K**] P. A. Kolgushkin classified stably simple symplectic singularities of parameterized curves (in the $\mathbb{C}$-analytic category). All stably simple symplectic singularities of curves are quasi-homogeneous too.

In [**DJZ2**] new symplectic invariants of singular quasi-homogeneous subsets of a symplectic space were explained by the algebraic restrictions of the symplectic form to these subsets. The algebraic restriction is an equivalence class of the following relation on the space of differential k-forms:

Differential k-forms ω_1 and ω_2 have the same **algebraic restriction** to a subset N if $\omega_1 - \omega_2 = \alpha + d\beta$, where α is a k-form vanishing on N and β is a $(k-1)$-form vanishing on N.

In [**DJZ2**] a generalization of the Darboux-Givental theorem ([**AG**]) to germs of arbitrary subsets of the symplectic space was obtained. This result reduces the problem of symplectic classification of germs of quasi-homogeneous subsets to the problem of classification of algebraic restrictions of symplectic forms to these subsets. For non-quasi-homogeneous subsets there is one more cohomological invariant apart of the algebraic restriction ([**DJZ2**], [**DJZ1**]). The dimension of the space of algebraic restrictions of closed 2-forms to a 1-dimensional quasi-homogeneous isolated complete intersection singularity C is equal to the multiplicity of C ([**DJZ2**]). In [**D**] it was proved that the space of algebraic restrictions of closed 2-forms to a 1-dimensional (singular) analytic variety is finite-dimensional.

In [**DJZ2**] the method of algebraic restrictions was applied to various classification problems in a symplectic space. In particular a complete symplectic classification of the 1-dimensional S_5 singularity was obtained. Most of different symplectic singularity classes were distinguished by new discrete symplectic invariants: the index of isotropy and the symplectic multiplicity.

In this paper we obtain the complete symplectic classification of the isolated complete intersection singularities S_μ for $\mu > 5$ using the method of algebraic restrictions (Theorem 4.1). The $S_\mu, \mu \geq 5$ are the first singularities appearing in the classification of simple 1-dimensional isolated complete intersection singularities in the space of dimension greater than 2 obtained by Giusti ([**G**], [**AVG**]). Isolated complete intersection singularities were intensively studied by many authors (e. g. see [**L**]), because of their interesting geometric, topological and algebraic properties. In this paper we study their symplectic invariants. The group of symplectomorphism-germs is not a geometric subgroup in the sense of Damon. Therefore symplectic classification problems are interesting and require new methods. We calculate discrete symplectic invariants for symplectic S_μ singularities (Theorems 4.6 and 4.4) and we present their geometric descriptions (Theorem 4.9).

In [**DT**] following ideas from [**A1**] and [**D**] new discrete symplectic invariants - the Lagrangian tangency orders were introduced and used to distinguish symplectic singularities of simple planar curves of type $A - D - E$, symplectic S_5 and T_7 singularities.

In this paper using Lagrangian tangency orders we are able to give detailed classification of the S_μ singularity for $\mu > 5$ (Theorem 4.6) and to present a geometric description of its symplectic orbits (Theorem 4.9).

The paper is organized as follows. In Section 2 we recall the method of algebraic restrictions. In Section 3 we present discrete symplectic invariants. Symplectic classification of the S_μ singularity is studied in Section 4.

2. The method of algebraic restrictions

In this section we present basic facts on the method of algebraic restrictions. The proofs of all results of this section can be found in [**DJZ2**].

Given a germ of a non-singular manifold M denote by $\Lambda^p(M)$ the space of all germs at 0 of differential p-forms on M. Given a subset $N \subset M$ introduce the following subspaces of $\Lambda^p(M)$:

$$\Lambda^p_N(M) = \{\omega \in \Lambda^p(M) : \quad \omega(x) = 0 \text{ for any } x \in N\};$$
$$\mathcal{A}^p_0(N, M) = \{\alpha + d\beta : \quad \alpha \in \Lambda^p_N(M), \ \beta \in \Lambda^{p-1}_N(M).\}$$

The relation $\omega(x) = 0$ means that the p-form ω annihilates any p-tuple of vectors in T_xM, i.e. all coefficients of ω in some (and then any) local coordinate system vanish at the point x.

Definition 2.1. Let N be the germ of a subset of M and let $\omega \in \Lambda^p(M)$. The **algebraic restriction** of ω to N is the equivalence class of ω in $\Lambda^p(M)$, where the equivalence is as follows: ω is equivalent to $\widetilde{\omega}$ if $\omega - \widetilde{\omega} \in \mathcal{A}^p_0(N, M)$.

Notation. The algebraic restriction of the germ of a p-form ω on M to the germ of a subset $N \subset M$ will be denoted by $[\omega]_N$. Writing $[\omega]_N = 0$ (or saying that ω has zero algebraic restriction to N) we mean that $[\omega]_N = [0]_N$, i.e. $\omega \in \mathcal{A}^p_0(N, M)$.

Let M and $\widetilde{M}$ be non-singular equal-dimensional manifolds and let $\Phi : \widetilde{M} \to M$ be a local diffeomorphism. Let N be a subset of M. It is clear that $\Phi^*\mathcal{A}^p_0(N, M) = \mathcal{A}^p_0(\Phi^{-1}(N), \widetilde{M})$. Therefore the action of the group of diffeomorphisms can be defined as follows: $\Phi^*([\omega]_N) = [\Phi^*\omega]_{\Phi^{-1}(N)}$, where ω is an arbitrary p-form on M.

Definition 2.2. Two algebraic restrictions $[\omega]_N$ and $[\widetilde{\omega}]_{\widetilde{N}}$ are called **diffeomorphic** if there exists the germ of a diffeomorphism $\Phi : \widetilde{M} \to M$ such that $\Phi(\widetilde{N}) = N$ and $\Phi^*([\omega]_N) = [\widetilde{\omega}]_{\widetilde{N}}$.

Remark 2.3. The above definition does not depend on the choice of ω and $\widetilde{\omega}$ since a local diffeomorphism maps forms with zero algebraic restriction to N to forms with zero algebraic restrictions to $\widetilde{N}$. If $M = \widetilde{M}$ and $N = \widetilde{N}$ then the definition of diffeomorphic algebraic restrictions reduces to the following one: two algebraic restrictions $[\omega]_N$ and $[\widetilde{\omega}]_N$ are diffeomorphic if there exists a local symmetry Φ of N (i.e. a local diffeomorphism preserving N) such that $[\Phi^*\omega]_N = [\widetilde{\omega}]_N$.

Definition 2.4. A subset N of $\mathbb{R}^m$ is quasi-homogeneous if there exists a coordinate system $(x_1, \cdots, x_m)$ on $\mathbb{R}^m$ and positive numbers $\lambda_1, \cdots, \lambda_n$ such that for any point $(y_1, \cdots, y_m) \in \mathbb{R}^m$ and any $t > 0$ if $(y_1, \cdots, y_m)$ belongs to N then the point $(t^{\lambda_1}y_1, \cdots, t^{\lambda_m}y_m)$ belongs to N.

The method of algebraic restrictions applied to singular quasi-homogeneous subsets is based on the following theorem.

THEOREM 2.5 (Theorem A in [**DJZ2**]). *Let N be the germ of a quasi-homogeneous subset of $\mathbb{R}^{2n}$. Let ω_0, ω_1 be germs of symplectic forms on $\mathbb{R}^{2n}$ with the same algebraic restriction to N. There exists a local diffeomorphism Φ such that $\Phi(x) = x$ for any $x \in N$ and $\Phi^*\omega_1 = \omega_0$.*

Two germs of quasi-homogeneous subsets N_1, N_2 of a fixed symplectic space $(\mathbb{R}^{2n}, \omega)$ are symplectomorphic if and only if the algebraic restrictions of the symplectic form ω to N_1 and N_2 are diffeomorphic.

Theorem 2.5 reduces the problem of symplectic classification of germs of singular quasi-homogeneous subsets to the problem of diffeomorphic classification of the algebraic restrictions of the germ of the symplectic form to the germs of singular quasi-homogeneous subsets.

The geometric meaning of the zero algebraic restriction is explained by the following theorem.

THEOREM 2.6 (Theorem **B** in [**DJZ2**]). *The germ of a quasi-homogeneous set N of a symplectic space $(\mathbb{R}^{2n}, \omega)$ is contained in a non-singular Lagrangian submanifold if and only if the symplectic form ω has zero algebraic restriction to N.*

PROPOSITION 2.7 (Lemma 2.20 in [**DJZ2**]). *Let $N \subset \mathbb{R}^m$. Let $W \subseteq T_0\mathbb{R}^m$ be the tangent space to some (and then any) non-singular submanifold containing N of minimal dimension within such submanifolds. If ω is the germ of a p-form with the zero algebraic restriction to N then $\omega|_W = 0$.*

The following result shows that the method of algebraic restrictions is a very powerful tool in symplectic classification of singular curves.

THEOREM 2.8 (Theorem 2 in [**D**]). *Let C be the germ of a $\mathbb{K}$-analytic curve (for $\mathbb{K} = \mathbb{R}$ or $\mathbb{K} = \mathbb{C}$). Then the space of algebraic restrictions of germs of closed 2-forms to C is a finite dimensional vector space.*

By a **$\mathbb{K}$-analytic curve** we understand a subset of $\mathbb{K}^m$ which is locally diffeomorphic to a 1-dimensional (possibly singular) $\mathbb{K}$-analytic subvariety of $\mathbb{K}^m$. Germs of $\mathbb{C}$-analytic parameterized curves can be identified with germs of irreducible $\mathbb{C}$-analytic curves.

We now recall basic properties of algebraic restrictions which are useful for a description of this subset ([**DJZ2**]).

First we can reduce the dimension of the manifold we consider due to the following propositions.

If the germ of a set $N \subset \mathbb{R}^m$ is contained in a non-singular submanifold $M \subset \mathbb{R}^m$ then the classification of the algebraic restrictions to N of p-forms on $\mathbb{R}^m$ reduces to the classification of the algebraic restrictions to N of p-forms on M. At first note that the algebraic restrictions $[\omega]_N$ and $[\omega|_{TM}]_N$ can be identified:

PROPOSITION 2.9. *Let N be the germ at 0 of a subset of $\mathbb{R}^m$ contained in a non-singular submanifold $M \subset \mathbb{R}^m$ and let ω_1, ω_2 be p-forms on $\mathbb{R}^m$. Then $[\omega_1]_N = [\omega_2]_N$ if and only if $[\omega_1|_{TM}]_N = [\omega_2|_{TM}]_N$.*

The following, less obvious statement, means that the *orbits* of the algebraic restrictions $[\omega]_N$ and $[\omega|_{TM}]_N$ also can be identified.

PROPOSITION 2.10. *Let N_1, N_2 be germs of subsets of $\mathbb{R}^m$ contained in equal-dimensional non-singular submanifolds M_1, M_2 respectively. Let ω_1, ω_2 be two germs of p-forms. The algebraic restrictions $[\omega_1]_{N_1}$ and $[\omega_2]_{N_2}$ are diffeomorphic if and only if the algebraic restrictions $\left[\omega_1|_{TM_1}\right]_{N_1}$ and $\left[\omega_2|_{TM_2}\right]_{N_2}$ are diffeomorphic.*

To calculate the space of algebraic restrictions of 2-forms we will use the following obvious properties.

PROPOSITION 2.11. *If $\omega \in \mathcal{A}_0^k(N, \mathbb{R}^{2n})$ then $d\omega \in \mathcal{A}_0^{k+1}(N, \mathbb{R}^{2n})$ and $\omega \wedge \alpha \in \mathcal{A}_0^{k+p}(N, \mathbb{R}^{2n})$ for any p-form α on $\mathbb{R}^{2n}$.*

The next step of our calculation is the description of the subspace of algebraic restrictions of closed 2-forms. The following proposition is very useful for this step.

PROPOSITION 2.12. *Let $a_1, \ldots, a_k$ be a basis of the space of algebraic restrictions of 2-forms to N satisfying the following conditions*

(1) $da_1 = \cdots = da_j = 0$,
(2) *the algebraic restrictions $da_{j+1}, \ldots, da_k$ are linearly independent.*

Then $a_1, \ldots, a_j$ is a basis of the space of algebraic restrictions of closed 2-forms to N.

Then we need to determine which algebraic restrictions of closed 2-forms are realizable by symplectic forms. This is possible due to the following fact.

PROPOSITION 2.13. *Let $N \subset \mathbb{R}^{2n}$. Let r be the minimal dimension of non-singular submanifolds of $\mathbb{R}^{2n}$ containing N. Let M be one of such r-dimensional submanifolds. The algebraic restriction $[\theta]_N$ of the germ of a closed 2-form θ is realizable by the germ of a symplectic form on $\mathbb{R}^{2n}$ if and only if $rank(\theta|_{T_0M}) \geq 2r - 2n$.*

Let us fix the following notations:

- $\left[\Lambda^2(\mathbb{R}^{2n})\right]_N$: the vector space consisting of the algebraic restrictions of germs of all 2-forms on $\mathbb{R}^{2n}$ to the germ of a subset $N \subset \mathbb{R}^{2n}$;
- $\left[Z^2(\mathbb{R}^{2n})\right]_N$: the subspace of $\left[\Lambda^2(\mathbb{R}^{2n})\right]_N$ consisting of the algebraic restrictions of germs of all closed 2-forms on $\mathbb{R}^{2n}$ to N;
- $\left[\mathrm{Symp}(\mathbb{R}^{2n})\right]_N$: the open set in $\left[Z^2(\mathbb{R}^{2n})\right]_N$ consisting of the algebraic restrictions of germs of all symplectic 2-forms on $\mathbb{R}^{2n}$ to N.

3. Discrete symplectic invariants.

We can use some discrete symplectic invariants to characterize symplectic singularity classes. They show how far a curve N is from the closest non-singular Lagrangian submanifold.

The first invariant is a symplectic multiplicity ([**DJZ2**]) introduced in [**IJ1**] as a symplectic defect of a curve.

Let N be the germ of a subvariety of $(\mathbb{R}^{2n}, \omega)$.

DEFINITION 3.1. The **symplectic multiplicity** $\mu_{sympl}(N)$ of N is the codimension of the symplectic orbit of N in the orbit of N with respect to the action of the group of diffeomorphism-germs.

To make the definition of the symplectic multiplicity precise we present some explanations (see [**DJZ2**] for details). Throughout the paper *by a variety in* $\mathbb{R}^{2n}$ *we mean the zero set of a* k*-generated ideal having the property of zeros,* $k \geq 1$. Denote by $\mathrm{Var}(k, 2n)$ the space of all germs at 0 of varieties described by k-generated ideals. We associate with the germ $N \in \mathrm{Var}(k, 2n)$ the map-germ $H : (\mathbb{R}^{2n}, 0) \to (\mathbb{R}^k, 0)$ whose k components are generators of the ideal of function-germs vanishing on N. We denote by (N) the orbit of N with respect to the group of diffeomorphism-germs. Then the orbit (N) can be identified with the V-orbit of H (see [**AVG**]). Recall from [**AVG**] that the V-equivalence of two map germs $H, \tilde{H} : (\mathbb{R}^{2n}, 0) \to (\mathbb{R}^k, 0)$ means the existence of a diffeomorphism-germ Φ and a germ M of a map from $\mathbb{R}^{2n}$ to the manifold of non-singular $k \times k$ matrices such that $\tilde{H} = M \cdot H(\Phi)$. The symplectic V-equivalence is defined in the same way as the V-equivalence; the only difference is that we require that Φ is a symplectomorphism-germ. The symplectic orbit of N can be identified with the symplectic V-orbit of H.

The codimension of the symplectic orbit of N in (N) is the codimension of the symplectic V-orbit of H in the V-orbit of H.

The second invariant is the index of isotropy [**DJZ2**].

DEFINITION 3.2. The **index of isotropy** $\iota(N)$ of N is the maximal order of vanishing of the 2-forms $\omega|_{TM}$ over all smooth submanifolds M containing N.

They can be described in terms of algebraic restrictions.

PROPOSITION 3.3 ([**DJZ2**]). *The symplectic multiplicity of the germ of a quasi-homogeneous variety* N *in a symplectic space is equal to the codimension of the orbit of the algebraic restriction* $[\omega]_N$ *with respect to the group of local diffeomorphisms preserving* N *in the space of the algebraic restrictions of closed* 2*-forms to* N.

PROPOSITION 3.4 ([**DJZ2**]). *The index of isotropy of the germ of a quasi-homogeneous variety* N *in a symplectic space* $(\mathbb{R}^{2n}, \omega)$ *is equal to the maximal order of vanishing of closed* 2*-forms representing the algebraic restriction* $[\omega]_N$.

One more discrete symplectic invariant was introduced in [**D**] following ideas from [**A1**]. It is defined specifically for a parameterized curve. This is the maximal tangency order of a curve $f : \mathbb{R} \to M$ with a smooth Lagrangian submanifold. If $H_1 = ... = H_n = 0$ define a smooth submanifold L in the symplectic space then the tangency order of a curve $f : \mathbb{R} \to M$ to L is the minimum of the orders of vanishing at 0 of functions $H_1 \circ f, \cdots, H_n \circ f$. We denote the tangency order of f with L by $t(f, L)$.

DEFINITION 3.5. The **Lagrangian tangency order** $Lt(f)$ **of a curve** f is the maximum of $t(f, L)$ over all smooth Lagrangian submanifolds L of the symplectic space.

The Lagrangian tangency order of a quasi-homogeneous curve in a symplectic space can also be expressed in terms of the algebraic restrictions.

PROPOSITION 3.6 ([**D**]). *Let* f *be the germ of a quasi-homogeneous curve such that the algebraic restriction of a symplectic form to it can be represented by a closed* 2*-form vanishing at* 0*. Then the Lagrangian tangency order of the germ of a quasi-homogeneous curve* f *is the maximum of the order of vanishing on* f *over all* 1*-forms* α *such that* $[\omega]_f = [d\alpha]_f$.

In [**DT**] the above invariant was generalized for germs of curves and multi-germs of curves which may be parameterized analytically since the Lagrangian tangency order is the same for every 'good' analytic parameterization of a curve.

Consider a multi-germ $(f_i)_{i\in\{1,\cdots,r\}}$ of analytically parameterized curves f_i. We have r-tuples $(t(f_1, L), \cdots, t(f_r, L))$ for any smooth submanifold L in the symplectic space.

DEFINITION 3.7. For any $I \subseteq \{1, \cdots, r\}$ we define **the tangency order of the multi-germ** $(f_i)_{i\in I}$ to L:

$$t[(f_i)_{i\in\, I}, L] = \min_{i\in\, I} t(f_i, L).$$

DEFINITION 3.8. The **Lagrangian tangency order** $Lt((f_i)_{i\in\, I})$ **of a multi-germ** $(f_i)_{i\in I}$ is the maximum of $t[(f_i)_{i\in\, I}, L]$ over all smooth Lagrangian submanifolds L of the symplectic space.

For multi-germs one can also define relative invariants according to selected branches or collections of branches [**DT**].

DEFINITION 3.9. For fixed $j \in I$ the **Lagrangian tangency order related to** f_j of a multi-germ $(f_i)_{i\in I}$ denoted by $Lt[(f_i)_{i\in I} : f_j]$ is the maximum of $t[(f_i)_{i\in I\setminus\{j\}}, L]$ over all smooth Lagrangian submanifolds L of the symplectic space for which $t(f_j, L) = Lt(f_j)$.

These invariants have geometric interpretation. If a branch f_i is contained in a smooth Lagrangian submanifold then $Lt(f_i) = \infty$. If all curves f_i for $i \in I$ are contained in the same non-singular Lagrangian submanifold then $Lt((f_i)_{i\in\, I}) = \infty$. (In the analytic category "if" can be replaced by "if and only if").

We may use these invariants for distinguishing symplectic singularities.

4. Symplectic S_μ-singularities

Denote by (S_μ) (for $\mu > 5$) the class of varieties in a fixed symplectic space $(\mathbb{R}^{2n}, \omega)$ which are diffeomorphic to

$$S_\mu = \{x \in \mathbb{R}^{2n\geq 4} : x_1^2 - x_2^2 - x_3^{\mu-3} = x_2x_3 = x_{\geq 4} = 0\}. \tag{4.1}$$

The S_μ, $\mu \geq 5$ are simple 1-dimensional isolated complete intersection singularities in the space of dimension greater than 2 ([**G**]). Let $N \in (S_\mu)$. Then N is the union of two 1-dimensional components invariant under the action of local diffeomorphisms preserving N: C_1 – diffeomorphic to the A_1 singularity and C_2 – diffeomorphic to the $A_{\mu-4}$ singularity. Here N is quasi-homogeneous with weights $w(x_1) = w(x_2) = \mu - 3$, $w(x_3) = 2$ when μ is an even number, or $w(x_1) = w(x_2) = (\mu - 3)/2$, $w(x_3) = 1$ when μ is an odd number. In our paper we often use the notation $r = \mu - 3$.

We will use the method of algebraic restrictions to obtain a complete classification of symplectic singularities in (S_μ) presented in the following theorem.

THEOREM 4.1. *Any submanifold of the symplectic space* $(\mathbb{R}^{2n}, \sum_{i=1}^n dp_i \wedge dq_i)$ *which is diffeomorphic to* S_μ *is symplectically equivalent to one and only one of the normal forms* $S_\mu^{i,j}$, $S_{\mu\ l}^{i,j}$ *listed below. The parameters* c_i *of the normal forms are moduli.*

$S_\mu^{0,0}$: $p_1^2 - p_2^2 - q_1^r = 0, \quad p_2q_1 = 0, \quad q_2 = c_1q_1 - c_2p_1, \quad p_{\geq 3} = q_{\geq 3} = 0;$

$S^{k,0}_{\mu\ 2}(1 \le k \le \mu-5): p_2^2 - p_1^2 - q_1^r = 0,\ \ p_1q_1 = 0,\ q_2 = c_3p_1 + \frac{c_{4+k}}{k+1}q_1^{k+1},$
$p_{\ge 3} = q_{\ge 3} = 0,\ c_{4+k} \ne 0;$
$S^{\mu-4,0}_{\mu\ 2}: p_2^2 - p_1^2 - q_1^r = 0,\ \ p_1q_1 = 0,\ q_2 = c_3p_1 + \frac{c_\mu}{r}q_1^r,\ p_{\ge 3} = q_{\ge 3} = 0,\ c_3c_\mu = 0;$
$S^{1+k,0}_{\mu\ r}(1 \le k \le \mu-6): p_1^2 - q_1^2 - q_2^r = 0,\ \ q_1q_2 = 0,\ p_2 = p_1q_2^k(c_{4+k} + c_{5+k}q_2),$
$p_{\ge 3} = q_{\ge 3} = 0,\ \ \ c_{4+k} \ne 0;$
$S^{\mu-4,0}_{\mu\ r}: p_1^2 - q_1^2 - q_2^r = 0,\ \ q_1q_2 = 0,\ p_2 = c_{\mu-1}p_1q_2^{r-2},\ p_{\ge 3} = q_{\ge 3} = 0;$
$S^{3,1}_\mu: p_1^2 - p_2^2 - p_3^r = 0,\ p_2p_3 = 0,\ q_1 = \frac{1}{2}p_3^2,\ q_2 = -c_4p_1p_3,\ p_{\ge 4} = q_{\ge 3} = 0;$
$S^{2+k,1}_\mu(2 \le k \le \mu-4): p_1^2 - p_2^2 - p_3^r = 0,\ p_2p_3 = 0,\ q_1 = \frac{c_{4+k}}{k+1}p_3^{k+1},\ \ q_2 = -p_1p_3,$
$p_{\ge 4} = q_{\ge 3} = 0,\ \ (c_{4+k} \ne 0 \textit{ for } 2 \le k \le \mu-5);$
$S^{3+k,k}_\mu(2 \le k \le \mu-4): p_1^2 - p_2^2 - p_3^r = 0,\ p_2p_3 = 0,\ q_1 = \frac{1}{k+1}p_3^{k+1},\ p_{\ge 4} = q_{\ge 2} = 0;$
$S^{\mu,\infty}_\mu: p_1^2 - p_2^2 - p_3^r = 0,\ p_2p_3 = 0,\ \ p_{\ge 4} = q_{\ge 1} = 0.$
(Here we wrote r for $\mu-3$).

In Section 4.1 we calculate the manifolds $[\mathrm{Symp}(\mathbb{R}^{2n})]_{S_\mu}$ and classify their algebraic restrictions. This allows us to decompose S_μ into symplectic singularity classes. In Section 4.2 we transfer the normal forms for the algebraic restrictions to the symplectic normal forms to obtain a proof of Theorem 4.1. In Section 4.3 we use the Lagrangian tangency orders to distinguish more symplectic singularity classes. In Section 4.4 we propose a geometric description of these singularities which confirms this more detailed classification. Some of the proofs are presented in Section 4.5.

4.1. Algebraic restrictions and their classification. One has the relations for the S_μ-singularities

$$[d(x_2x_3)]_{S_\mu} = [x_2dx_3 + x_3dx_2]_{S_\mu} = 0, \tag{4.2}$$

$$[d(x_1^2 - x_2^2 - x_3^{\mu-3})]_{S_\mu} = [2x_1dx_1 - 2x_2dx_2 - (\mu-3)x_3^{\mu-4}dx_3]_{S_\mu} = 0. \tag{4.3}$$

Multiplying these relations by suitable 1-forms we obtain the relations in Table 1.

TABLE 1. Relations towards calculating $[\Lambda^2(\mathbb{R}^{2n})]_N$ for $N = S_\mu$

	relations	proof
1.	$[x_2dx_2 \wedge dx_3]_N = 0$	(4.2)$\wedge\, dx_2$
2.	$[x_3dx_2 \wedge dx_3]_N = 0$	(4.2)$\wedge\, dx_3$
3.	$[x_1dx_1 \wedge dx_2]_N = 0$	(4.3)$\wedge\, dx_2$ and row 2.
4.	$[x_1dx_1 \wedge dx_3]_N = 0$	(4.3)$\wedge\, dx_3$ and row 1.
5.	$[x_3dx_1 \wedge dx_2]_N = [x_2dx_3 \wedge dx_1]_N$	(4.2)$\wedge\, dx_1$
6.	$[2x_2dx_1 \wedge dx_2]_N = (\mu-3)[x_3^{\mu-4}dx_3 \wedge dx_1]_N$	(4.3)$\wedge\, dx_1$
7.	$[x_1^2dx_2 \wedge dx_3]_N = 0$	rows 1. and 2. and $[x_1^2]_N = [x_2^2 + x_3^{\mu-3}]_N$
8.	$[x_3^2dx_1 \wedge dx_2]_N = 0$	(4.2)$\wedge\, x_3dx_1$ and $[x_2x_3]_N = 0$
9.	$[x_2^2dx_1 \wedge dx_2]_N = 0$	(4.3)$\wedge\, x_2dx_1$ and $[x_2x_3]_N = 0$

Table 1 and Proposition 2.11 easily imply the following proposition:

PROPOSITION 4.2. *The space* $[\Lambda^2(\mathbb{R}^{2n})]_{S_\mu}$ *is a* $\mu+1$*-dimensional vector space spanned by the algebraic restrictions to* S_μ *of the* 2*-forms*

$$\theta_1 = dx_1 \wedge dx_3, \quad \theta_2 = dx_2 \wedge dx_3, \quad \theta_3 = dx_1 \wedge dx_2,$$
$$\sigma_1 = x_3 dx_1 \wedge dx_2, \quad \sigma_2 = x_1 dx_2 \wedge dx_3,$$
$$\theta_{4+k} = x_3^k dx_1 \wedge dx_3, \quad \text{for } 1 \leq k \leq \mu - 4.$$

Proposition 4.2 and results of Section 2 imply the following description of the space $[Z^2(\mathbb{R}^{2n})]_{S_\mu}$ and the manifold $[\mathrm{Symp}(\mathbb{R}^{2n})]_{S_\mu}$.

PROPOSITION 4.3. *The space* $[Z^2(\mathbb{R}^{2n})]_{S_\mu}$ *has dimension* μ. *It is spanned by the algebraic restrictions to* S_μ *of the* 2*-forms*

$$\theta_1, \theta_2, \theta_3, \quad \theta_4 = \sigma_1 - \sigma_2, \quad \theta_{4+k} = x_3^k dx_1 \wedge dx_3, \quad for \ \ 1 \leq k \leq \mu - 4.$$

If $n \geq 3$ *then* $[\mathrm{Symp}(\mathbb{R}^{2n})]_{S_\mu} = [Z^2(\mathbb{R}^{2n})]_{S_\mu}$. *The manifold* $[\mathrm{Symp}(\mathbb{R}^4)]_{S_\mu}$ *is an open part of the* μ*-space* $[Z^2(\mathbb{R}^4)]_{S_\mu}$ *consisting of the algebraic restrictions of the form* $[c_1\theta_1 + \cdots + c_\mu\theta_\mu]_{S_\mu}$ *such that* $(c_1, c_2, c_3) \neq (0, 0, 0)$.

THEOREM 4.4.

(i) Any algebraic restriction in $[Z^2(\mathbb{R}^{2n})]_{S_\mu}$ *can be brought by a symmetry of* S_μ *to one of the normal forms* $[S_\mu]^{i,j}$ *given in the second column of Table* 2.

(ii) The singularity classes corresponding to the normal forms are disjoint.

(iii) The parameters c_i *of the normal forms* $[S_\mu]^{i,j}$ *are moduli.*

(iv) The codimension in $[Z^2(\mathbb{R}^{2n})]_{S_\mu}$ *of the singularity class corresponding to the normal form* $[S_\mu]^{i,j}$ *is equal to* i *and the index of isotropness is equal to* j.

TABLE 2. Classification of symplectic S_μ singularities:
cod – codimension of the classes; μ^{sym}– symplectic multiplicity;
ind – index of isotropy.

Symplectic class	Normal forms for algebraic restrictions	cod	μ^{sym}	ind
$(S_\mu)^{0,0}$ $(2n \geq 4)$	$[S_\mu]^{0,0} : [\theta_1 + c_2\theta_2 + c_2\theta_3]_{S_\mu}$	0	2	0
$(S_\mu)_2^{k,0}$ $(2n \geq 4)$ for $1 \leq k \leq \mu - 5$	$[S_\mu]_2^{k,0} : [\theta_2 + c_3\theta_3 + c_{4+k}\theta_{4+k}]_{S_\mu}$ $c_{4+k} \neq 0$	k	$k+2$	0
$(S_\mu)_2^{\mu-4,0}$ $(2n \geq 4)$	$[S_\mu]_2^{\mu-4,0} : [\theta_2 + c_3\theta_3 + c_\mu\theta_\mu]_{S_\mu}, \ c_3c_\mu = 0$	$\mu - 4$	$\mu - 3$	0
$(S_\mu)_r^{1+k,0}$ $(2n \geq 4)$ for $1 \leq k \leq \mu - 6$	$[S_\mu]_r^{1+k,0} : [\theta_3 + c_{4+k}\theta_{4+k} + c_{5+k}\theta_{5+k}]_{S_\mu}$ $c_{4+k} \neq 0$	$k+1$	$k+3$	0
$(S_\mu)_r^{\mu-4,0}$ $(2n \geq 4)$	$[S_\mu]_r^{\mu-4,0} : [\theta_3 + c_{\mu-1}\theta_{\mu-1}]_{S_\mu}$	$\mu - 4$	$\mu - 3$	0
$(S_\mu)^{3,1}$ $(2n \geq 6)$	$[S_\mu]^{3,1} : [c_4\theta_4 + \theta_5]_{S_\mu}$	3	4	1
$(S_\mu)^{2+k,1}$ $(2n \geq 6)$ for $2 \leq k \leq \mu - 4$	$[S_\mu]^{2+k,1} : [\theta_4 + c_{4+k}\theta_{4+k}]_{S_\mu}$ $c_{4+k} \neq 0$ for $2 \leq k \leq \mu - 5$	$k+2$	$k+3$	1
$(S_\mu)^{3+k,k}$ $(2n \geq 6)$ for $2 \leq k \leq \mu - 4$	$[S_\mu]^{3+k,k} : [\theta_{4+k}]_{S_\mu}$ for $2 \leq k \leq \mu - 4$	$k+3$	$k+3$	k
$(S_\mu)^{\mu,\infty}$ $(2n \geq 6)$	$[S_\mu]^{\mu,\infty} : [0]_{S_\mu}$	μ	μ	∞

The proof of Theorem 4.4 is presented in Section 4.5.

In the first column of Table 2 by $(S_\mu)^{i,j}$ we denote a subclass of (S_μ) consisting of $N \in (S_\mu)$ such that the algebraic restriction $[\omega]_N$ is diffeomorphic to some algebraic restriction of the normal form $[S_\mu]^{i,j}$ where i is the codimension of the class and j denotes index of isotropy of the class. Classes $(S_\mu)_2^{i,0}$ and $(S_\mu)_r^{i,0}$ can be distinguished geometrically (see Section 4.4) and by relative Lagrangian tangency order $L_{2:1}$ defined in Section 4.3 (Remark 4.8). The classes $(S_\mu)_2^{i,0}$ have $L_{2:1} = \frac{2}{\lambda_\mu}$ and the classes $(S_\mu)_r^{i,0}$ have $L_{2:1} = \frac{r}{\lambda_\mu}$ where $\lambda_\mu = 1$ for even μ and $\lambda_\mu = 2$ for odd μ.

Theorem 2.5, Theorem 4.4 and Proposition 4.3 imply the following statement.

PROPOSITION 4.5. *The classes $(S_\mu)^{i,j}$ are symplectic singularity classes, i.e. they are closed with respect to the action of the group of symplectomorphisms. The class (S_μ) is the disjoint union of the classes $(S_\mu)^{i,j}$. The classes $(S_\mu)^{0,0}$, $(S_\mu)_2^{i,0}$ and $(S_\mu)_r^{i,0}$ for $1 \leq i \leq \mu - 4$ are non-empty for any dimension $2n \geq 4$ of the symplectic space; the classes $(S_\mu)^{i,1}$ for $3 \leq i \leq \mu - 2$ and $(S_\mu)^{i,i-3}$ for $5 \leq i \leq \mu - 1$ and $(S_\mu)^{\mu,\infty}$ are empty if $n = 2$ and not empty if $n \geq 3$.*

4.2. Symplectic normal forms. Proof of Theorem 4.1. Let us transfer the normal forms $[S_\mu]^{i,j}$ to symplectic normal forms using Theorem 2.12, i.e. realizing the algorithm in Section 2. Fix a family $\omega^{i,j}$ of symplectic forms on $\mathbb{R}^{2n}$ realizing the family $[S_\mu]^{i,j}$ of algebraic restrictions. We can fix, for example

$\omega^{0,0} = \theta_1 + c_2\theta_2 + c_3\theta_3 + dx_2 \wedge dx_4 + \sum_{i=3}^n dx_{2i-1} \wedge dx_{2i}$;

$\omega_2^{k,0} = \theta_2 + c_3\theta_3 + c_{4+k}\theta_{4+k} + dx_1 \wedge dx_4 + \sum_{i=3}^n dx_{2i-1} \wedge dx_{2i}$, $c_{4+k} \neq 0$, $1 \leq k \leq \mu - 5$;

$\omega_2^{\mu-4,0} = \theta_2 + c_3\theta_3 + c_\mu\theta_\mu + dx_1 \wedge dx_4 + \sum_{i=3}^n dx_{2i-1} \wedge dx_{2i}$, $c_3 c_\mu = 0$;

$\omega_r^{1+k,0} = \theta_3 + c_{4+k}\theta_{4+k} + c_{5+k}\theta_{5+k} + \sum_{i=2}^n dx_{2i-1} \wedge dx_{2i}$, $c_{4+k} \neq 0$, $1 \leq k \leq \mu - 6$;

$\omega_r^{\mu-4,0} = \theta_3 + c_{\mu-1}\theta_{\mu-1} + dx_4 \wedge dx_3 + \sum_{i=3}^n dx_{2i-1} \wedge dx_{2i}$;

$\omega^{3,1} = c_4\theta_4 + \theta_5 + \sum_{i=1}^3 dx_i \wedge dx_{i+3} + \sum_{i=4}^n dx_{2i-1} \wedge dx_{2i}$;

$\omega^{2+k,1} = \theta_4 + c_{4+k}\theta_{4+k} + \sum_{i=1}^3 dx_i \wedge dx_{i+3} + \sum_{i=4}^n dx_{2i-1} \wedge dx_{2i}$, $2 \leq k \leq \mu - 4$;

$\omega^{3+k,k} = \theta_{4+k} + \sum_{i=1}^3 dx_i \wedge dx_{i+3} + \sum_{i=4}^n dx_{2i-1} \wedge dx_{2i}$, $2 \leq k \leq \mu - 4$;

$\omega^{\mu,\infty} = \sum_{i=1}^3 dx_i \wedge dx_{i+3} + \sum_{i=4}^n dx_{2i-1} \wedge dx_{2i}$.

Let $\omega = \sum_{i=1}^m dp_i \wedge dq_i$, where $(p_1, q_1, \cdots, p_n, q_n)$ is the coordinate system on $\mathbb{R}^{2n}, n \geq 3$ (resp. $n = 2$). Fix a family $\Phi^{i,j}$ of local diffeomorphisms which bring the family of symplectic forms $\omega^{i,j}$ to the symplectic form ω: $(\Phi^{i,j})^*\omega^{i,j} = \omega$. Consider the families $S_\mu^{i,j} = (\Phi^{i,j})^{-1}(S_\mu)$. Any stratified submanifold of the symplectic space $(\mathbb{R}^{2n}, \omega)$ which is diffeomorphic to S_μ is symplectically equivalent to one and only one of the normal forms $S_\mu^{i,j}$ presented in Theorem 4.1. By Theorem 4.4 we obtain that the parameters c_i of the normal forms are moduli.

4.3. Distinguishing symplectic classes of S_μ by Lagrangian tangency orders. Lagrangian tangency orders will be used to obtain a more detailed classification of (S_μ). A curve $N \in (S_\mu)$ may be described as a union of two invariant components C_1 and C_2. The curve C_1 is diffeomorphic to the A_1 singularity and consists of two parameterized branches B_{1+} and B_{1-}. The curve C_2 is diffeomorphic to the $A_{\mu-4}$ singularity and consists of one parameterized branch if μ is even and consists of two branches B_{2+} and B_{2-} if μ is odd. The parametrization of these branches is given in the second column of Table 3 or Table 4. To distinguish the classes of this singularity completely we need following three invariants:

- $Lt(N) = Lt(C_1, C_2)$
- $L_1 = Lt(C_1) = \max_L(\min\{t(B_{1+}, L), t(B_{1-}, L)\})$
- $L_2 = Lt(C_2)$

where L is a smooth Lagrangian submanifold of the symplectic space.

Considering the triples $(Lt(N), L_1, L_2)$ we obtain a detailed classification of symplectic singularities of S_μ. Some subclasses (see Table 3 and 4) have a natural geometric interpretation (Table 5).

THEOREM 4.6. *A stratified submanifold $N \in (S_\mu)$ of a symplectic space $(\mathbb{R}^{2n}, \omega)$ with the canonical coordinates $(p_1, q_1, \cdots, p_n, q_n)$ is symplectically equivalent to one and only one of the curves presented in the second column of Table 3 or 4. The parameters c_i are moduli. The Lagrangian tangency orders of the curve are presented in the fifth, sixth and seventh columns of these tables and the codimension of the classes is given in the fourth column.*

TABLE 3. Lagrangian tangency orders for symplectic classes of S_μ singularity (μ even)

Class	Parametrization of branches $B_{1\pm}$ and C_2	Conditions for subclasses	cod	$Lt(N)$	L_1	L_2
$(S_\mu)^{0,0}$	$(t, 0, \pm t, -c_3 t, 0, \cdots)$	$c_3 \neq 0$	0	1	1	r
$2n \geq 4$	$(t^r, t^2, 0, c_2 t^2 - c_3 t^r, 0, \cdots)$	$c_3 = 0$	1	2	∞	r
$(S_\mu)_2^{k,0}$	$(t, 0, \pm t, c_3 t, 0, \cdots)$	$c_{4+k} \cdot c_3 \neq 0$	k	1	1	$r+2k$
$2n \geq 4$	$(0, t^2, t^r, \frac{c_{4+k}}{k+1} t^{2+2k}, 0, \cdots)$	$c_3 = 0, c_{4+k} \neq 0$	$k+1$	2	∞	$r+2k$
$(S_\mu)_2^{\mu-4,0}$	$(t, 0, \pm t, c_3 t, 0, \cdots)$	$c_3 \neq 0$	$\mu-4$	1	1	∞
$2n \geq 4$	$(0, t^2, t^r, \frac{c_\mu}{r} t^{2r}, 0, \cdots)$	$c_3 = 0$	$\mu-3$	2	∞	∞
$(S_\mu)_r^{1+k,0}$	$(t, \pm t, 0, 0, \cdots)$	$c_{k+4} \neq 0$	$k+1$	1	1	$r+2k$
$2n \geq 4$	$(t^r, 0, (c_{4+k} + c_{5+k} t^2) t^{r+2k}, t^2, 0, \cdots)$	$1 \leq k \leq \mu - 6$				
$(S_\mu)_r^{\mu-4,0}$	$(t, \pm t, 0, 0, \cdots)$	$c_{\mu-1} \neq 0$	$\mu-4$	1	1	$3r-4$
$2n \geq 4$	$(t^r, 0, c_{\mu-1} t^{3r-4}, t^2, 0, \cdots)$	$c_{\mu-1} = 0$	$\mu-3$	1	1	∞
$(S_\mu)^{3,1}$	$(t, 0, \pm t, 0, 0, \cdots)$		3	$r+2$	∞	$r+2$
$2n \geq 6$	$(t^r, \frac{1}{2} t^4, 0, -c_4 t^{r+2}, t^2, 0, \cdots)$					
$(S_\mu)^{2+k,1}$	$(t, 0, \pm t, 0, 0, 0, \cdots)$	$c_{4+k} \neq 0$	$k+2$	$r+2$	∞	$r+2k$
$2n \geq 6$	$(t^r, \frac{c_{4+k} t^{2(k+1)}}{k+1}, 0, -t^{r+2}, t^2, 0, \cdots)$	$2 \leq k \leq \mu - 5$				
		$k = \mu - 4$	$\mu-2$	$r+2$	∞	∞
$(S_\mu)^{3+k,k}$	$(t, 0, \pm t, 0, 0, 0, \cdots)$	$2 \leq k \leq \mu - 5$	$k+3$	$r+2k$	∞	$r+2k$
$2n \geq 6$	$(t^r, \frac{t^{2(k+1)}}{k+1}, 0, 0, t^2, 0, 0, \cdots)$	$k = \mu - 4$	$\mu-1$	$3r-2$	∞	∞
$(S_\mu)^{\mu,\infty}$	$(t, 0, \pm t, 0, 0, \cdots)$		μ	∞	∞	∞
$2n \geq 6$	$(t^r, 0, 0, 0, t^2, 0, \cdots)$					

TABLE 4. Lagrangian tangency orders for symplectic classes of S_μ singularity (μ odd)

Class	Parametrization of branches $B_{1\pm}$ and $B_{2\pm}$	Conditions for subclasses	cod	$Lt(N)$	L_1	L_2
$(S_\mu)^{0,0}$	$(t,0,\pm t,-c_3t,0,\cdots)$	$c_3\neq 0$	0	1	1	$\frac{r}{2}$
$2n\geq 4$	$(\pm t^{\frac{r}{2}},t,0,c_2t^2\mp c_3t^{\frac{r}{2}},0,\cdots)$	$c_3=0$	1	1	∞	$\frac{r}{2}$
$(S_\mu)_2^{k,0}$	$(t,0,\pm t,c_3t,0,\cdots)$	$c_{4+k}\cdot c_3\neq 0$	k	1	1	$\frac{r}{2}+k$
$2n\geq 4$	$(0,t,\pm t^{\frac{r}{2}},\frac{c_{4+k}}{k+1}t^{1+k},0,\cdots)$	$c_3=0, c_{4+k}\neq 0$	$k+1$	1	∞	$\frac{r}{2}+k$
$(S_\mu)_2^{\mu-4,0}$	$(t,0,\pm t,c_3t,0,\cdots)$	$c_3\neq 0$	$\mu-4$	1	1	∞
$2n\geq 4$	$(0,t,t^{\frac{r}{2}},\frac{c_\mu}{r}t^r,0,\cdots)$	$c_3=0$	$\mu-3$	1	∞	∞
$(S_\mu)_r^{1+k,0}$	$(t,\pm t,0,0,\cdots)$	$c_{k+4}\neq 0$	$k+1$	1	1	$\frac{r}{2}+k$
$2n\geq 4$	$(\pm t^{\frac{r}{2}},0,\pm(c_{4+k}+c_{5+k}t)t^{\frac{r}{2}+k},t,0,\cdots)$	$1\leq k\leq\mu-6$				
$(S_\mu)_r^{\mu-4,0}$	$(t,\pm t,0,0,\cdots)$	$c_{\mu-1}\neq 0$	$\mu-4$	1	1	$\frac{3r}{2}-2$
$2n\geq 4$	$(\pm t^{\frac{r}{2}},0,\pm c_{\mu-1}t^{\frac{3r-4}{2}},t,0,\cdots)$	$c_{\mu-1}=0$	$\mu-3$	1	1	∞
$(S_\mu)^{3,1}$	$(t,0,\pm t,0,0,\cdots)$		3	$\frac{r}{2}+1$	∞	$\frac{r}{2}+1$
$2n\geq 6$	$(\pm t^{\frac{r}{2}},\frac{1}{2}t^2,0,\mp c_4t^{\frac{r+2}{2}},t,0,\cdots)$					
$(S_\mu)^{2+k,1}$	$(t,0,\pm t,0,0,0,\cdots)$	$c_{4+k}\neq 0$	$k+2$	$\frac{r}{2}+1$	∞	$\frac{r}{2}+k$
$2n\geq 6$	$(\pm t^{\frac{r}{2}},\frac{c_{4+k}t^{k+1}}{k+1},0,\mp t^{\frac{r+2}{2}},t,0,\cdots)$	$2\leq k\leq\mu-5$				
		$k=\mu-4$	$\mu-2$	$\frac{r}{2}+1$	∞	∞
$(S_\mu)^{3+k,k}$	$(t,0,\pm t,0,0,0,\cdots)$	$2\leq k\leq\mu-5$	$k+3$	$\frac{r}{2}+k$	∞	$\frac{r}{2}+k$
$2n\geq 6$	$(\pm t^{\frac{r}{2}},\frac{t^{k+1}}{k+1},0,0,t,0,0,\cdots)$	$k=\mu-4$	$\mu-1$	$\frac{3}{2}r-1$	∞	∞
$(S_\mu)^{\mu,\infty}$	$(t,0,\pm t,0,0,\cdots)$		μ	∞	∞	∞
$2n\geq 6$	$(\pm t^{\frac{r}{2}},0,0,0,t,0,\cdots)$					

REMARK 4.7. The numbers L_1 and L_2 can be easily calculated knowing Lagrangian tangency orders for the A_1 and $A_{\mu-4}$ singularities (see Table 2 in [**DT**]) or by applying directly the definition of the Lagrangian tangency order and finding a Lagrangian submanifold nearest to the components. Next we calculate $Lt(N)$ from the definition knowing that it can not be greater than $\min(L_1,L_2)$.

We can compute L_1 using the algebraic restrictions $[\omega^{i,j}]_{C_1}$ where the space $[Z^2(\mathbb{R}^{2n})]_{C_1}$ is spanned only by the algebraic restriction to C_1 of the 2-form θ_3. For example for the class $(S_\mu)^{0,0}$ we have $[\theta_1+c_2\theta_2+c_3\theta_3]_{C_1}=[c_3\theta_3]_{C_1}$ and thus $L_1=1$ when $c_3\neq 0$ and $L_1=\infty$ when $c_3=0$.

We can compute L_2 using the algebraic restrictions $[\omega^{i,j}]_{C_2}$ where the space $[Z^2(\mathbb{R}^{2n})]_{C_2}$ is spanned only by the algebraic restrictions to C_2 of the 2-forms θ_1, θ_{4+k} for $k=1,2,\ldots,\theta_{\mu-1}$. For example for the class $(S_\mu)^{0,0}$ we have $[\theta_1+c_2\theta_2+c_3\theta_3]_{C_2}=[\theta_1]_{C_2}$ and thus $L_2=\mu-3$ if μ is an even number and $L_2=\frac{\mu-3}{2}$ if μ is an odd number.

$Lt(N)\leq 1=\min(L_1,L_2)$ when $c_3\neq 0$. Applying the definition of $Lt(N)$ we find the smooth Lagrangian submanifold L described by the conditions $p_i=0$, $i\in\{1,\ldots,n\}$ and we get $Lt(N)\geq t(N,L)=1$ in this case.

If $c_3=0$ then $Lt(N)\leq L_2=\min(L_1,L_2)$, but applying the definition of $Lt(N)$ we have $t(N,L)\leq 2$ (resp. $t(N,L)\leq 1$) for all Lagrangian submanifolds L. For L

described by the conditions $q_i = 0,\ i \in \{1, \ldots, n\}$ we get $Lt(N) = t(N, L) = 2$ if μ is even and $Lt(N) = t(N, L) = 1$ if μ is odd.

REMARK 4.8. We are not able to distinguish some classes $(S_\mu)_2^{i,0}$ and $(S_\mu)_r^{i,0}$ by the triples $(Lt(N), L_1, L_2)$ but we can do this using relative Lagrangian tangency orders.

We define $L_{2:1} = Lt[C_2 : B_{1\pm}] = \max(Lt[C_2 : B_{1+}], Lt[C_2 : B_{1-}])$.

Since branches B_{1+} and B_{1-} are smooth curves then $Lt(B_{1+}) = Lt(B_{1-}) = \infty$ and $L_{2:1} = \max_L(t(C_2, L))$ where L is a smooth Lagrangian submanifold containing B_{1+} or B_{1-}.

Considering such smooth Lagrangian submanifolds we obtain $L_{2:1} = \frac{2}{\lambda_\mu}$ for the classes $(S_\mu)_2^{i,0}$ and $L_{2:1} = \frac{\mu-3}{\lambda_\mu}$ for the classes $(S_\mu)_r^{i,0}$ ($\lambda_\mu = 1$ for even μ and $\lambda_\mu = 2$ for odd μ).

4.4. Geometric conditions for the classes $(S_\mu)^{i,j}$. The classes $(S_\mu)^{i,j}$ can be distinguished geometrically, without using any local coordinate system.

Let $N \in (S_\mu)$. Then N is the union of two singular 1-dimensional irreducible components diffeomorphic to the A_1 and $A_{\mu-4}$ singularities. In local coordinates they have the form

$$\mathcal{C}_1 = \{x_1^2 - x_2^2 = 0,\ x_{\geq 3} = 0\},$$
$$\mathcal{C}_2 = \{x_1^2 - x_3^{\mu-3} = 0,\ x_2 = x_{\geq 4} = 0\}.$$

Denote by ℓ_{1+}, ℓ_{1-} the tangent lines at 0 to the branches $\mathcal{B}_{1+}$ and $\mathcal{B}_{1-}$ respectively. These lines span a 2-space P_1. Denote by ℓ_2 the tangent line at 0 to the component $\mathcal{C}_2$ and let P_2 be the 2-space tangent at 0 to the component $\mathcal{C}_2$. Define the line $\ell_3 = P_1 \cap P_2$. The lines $\ell_{1\pm}, \ell_2$ span a 3-space $W = W(N)$. Equivalently W is the tangent space at 0 to some (and then any) non-singular 3-manifold containing N. The classes $(S_\mu)^{i,j}$ satisfy special conditions in terms of the restriction $\omega|_W$, where ω is the symplectic form. For $N = S_\mu =$(4.1) it is easy to calculate

$$\ell_{1\pm} = \mathrm{span}(\partial/\partial x_1 \pm \partial/\partial x_2),\ \ell_2 = \mathrm{span}(\partial/\partial x_3),\ \ell_3 = \mathrm{span}(\partial/\partial x_1). \tag{4.4}$$

THEOREM 4.9. *A stratified submanifold $N \in (S_\mu)$ of a symplectic space $(\mathbb{R}^{2n}, \omega)$ belongs to the class $(S_\mu)^{i,j}$ if and only if the couple (N, ω) satisfies the corresponding conditions in the last column of Table 5.*

PROOF OF THEOREM 4.9. The conditions on the pair (ω, N) in the last column of Table 5 are disjoint. It suffices to prove that these conditions in the row of $(S_\mu)^{i,j}$, are satisfied for any $N \in (S_\mu)^{i,j}$. This is a corollary of the following claims:

1. Each of the conditions in the last column of Table 5 is invariant with respect to the action of the group of diffeomorphisms in the space of pairs (ω, N);
2. Each of these conditions depends only on the algebraic restriction $[\omega]_N$;
3. Take the simplest 2-forms $\omega^{i,j}$ representing the normal forms $[S_\mu]^{i,j}$ for the algebraic restrictions. The pair $(\omega = \omega^{i,j}, S_\mu)$ satisfies the condition in the last column of Table 5, the row of $(S_\mu)^{i,j}$.

The first statement is obvious, the second one follows from Lemma 2.7.

TABLE 5. Geometric interpretation of singularity classes of S_μ: W is the tangent space to a non-singular 3-dimensional manifold in $(\mathbb{R}^{2n\geq 4}, \omega)$ containing $N \in (S_\mu)$, $\lambda_\mu = 1$ for even μ and $\lambda_\mu = 2$ for odd μ.

Class	Normal form	Geometric conditions
$(S_\mu)^{0,0}$		$\omega\vert_{\ell_2+\ell_3} \neq 0$
	$[S_\mu]^{0,0}_{L_1=1} : [\theta_1 + c_2\theta_2 + c_3\theta_3]_{S_\mu}$ $c_3 \neq 0$	$\omega\vert_{\ell_{1+}+\ell_{1-}} \neq 0$ and none of the components is contained in a Lagrangian submanifold
	$[S_\mu]^{0,0}_{L_1=\infty} : [\theta_1 + c_2\theta_2]_{S_\mu}$	$\omega\vert_{\ell_{1+}+\ell_{1-}} = 0$ (so component C_1 is contained in a Lagrangian submanifold)
$(S_\mu)^{i,0}_2$		$\omega\vert_{\ell_2+\ell_3} = 0$ but $\omega\vert_{\ell_{1\pm}+\ell_2} \neq 0$
$(S_\mu)^{k,0}_2$	$[S_\mu]^{k,0}_{2,L_1=1}$: $[\theta_2+c_3\theta_3+c_{4+k}\theta_{4+k}]_{S_\mu}$ $c_3 \cdot c_{4+k} \neq 0$ for $1 \leq k \leq \mu - 5$	$\omega\vert_{\ell_{1+}+\ell_{1-}} \neq 0$ and $L_2 = \frac{r+2k}{\lambda_\mu}$
	$[S_\mu]^{k,0}_{2,L_1=\infty}$: $[\theta_2+c_{4+k}\theta_{4+k}]_{S_\mu}$ $c_{4+k} \neq 0$ for $1 \leq k \leq \mu - 5$	$\omega\vert_{\ell_{1+}+\ell_{1-}} = 0$ (so component C_1 is contained in a Lagrangian submanifold) and $L_2 = \frac{r+2k}{\lambda_\mu}$
$(S_\mu)^{\mu-4,0}_2$	$[S_\mu]^{\mu-4,0}_2 : [\theta_2 + c_3\theta_3]_{S_\mu}$, $c_3 \neq 0$	$\omega\vert_{\ell_{1+}+\ell_{1-}} \neq 0$ and component C_2 is contained in a Lagrangian submanifold
	$[S_\mu]^{\mu-3,0}_2 : [\theta_2 + c_\mu\theta_\mu]_{S_\mu}$	$\omega\vert_{\ell_{1+}+\ell_{1-}} = 0$, both components are contained in Lagrangian submanifolds
$(S_\mu)^{i,0}_r$		$\omega\vert_{\ell_2+\ell_3} = 0$ and $\omega\vert_{\ell_{1\pm}+\ell_2} = 0$ but $\omega\vert_{\ell_{1+}+\ell_{1-}} \neq 0$
$(S_\mu)^{1+k,0}_r$	$[S_\mu]^{1+k,0}_r$: $[\theta_3+c_{4+k}\theta_{4+k}+c_{5+k}\theta_{5+k}]_{S_\mu}$ $c_{4+k} \neq 0$ for $1 \leq k \leq \mu - 6$	none of the components is contained in a Lagrangian submanifold and $L_2 = \frac{r+2k}{\lambda_\mu}$
$(S_\mu)^{\mu-4,0}_r$	$[S_\mu]^{\mu-4,0}_r : [\theta_3 + c_{\mu-1}\theta_{\mu-1}]_{S_\mu}$ $c_{\mu-1} \neq 0$	none of the components is contained in a Lagrangian submanifold and $L_2 = \frac{3r-4}{\lambda_\mu}$
	$[S_\mu]^{\mu-3,0}_r : [\theta_3]_{S_\mu}$	component C_2 is contained in Lagrangian submanifolds
		$\omega\vert_W = 0$ and component C_1 is contained in a Lagrangian submanifold
$(S_\mu)^{3,1}$	$[S_\mu]^{3,1} : [c_4\theta_4 + \theta_5]_{S_\mu}$	$L_2 = \frac{r+2}{\lambda_\mu}$ and $Lt(N) = \frac{r+2}{\lambda_\mu}$
$(S_\mu)^{2+k,1}$	$[S_\mu]^{2+k,1} : [\theta_4 + c_{4+k}\theta_{4+k}]_{S_\mu}$ $c_{4+k} \neq 0$ and $2 \leq k \leq \mu - 5$	$L_2 = \frac{r+2k}{\lambda_\mu}$ and $Lt(N) = \frac{r+2}{\lambda_\mu}$
	$[S_\mu]^{\mu-2,1} : [\theta_4 + c\theta_\mu]_{S_\mu}$	both components are contained in Lagrangian submanifolds and $Lt(N) = \frac{r+2}{\lambda_\mu}$
$(S_\mu)^{3+k,k}$	$[S_\mu]^{3+k,k} : [\theta_{4+k}]_{S_\mu}$ $2 \leq k \leq \mu - 5$	$L_2 = \frac{r+2k}{\lambda_\mu}$ and $Lt(N) = \frac{r+2k}{\lambda_\mu}$
	$[S_\mu]^{\mu-1,\mu-4} : [\theta_\mu]_{S_\mu}$	both components are contained in Lagrangian submanifolds and $Lt(N) = \frac{3r-2}{\lambda_\mu}$
$(S_\mu)^{\mu,\infty}$	$[S_\mu]^{\mu,\infty} : [0]_{S_\mu}$	both components are contained in the same Lagrangian submanifold

To prove the third statement we note that in the case $N = S_\mu = (4.1)$ one has $\ell_{1\pm} = \mathrm{span}(\partial/\partial x_1 \pm \partial/\partial x_2)$, $\ell_2 = \mathrm{span}(\partial/\partial x_3)$, $\ell_3 = \mathrm{span}(\partial/\partial x_1)$ and $W = \mathrm{span}(\partial/\partial x_1, \partial/\partial x_2, \partial/\partial x_3)$. By simple calculation and observation of the Lagrangian tangency orders we obtain that the conditions in the last column of Table 5, the row of $(S_\mu)^{i,j}$ are satisfied. □

4.5. Proof of Theorem 4.4.

PROOF. In our proof we use vector fields tangent to $N \in S_\mu$. Any vector fields tangent to $N \in S_\mu$ may be described as $V = g_1 E + g_2 \mathcal{H}$ where E is the Euler vector field and $\mathcal{H}$ is a Hamiltonian vector field and g_1, g_2 are functions. It was shown in [**DT**] (Prop. 6.13) that the action of a Hamiltonian vector field on any 1-dimensional complete intersection is trivial.

The germ of a vector field tangent to S_μ of non trivial action on algebraic restrictions of closed 2-forms to S_μ may be described as a linear combination germs of the following vector fields: $X_0 = E$, $X_1 = x_1 E$, $X_2 = x_2 E$, $X_3 = x_3 E$, $X_{l+2} = x_3^l E$ for $1 < l < \mu - 3$, where E is the Euler vector field $E = \sum_{i=1}^{3} \lambda_i x_i \partial/\partial x_i$ and the λ_i are the weights for x_i.

PROPOSITION 4.10. *When μ is an even number then the infinitesimal action of germs of quasi-homogeneous vector fields tangent to N on the basis of the vector space of the algebraic restrictions of closed 2-forms to N is presented in Table 6.*

TABLE 6. Infinitesimal actions on algebraic restrictions of closed 2-forms to S_μ. $E = (\mu-3)x_1\partial/\partial x_1 + (\mu-3)x_2\partial/\partial x_2 + 2x_3\partial/\partial x_3$

$\mathcal{L}_{X_i}[\theta_j]$	$[\theta_1]$	$[\theta_2]$	$[\theta_3]$	$[\theta_4]$	$[\theta_{4+k}]$ *for* $0<k<r$
$X_0 = E$	$(r+2)[\theta_1]$	$(r+2)[\theta_2]$	$2r[\theta_3]$	$(2r+2)[\theta_4]$	$(r+2(k+1))[\theta_{4+k}]$
$X_1 = x_1 E$	$[0]$	$-(r+2)[\theta_4]$	$[0]$	$[0]$	$[0]$
$X_2 = x_2 E$	$-r[\theta_4]$	$[0]$	$\frac{-3r^2}{2}[\theta_\mu]$	$[0]$	$[0]$
$X_3 = x_3 E$	$(r+4)[\theta_5]$	$[0]$	$r[\theta_4]$	$[0]$	$(r+2(k+2))[\theta_{5+k}]$
$X_{l+2} = x_3^l E$ $l < r-k$	$(r+2l+2)[\theta_{4+l}]$	$[0]$	$[0]$	$[0]$	$(r+2(k+l+1))[\theta_{4+k+l}]$
$X_{l+2} = x_3^l E$ $r-k \le l \le r-1$	$(r+2l+2)[\theta_{4+l}]$	$[0]$	$[0]$	$[0]$	$[0]$

REMARK 4.11. When μ is odd we obtain a very similar table, we only have to divide by 2 all coefficients in Table 6. The next part of the proof is written for even μ. In the case of odd μ we repeat the same scheme.

Let $\mathcal{A} = [\sum_{l=1}^{\mu} c_l \theta_l]_{S_\mu}$ be the algebraic restriction of a symplectic form ω.

The first statement of Theorem 4.4 follows from the following lemmas.

LEMMA 4.12. *If $c_1 \neq 0$ then the algebraic restriction $\mathcal{A} = [\sum_{l=1}^{\mu} c_l \theta_l]_{S_\mu}$ can be reduced by a symmetry of S_μ to an algebraic restriction $[\theta_1 + \widetilde{c}_2\theta_2 + \widetilde{c}_3\theta_3]_{S_\mu}$.*

PROOF OF LEMMA 4.12. We use the homotopy method to prove that $\mathcal{A}$ is diffeomorphic to $[\theta_1+\widetilde{c}_2\theta_2+\widetilde{c}_3\theta_3]_{S_\mu}$. Let $\mathcal{B}_t = [c_1\theta_1+c_2\theta_2+c_3\theta_3+(1-t)\sum_{l=4}^{\mu} c_l\theta_l]_{S_\mu}$ for $t \in [0;1]$. Then $\mathcal{B}_0 = \mathcal{A}$ and $\mathcal{B}_1 = [c_1\theta_1 + c_2\theta_2 + c_3\theta_3]_{S_\mu}$. We prove that there exists a family $\Phi_t \in Symm(S_\mu)$, $t \in [0;1]$ such that

$$\Phi_t^*\mathcal{B}_t = \mathcal{B}_0, \ \Phi_0 = id. \tag{4.5}$$

Let V_t be a vector field defined by $\frac{d\Phi_t}{dt} = V_t(\Phi_t)$. Then differentiating (4.5) we obtain

$$\mathcal{L}_{V_t}\mathcal{B}_t = [\sum_{l=4}^{\mu} c_l\theta_l]. \tag{4.6}$$

We are looking for V_t in the form $V_t = \sum_{k=1}^{\mu-2} b_k(t)X_k$ where the $b_k(t)$ for $k = 1,\dots,\mu-2$ are smooth functions $b_k : [0;1] \to \mathbb{R}$. Then by Proposition 4.10 equation (4.6) has a form

(4.7)
$$\begin{bmatrix} -(r+2)c_2 & -rc_1 & rc_3 & 0 & 0 & 0 & 0 \\ 0 & 0 & (r+4)c_1 & 0 & 0 & 0 & 0 \\ 0 & 0 & (1-t)(r+6)c_5 & (r+6)c_1 & 0 & 0 & 0 \\ \vdots & \vdots & \vdots & \ddots & 0 & 0 & 0 \\ 0 & 0 & (1-t)(r+2k)c_{k+2} & \cdots & (r+2k)c_1 & 0 & 0 \\ 0 & 0 & \vdots & \cdots & \vdots & \ddots & 0 \\ 0 & -\frac{3r^2}{2}c_3 & 3(1-t)rc_{\mu-1} & \cdots & 3(1-t)rc_{\mu-k+1} & \cdots & 3rc_1 \end{bmatrix} \begin{bmatrix} b_1(t) \\ b_2(t) \\ b_3(t) \\ \vdots \\ b_{k+1}(t) \\ \vdots \\ b_{\mu-2}(t) \end{bmatrix} = \begin{bmatrix} c_4 \\ c_5 \\ c_6 \\ \vdots \\ c_{k+3} \\ \vdots \\ c_\mu \end{bmatrix}$$

If $c_1 \neq 0$ we can solve (4.7).
We obtain $b_3(t) = \frac{c_5}{c_1(r+4)}$ and we may choose any b_1.
Other functions b_k are determined by that choice.
Let $b_1(t) = 0$. This implies $b_2(t) = \frac{rc_3b_3(t)-c_4}{rc_1} = \frac{c_3c_5}{(r+4)c_1^2} - \frac{c_4}{rc_1}$.
Next $b_4(t) = \frac{c_6}{(r+6)c_1} - \frac{(1-t)}{c_1}c_5b_3(t)$, $b_5(t) = \frac{c_7}{(r+8)c_1} - \frac{(1-t)}{c_1}(c_6b_3(t) + c_5b_4(t))$,
consequently $b_{k+1}(t) = \frac{c_{k+3}}{(r+2k)c_1} - \frac{(1-t)}{c_1}\sum_{l=3}^{k} c_{k+5-l}b_l(t)$ for $k < \mu - 3$,
and eventually $b_{\mu-2}(t) = \frac{c_\mu}{3rc_1} + \frac{r}{2c_1}c_3b_2(t) - \frac{(1-t)}{c_1}\sum_{l=3}^{\mu-3} c_{\mu+2-l}b_l(t)$.

Diffeomorphisms Φ_t may be obtained as a flow of the vector field V_t. The family Φ_t preserves S_μ, because V_t is tangent to S_μ and $\Phi_t^*\mathcal{B}_t = \mathcal{A}$. Using the homotopy arguments we have $\mathcal{A}$ diffeomorphic to $\mathcal{B}_1 = [c_1\theta_1+c_2\theta_2+c_3\theta_3]_{S_\mu}$. By the condition $c_1 \neq 0$ we have a diffeomorphism $\Psi \in Symm(S_\mu)$ of the form

$$\Psi : (x_1, x_2, x_3) \mapsto (c_1^{-\frac{r}{r+2}}x_1, c_1^{-\frac{r}{r+2}}x_2, c_1^{-\frac{2}{r+2}}x_3), \tag{4.8}$$

and we obtain

$$\Psi^*(\mathcal{B}_1) = [\theta_1 + \frac{c_2}{c_1}\theta_2 + c_3c_1^{-\frac{2r}{r+2}}\theta_3]_{S_\mu} = [\theta_1 + \widetilde{c}_2\theta_2 + \widetilde{c}_3\theta_3]_{S_\mu}.$$

□

LEMMA 4.13. *If $c_1 = 0$ and $c_2 \neq 0$ and $c_{4+k} \neq 0$ and $c_l = 0$ for $5 \leq l < 4 + k$, then the algebraic restriction $\mathcal{A} = [\sum_{l=1}^{\mu} c_l\theta_l]_{S_\mu}$ can be reduced by a symmetry of S_μ to an algebraic restriction $[\theta_2 + \widetilde{c}_3\theta_3 + +\widetilde{c}_{4+k}\theta_{4+k}]_{S_\mu}$.*

PROOF OF LEMMA 4.13. If $c_1 = 0$ and $c_2 \neq 0$ and $c_{4+k} \neq 0$ and $c_l = 0$ for $5 \leq l < 4+k$, then $\mathcal{A} = [c_2\theta_2 + c_3\theta_3 + c_4\theta_4 + \sum_{l=4+k}^{\mu} c_l\theta_l]_{S_\mu}$.
Let $\mathcal{B}_t = [c_2\theta_2 + c_3\theta_3 + (1-t)c_4\theta_4 + c_{4+k}\theta_{4+k} + (1-t)\sum_{l=5+k}^{\mu} c_l\theta_l]_{S_\mu}$ for $t \in [0;1]$. Then $\mathcal{B}_0 = \mathcal{A}$ and $\mathcal{B}_1 = [c_2\theta_2 + c_3\theta_3 + c_{4+k}\theta_{4+k}]_{S_\mu}$. We prove that there exists a family $\Phi_t \in Symm(S_\mu)$, $t \in [0;1]$ such that

$$\Phi_t^*\mathcal{B}_t = \mathcal{B}_0, \ \Phi_0 = id. \tag{4.9}$$

Let V_t be a vector field defined by $\frac{d\Phi_t}{dt} = V_t(\Phi_t)$. Then differentiating (4.9) we obtain

$$\mathcal{L}_{V_t}\mathcal{B}_t = [c_4\theta_4 + \sum_{l=5+k}^{\mu} c_l\theta_l]_{S_\mu}. \tag{4.10}$$

We are looking for V_t in the form $V_t = \sum_{k=1}^{\mu-2} b_k(t)X_k$ where the $b_k(t)$ are smooth functions $b_k : [0;1] \to \mathbb{R}$ for $k = 1, \ldots, \mu-2$. Then by Proposition 4.10 equation (4.10) has a form

$$\begin{bmatrix} -(r+2)c_2 & 0 & rc_3 & 0 & \cdots & \cdots & 0 \\ 0 & 0 & (r+2k+4)c_{k+4} & 0 & \cdots & \cdots & 0 \\ 0 & 0 & \vdots & \ddots & 0 & \cdots & 0 \\ 0 & -\frac{3r^2}{2}c_3 & 3(1-t)rc_{\mu-1} & \cdots & 3rc_{k+4} & 0\cdots & 0 \end{bmatrix} \begin{bmatrix} b_1(t) \\ b_2(t) \\ b_3(t) \\ \vdots \\ b_{\mu-2}(t) \end{bmatrix} = \begin{bmatrix} c_4 \\ c_{k+5} \\ \vdots \\ c_\mu \end{bmatrix} \tag{4.11}$$

If $c_2 \neq 0$ we can solve (4.11). Diffeomorphisms Φ_t may be obtained as a flow of the vector field V_t. The family Φ_t preserves S_μ, because V_t is tangent to S_μ and $\Phi_t^*\mathcal{B}_t = \mathcal{A}$. Using the homotopy arguments we have that $\mathcal{A}$ is diffeomorphic to $\mathcal{B}_1 = [c_2\theta_2 + c_3\theta_3 + c_{4+k}\theta_{4+k}]_{S_\mu}$. By the condition $c_2 \neq 0$ we have a diffeomorphism $\Psi \in Symm(S_\mu)$ of the form

$$\Psi : (x_1, x_2, x_3) \mapsto (c_2^{-\frac{r}{r+2}}x_1, c_2^{-\frac{r}{r+2}}x_2, c_2^{-\frac{2}{r+2}}x_3), \tag{4.12}$$

and we obtain

$$\Psi^*(\mathcal{B}_1) = [\theta_2 + c_3c_2^{-\frac{2r}{r+2}}\theta_3 + c_{4+k}c_2^{-(1+\frac{2k}{r+2})}\theta_{4+k}]_{S_\mu} = [\theta_2 + \widetilde{c}_3\theta_3 + \widetilde{c}_{4+k}\theta_{4+k}]_{S_\mu}.$$

□

LEMMA 4.14. *If $c_1 = 0$ and $c_2 \neq 0$ and $c_{4+k} = 0$ for $k \in \{1, ..., \mu-5\}$, then the algebraic restriction $\mathcal{A} = [\sum_{l=1}^{\mu} c_l\theta_l]_{S_\mu}$ can be reduced by a symmetry of S_μ to an algebraic restriction $[\theta_2 + \widetilde{c}_3\theta_3 + \widetilde{c}_\mu\theta_\mu]_{S_\mu}$ where $\widetilde{c}_3\widetilde{c}_\mu = 0$.*

PROOF OF LEMMA 4.14. We use methods similar to those in the proof of the previous lemma. Now $\mathcal{A} = [c_2\theta_2 + c_3\theta_3 + c_4\theta_4 + c_\mu\theta_\mu]_{S_\mu}$.
When $c_3 \neq 0$ let $\mathcal{B}_t = [c_2\theta_2 + c_3\theta_3 + (1-t)c_4\theta_4 + (1-t)c_\mu\theta_\mu]_{S_\mu}$ for $t \in [0;1]$. Then $\mathcal{B}_0 = \mathcal{A}$ and $\mathcal{B}_1 = [c_2\theta_2 + c_3\theta_3]_{S_\mu}$. We prove that there exists a family $\Phi_t \in Symm(S_\mu)$, $t \in [0;1]$ such that

$$\Phi_t^*\mathcal{B}_t = \mathcal{B}_0, \ \Phi_0 = id. \tag{4.13}$$

Let V_t be a vector field defined by $\frac{d\Phi_t}{dt} = V_t(\Phi_t)$. Then differentiating (4.13) we obtain

$$\mathcal{L}_{V_t}\mathcal{B}_t = [c_4\theta_4 + c_\mu\theta_\mu]_{S_\mu}. \tag{4.14}$$

We are looking for V_t in the form $V_t = \sum_{k=1}^{3} b_k X_k$ where the $b_k \in \mathbb{R}$. Then by Proposition 4.10 equation (4.14) has a form

$$\begin{bmatrix} -(r+2)c_2 & 0 & rc_3 \\ 0 & -\frac{3r^2}{2}c_3 & 0 \end{bmatrix} \begin{bmatrix} b_1 \\ b_2 \\ b_3 \end{bmatrix} = \begin{bmatrix} c_4 \\ c_\mu \end{bmatrix} \tag{4.15}$$

If $c_3 \neq 0$ we can solve (4.15) and Φ_t may be obtained as a flow of the vector field V_t. The family Φ_t preserves S_μ, because V_t is tangent to S_μ and $\Phi_t^*\mathcal{B}_t = \mathcal{A}$. Using the homotopy arguments we have that $\mathcal{A}$ is diffeomorphic to $\mathcal{B}_1 = [c_2\theta_2 + c_3\theta_3]_{S_\mu}$. By the condition $c_2 \neq 0$ we have a diffeomorphism $\Psi \in Symm(S_\mu)$ of the form

$$\Psi : (x_1, x_2, x_3) \mapsto (c_2^{-\frac{r}{r+2}} x_1, c_2^{-\frac{r}{r+2}} x_2, c_2^{-\frac{2}{r+2}} x_3), \tag{4.16}$$

and we obtain

$$\Psi^*(\mathcal{B}_1) = [\theta_2 + c_3 c_2^{-\frac{2r}{r+2}} \theta_3]_{S_\mu} = [\theta_2 + \widetilde{c}_3\theta_3]_{S_\mu}.$$

In the case $c_3 = 0$ we take $\mathcal{B}_t = [c_2\theta_2 + (1-t)c_4\theta_4 + c_\mu\theta_\mu]_{S_\mu}$ for $t \in [0;1]$ and we can solve only the first equation of (4.15). Using the homotopy arguments we have that $\mathcal{A}$ is diffeomorphic to $\mathcal{B}_1 = [c_2\theta_2 + c_\mu\theta_\mu]_{S_\mu}$. Using the diffeomorphism (4.16) we obtain

$$\Psi^*(\mathcal{B}_1) = [\theta_2 + c_\mu c_2^{-\frac{3r}{r+2}} \theta_\mu]_{S_\mu} = [\theta_2 + \widetilde{c}_\mu\theta_\mu]_{S_\mu}.$$

□

LEMMA 4.15. *If $c_1 = 0$ and $c_2 = 0$ and $c_3 c_{4+k} \neq 0$ and $c_l = 0$ for $5 \leq l < 4+k$, then the algebraic restriction $\mathcal{A} = [\sum_{l=1}^{\mu} c_l\theta_l]_{S_\mu}$ can be reduced by a symmetry of S_μ to an algebraic restriction $[\theta_3 + \widetilde{c}_{4+k}\theta_{4+k} + +\widetilde{c}_{5+k}\theta_{5+k}]_{S_\mu}$.*

PROOF OF LEMMA 4.15. If $c_1 = 0, c_2 = 0$ and $c_3 \neq 0$ and $c_{4+k} \neq 0$ and $c_l = 0$ for $5 \leq l < 4+k$, then $\mathcal{A} = [c_3\theta_3 + c_4\theta_4 + \sum_{l=4+k}^{\mu} c_l\theta_l]_{S_\mu}$.
Let $\mathcal{B}_t = [c_3\theta_3 + (1-t)c_4\theta_4 + c_{4+k}\theta_{4+k} + \sum_{l=5+k}^{\mu} \widetilde{c}_l(t)\theta_l]_{S_\mu}$ for $t \in [0;1]$ where the $\widetilde{c}_l(t)$ are smooth functions $\widetilde{c}_l(t) : [0;1] \to \mathbb{R}$ such that $\widetilde{c}_l(0) = c_l$. Then $\mathcal{B}_0 = \mathcal{A}$ and $\mathcal{B}_1 = [c_3\theta_3 + c_{4+k}\theta_{4+k} + \sum_{l=5+k}^{\mu} \widetilde{c}_l(1)\theta_l]_{S_\mu}$.

Let Φ_t, $t \in [0;1]$, be the flow of the vector field $V = \frac{c_4}{rc_3}X_3$. We show that there exist functions $\widetilde{c}_l$ such that

$$\Phi_t^*\mathcal{B}_t = \mathcal{B}_0, \ \Phi_0 = id. \tag{4.17}$$

Then differentiating (4.17) we obtain

$$\mathcal{L}_V\mathcal{B}_t = [c_4\theta_4 - \sum_{l=5+k}^{\mu} \frac{d\widetilde{c}_l}{dt}\theta_l]_{S_\mu}. \tag{4.18}$$

We can find the $\widetilde{c}_l$ as solutions of the system of first order linear ODEs defined by (4.18) with the initial data $\widetilde{c}_l(0) = c_l$ for $l = 5+k, \ldots, \mu$. This implies that $\mathcal{B}_0 = \mathcal{A}$ and $\mathcal{B}_1 = [c_3\theta_3 + c_{4+k}\theta_{4+k} + \sum_{l=5+k}^{\mu} \widetilde{c}_l(1)\theta_l]_{S_\mu}$ are diffeomorphic. Denote $\hat{c}_l = \widetilde{c}_l(1)$ for $l = 5+k, \ldots, \mu$.

Next let $\mathcal{C}_t = [c_3\theta_3 + c_{4+k}\theta_{4+k} + \hat{c}_{5+k}\theta_{5+k} + (1-t)\sum_{l=6+k}^{\mu} \hat{c}_l\theta_l]_{S_\mu}$ for $t \in [0;1]$. Then $\mathcal{C}_0 = \mathcal{B}_1$ and $\mathcal{C}_1 = [c_3\theta_3 + c_{4+k}\theta_{4+k} + \hat{c}_{5+k}\theta_{5+k}]_{S_\mu}$.

We prove that there exists a family $\Upsilon_t \in Symm(S_\mu)$, $t \in [0;1]$ such that

$$\Upsilon_t^*\mathcal{C}_t = \mathcal{C}_0, \ \Upsilon_0 = id. \tag{4.19}$$

Let V_t be a vector field defined by $\frac{d\Upsilon_t}{dt} = V_t(\Upsilon_t)$. Then differentiating (4.19) we obtain

$$\mathcal{L}_{V_t}\mathcal{B}_t = [\sum_{l=6+k}^{\mu} \hat{c}_l\theta_l]_{S_\mu}. \tag{4.20}$$

We are looking for V_t in the form $V_t = \sum_{k=4}^{\mu-2} b_k(t)X_k$ where the $b_k(t)$ are smooth functions $b_k : [0;1] \to \mathbb{R}$ for $k = 4,\dots,\mu-2$. Then by Proposition 4.10 equation (4.20) has a form

$$\begin{bmatrix} (r+2k+6)c_{k+4} & 0 & 0 & \cdots & \cdots & 0 \\ (r+2k+8)\hat{c}_{k+5} & (r+2k+8)c_{k+4} & 0 & \cdots & \cdots & 0 \\ \vdots & \vdots & \ddots & 0 & \cdots & 0 \\ 3r(1-t)\hat{c}_{\mu-1} & 3r\hat{c}_{\mu-2}(1-t) & \cdots & 3rc_{k+4} & 0\cdots & 0 \end{bmatrix} \begin{bmatrix} b_4(t) \\ b_5(t) \\ \vdots \\ b_{\mu-2}(t) \end{bmatrix} = \begin{bmatrix} \hat{c}_{k+6} \\ \hat{c}_{k+7} \\ \vdots \\ \hat{c}_\mu \end{bmatrix} \tag{4.21}$$

If $c_{4+k} \neq 0$ we can solve (4.21) and Υ_t may be obtained as a flow of the vector field V_t. The family Υ_t preserves S_μ, because V_t is tangent to S_μ and $\Upsilon_t^*\mathcal{C}_t = \mathcal{C}_0 = \mathcal{B}_1$. Using the homotopy arguments we have that $\mathcal{A}$ is diffeomorphic to $\mathcal{B}_1$ and $\mathcal{B}_1$ is diffeomorphic to $\mathcal{C}_1$. By the condition $c_3 \neq 0$ we have a diffeomorphism $\Psi \in Symm(S_\mu)$ of the form

$$\Psi : (x_1,x_2,x_3) \mapsto (|c_3|^{-\frac{1}{2}}x_1, |c_3|^{-\frac{1}{2}}x_2, |c_3|^{-\frac{1}{r}}x_3), \tag{4.22}$$

and we obtain

$$\Psi^*(\mathcal{C}_1) = [\frac{c_3}{|c_3|}\theta_3 + \widetilde{c}_{4+k}\theta_{4+k} + \widetilde{c}_{5+k}\theta_{5+k}]_{S_\mu} = [sgn(c_3)\theta_3 + \widetilde{c}_{4+k}\theta_{4+k} + \widetilde{c}_{5+k}\theta_{5+k}]_{S_\mu}.$$

By the following symmetry of S_μ: $(x_1,x_2,x_3) \mapsto (-x_1,x_2,x_3)$, we have that $[-\theta_3 + \widetilde{c}_{4+k}\theta_{4+k} + \widetilde{c}_{5+k}\theta_{5+k}]_{S_\mu}$ is diffeomorphic to $[\theta_3 - \widetilde{c}_{4+k}\theta_{4+k} - \widetilde{c}_{5+k}\theta_{5+k}]_{S_\mu}$.

□

LEMMA 4.16. *If $c_1 = 0$ and $c_2 = 0$ and $c_3 \neq 0$ and $c_l = 0$ for $5 \le l < \mu - 1$, then the algebraic restriction $\mathcal{A} = [\sum_{l=1}^{\mu} c_l\theta_l]_{S_\mu}$ can be reduced by a symmetry of S_μ to an algebraic restriction $[\theta_3 + \widetilde{c}_{\mu-1}\theta_{\mu-1}]_{S_\mu}$.*

PROOF OF LEMMA 4.16. The proof of this lemma is very similar to the previous case. It suffices to notice that if $c_3 \neq 0$ we can solve the equation

$$\begin{bmatrix} 0 & rc_3 \\ -\frac{3r^2}{2}c_3 & 3rc_{\mu-1} \end{bmatrix} \begin{bmatrix} b_2 \\ b_3 \end{bmatrix} = \begin{bmatrix} c_4 \\ c_\mu \end{bmatrix} \tag{4.23}$$

□

LEMMA 4.17. *If $c_1 = c_2 = c_3 = 0$ and $c_{4+k} \neq 0$ and $c_l = 0$ for $5 \le l < 4+k$, then the algebraic restriction $\mathcal{A} = [\sum_{l=1}^{\mu} c_l\theta_l]_{S_\mu}$ can be reduced by a symmetry of S_μ to an algebraic restriction $[c_4\theta_4 + c_{4+k}\theta_{4+k}]_{S_\mu}$.*

PROOF OF LEMMA 4.17. We use similar methods as above to prove this lemma. In this case $\mathcal{A} = [c_4\theta_4 + \sum_{l=4+k}^{\mu} c_l\theta_l]_{S_\mu}$. Let $\mathcal{B}_t = [c_4\theta_4 + c_{4+k}\theta_{4+k} + (1-t)\sum_{l=5+k}^{\mu} c_l\theta_l]_{S_\mu}$ for $t \in [0;1]$. Then $\mathcal{B}_0 = \mathcal{A}$ and $\mathcal{B}_1 = [c_4\theta_4 + c_{4+k}\theta_{4+k}]_{S_\mu}$. We prove that there exists a family $\Phi_t \in Symm(S_\mu)$, $t \in [0;1]$ such that

$$\Phi_t^*\mathcal{B}_t = \mathcal{B}_0,\ \Phi_0 = id. \tag{4.24}$$

Let V_t be a vector field defined by $\frac{d\Phi_t}{dt} = V_t(\Phi_t)$. Then differentiating (4.24) we obtain

$$\mathcal{L}_{V_t}\mathcal{B}_t = [\sum_{l=5+k}^{\mu} c_l\theta_l]_{S_\mu}. \tag{4.25}$$

We are looking for V_t in the form $V_t = \sum_{k=3}^{\mu-2} b_k(t)X_k$ where the $b_k(t)$ are smooth functions $b_k : [0;1] \to \mathbb{R}$ for $k = 3,\dots,\mu-2$. Then by Proposition 4.10 equation (4.25) has a form

$$\begin{bmatrix} (r+2k+4)c_{k+4} & 0 & 0 & \cdots & \cdots & 0 \\ (r+2k+6)c_{k+5}(1-t) & (r+2k+6)c_{k+4} & 0 & \cdots & \cdots & 0 \\ \vdots \quad \ddots & & \ddots & \ddots & 0 & \cdots\ 0 \\ 3rc_{\mu-1}(1-t) & 3rc_{\mu-2}(1-t) & \cdots & 3rc_{k+4} & 0 \cdots & 0 \end{bmatrix} \begin{bmatrix} b_3(t) \\ \vdots \\ b_{\mu-2}(t) \end{bmatrix} = \begin{bmatrix} c_{k+5} \\ c_{k+6} \\ \vdots \\ c_\mu \end{bmatrix} \tag{4.26}$$

If $c_{4+k} \neq 0$ we can solve (4.26) and Φ_t may be obtained as a flow of the vector field V_t. The family Φ_t preserves S_μ, because V_t is tangent to S_μ and $\Phi_t^*\mathcal{B}_t = \mathcal{A}$. Using the homotopy arguments we have that $\mathcal{A}$ is diffeomorphic to $\mathcal{B}_1 = [c_4\theta_4 + c_{4+k}\theta_{4+k}]_{S_\mu}$.

When $c_4 \neq 0$ we have a diffeomorphism $\Psi \in Symm(S_\mu)$ of the form

$$\Psi : (x_1,x_2,x_3) \mapsto (|c_4|^{-\frac{r}{2r+2}}x_1, |c_4|^{-\frac{r}{2r+2}}x_2, |c_4|^{-\frac{2}{2r+2}}x_3), \tag{4.27}$$

and we obtain

$$\Psi^*(\mathcal{B}_1) = [sgn(c_4)\theta_4 + c_{4+k}|c_4|^{-(\frac{2k+r+2}{2r+2})}\theta_{4+k}]_{S_\mu} = [\pm\theta_4 + \widetilde{c}_{4+k}\theta_{4+k}]_{S_\mu}.$$

By the following symmetry of S_μ: $(x_1,x_2,x_3) \mapsto (-x_1,x_2,x_3)$, we have that $[-\theta_4 + \widetilde{c}_{4+k}\theta_{4+k}]_{S_\mu}$ is diffeomorphic to $[\theta_4 - \widetilde{c}_{4+k}\theta_{4+k}]_{S_\mu}$.

When $c_{4+k} \neq 0$ then we may use a diffeomorphism $\Psi_1 \in Symm(S_\mu)$ of the form

$$\Psi_1 : (x_1,x_2,x_3) \mapsto (c_{4+k}^{-\frac{r}{2k+r+2}}x_1, c_{4+k}^{-\frac{r}{2k+r+2}}x_2, c_{4+k}^{-\frac{2}{2k+r+2}}x_3), \tag{4.28}$$

and we obtain

$$\Psi_1^*(\mathcal{B}_1) = [c_4 c_{4+k}^{-(\frac{2r+2}{2k+r+2})}\theta_4 + \theta_{4+k}]_{S_\mu} = [\widetilde{c}_4\theta_4 + \theta_{4+k}]_{S_\mu}.$$

□

Statement (ii) of Theorem 4.4 follows from Theorem 4.9.

(iii) Now we prove that the parameters c_i are moduli in the normal forms. The proofs are very similar in all cases. We consider as an example the normal form with two parameters $[\theta_1 + c_2\theta_2 + c_3\theta_3]_{S_\mu}$. From Table 6 we see that the tangent space to the orbit of $[\theta_1 + c_2\theta_2 + c_3\theta_3]_{S_\mu}$ at $[\theta_1 + c_2\theta_2 + c_3\theta_3]_{S_\mu}$ is spanned by the linearly independent algebraic restrictions $[r\theta_1 + rc_2\theta_2 + 2c_3\theta_3]_{S_\mu}$, $[\theta_4]_{S_\mu}$, $[\theta_5]_{S_\mu}, \dots, [\theta_\mu]_{S_\mu}$. Hence the algebraic restrictions $[\theta_2]_{S_\mu}$ and $[\theta_3]_{S_\mu}$ do not belong to it. Therefore the parameters c_2 and c_3 are independent moduli in the normal form $[\theta_1+c_2\theta_2+c_3\theta_3]_{S_\mu}$.

Statement (iv) of Theorem 4.4 follows from the conditions in the proof of part (i) (codimension) and from Theorem 2.6 and Propositions 3.4 and 2.7 (index of isotropy).

□

References

[A1] V. I. Arnold, *First step of local symplectic algebra*, Differential topology, infinite-dimensional Lie algebras, and applications. D. B. Fuchs' 60th anniversary collection. Providence, RI: American Mathematical Society. Transl., Ser. 2, Am. Math. Soc. 194(44), 1999,1-8. MR1729356 (2001h:58055)

[AG] V. I. Arnold, A. B. Givental *Symplectic geometry*, in Dynamical systems, IV, 1-138, Encyclopedia of Matematical Sciences, vol. 4, Springer, Berlin, 2001. MR1866631

[AVG] V. I. Arnold, S. M. Gusein-Zade, A. N. Varchenko, *Singularities of Differentiable Maps*, Vol. 1, Birhauser, Boston, 1985. MR777682 (86f:58018)

[D] W. Domitrz, *Local symplectic algebra of quasi-homogeneous curves*, Fundamentae Mathematicae 204 (2009), 57-86. MR2507690 (2010j:58092)

[DJZ1] W. Domitrz, S. Janeczko, M. Zhitomirskii, *Relative Poincare lemma, contractibility, quasi-homogeneity and vector fields tangent to a singular variety*, Ill. J. Math. 48, No.3 (2004), 803-835. MR2114253 (2006b:32010)

[DJZ2] W. Domitrz, S. Janeczko, M. Zhitomirskii, *Symplectic singularities of varietes: the method of algebraic restrictions*, J. reine und angewandte Math. 618 (2008), 197-235. MR2404750 (2009e:53098)

[DR] W. Domitrz, J. H. Rieger, *Volume preserving subgroups of $\mathcal{A}$ and $\mathcal{K}$ and singularities in unimodular geometry*, Mathematische Annalen 345 (2009), 783-817. MR2545866 (2011d:58096)

[DT] W. Domitrz, Ż. Trębska, *Symplectic T_7, T_8 singularities and Lagrangian tangency orders*, to appear in Proceedings of the Edinburgh Mathematical Society.

[G] M. Giusti, *Classification des singularités isolées d'intersections complètes simples*, C. R. Acad. Sci., Paris, Sér. A 284 (1977), 167-170 . MR0440067 (55:12948)

[IJ1] G. Ishikawa, S. Janeczko, *Symplectic bifurcations of plane curves and isotropic liftings*, Q. J. Math. **54**, No.1 (2003), 73-102. MR1967071 (2004b:58063)

[IJ2] G. Ishikawa, S. Janeczko, *Symplectic singularities of isotropic mappings*, Geometric singularity theory, Banach Center Publications **65** (2004), 85-106. MR2104340 (2005h:58073)

[K] P. A. Kolgushkin, *Classification of simple multigerms of curves in a space endowed with a symplectic structure*, St. Petersburg Math. J. **15** (2004), no. 1, 103-126. MR1979720 (2004g:58054)

[L] E. J. M. Looijenga *Isolated Singular Points on Complete Intersections*, London Mathematical Society Lecture Note Series 77, Cambridge University Press 1984. MR747303 (86a:32021)

[W] C. T. C. Wall, *Singular points of plane curves*, London Mathematical Society Student Texts, 63, Cambridge University Press, Cambridge, 2004. MR2107253 (2005i:14031)

[Z] M. Zhitomirskii, *Relative Darboux theorem for singular manifolds and local contact algebra*, Can. J. Math. **57**, No.6 (2005), 1314-1340. MR2178564 (2006f:53116)

WARSAW UNIVERSITY OF TECHNOLOGY, FACULTY OF MATHEMATICS AND INFORMATION SCIENCE, PLAC POLITECHNIKI 1, 00-661 WARSAW, POLAND

E-mail address: domitrz@mini.pw.edu.pl

WARSAW UNIVERSITY OF TECHNOLOGY, FACULTY OF MATHEMATICS AND INFORMATION SCIENCE, PLAC POLITECHNIKI 1, 00-661 WARSAW, POLAND

E-mail address: ztrebska@mini.pw.edu.pl

Contemporary Mathematics
Volume **569**, 2012
http://dx.doi.org/10.1090/conm/569/11242

Topology of the real Milnor fiber for isolated singularities

R. Araújo dos Santos, D. Dreibelbis, and N. Dutertre

Abstract. Let $\psi : (\mathbb{R}^n, 0) \to (\mathbb{R}^p, 0)$ be a real analytic map-germ with an isolated singularity at the origin. In these notes we use tools from singularity theory and Morse theory to provide a formula for the Euler characteristic of the fiber of Milnor fibrations. As application, we use this to show a topological/geometric description of the Milnor fiber for mapping from $(\mathbb{R}^4, 0)$ to $(\mathbb{R}^2, 0)$.

1. Introduction

It is well known that the existence of the local Milnor fibrations for real and complex singularities is a major tool in the study of the topology of singularities. In [**Ki**] H. King showed that a necessary and sufficient condition for two real analytic map-germs with isolated singularities at the origin to be $\mathcal{C}^0$-$\mathcal{A}$-equivalent is that their respective links are invertible cobordant and the local Milnor fibrations are isomorphic. It means that, in order to do a classification in the space of real analytic map germs, we can study their respective Milnor fibrations.

Although in the complex setting the Milnor fibration and several results about the topology of Milnor's fiber are well understood, the same cannot be said in the real case. For instance, only recently some results about the existence of Milnor's fibration for non-isolated singularities were obtained, see for instance [**AT, TYA, GA, CA, RS**] and references. However, in some cases there are interesting formulas for the Euler characteristic of the real Milnor fiber, see [**Du3**] and references. Let us see next a short account on some formulas about the topology of the real Milnor fiber for real functions and mappings.

Let $f : (\mathbb{R}^n, 0) \to (\mathbb{R}, 0)$ be an analytic function-germ with an isolated critical point at the origin. Khimshiashvili proved in [**Kh**] the following formula:

$$\chi\big(f^{-1}(\delta) \cap B_\varepsilon\big) = 1 - \mathrm{sign}(-\delta)^n \deg_0 \nabla f,$$

where $0 < |\delta| \ll \varepsilon \ll 1$, ∇f is the gradient of f and $\deg_0 \nabla f$ is the topological degree of the mapping $\frac{\nabla f}{|\nabla f|} : S_\varepsilon \to S_1$.

2010 *Mathematics Subject Classification.* 14P25, 58K15, 58K65.

Key words and phrases. Real Milnor fibration, geometry and topology of the real fibers, topology of real singularities.

The first author is thankful for the support from the Brazilian CNPq, PQ-2, proc. num. 305183/2009-5 and FAPESP/São Paulo proc. num. 2009/14383-3.

The third author is supported by Agence Nationale de la Recherche (reference ANR-08-JCJC-0118-01).

Now let $\psi = (f_1, \dots, f_k) : (\mathbb{R}^n, 0) \to (\mathbb{R}^k, 0)$, $2 \leq k \leq n$, be an analytic map-germ and let us denote by ϕ the map-germ $(f_1, \dots, f_{k-1}) : (\mathbb{R}^n, 0) \to (\mathbb{R}^{k-1}, 0)$. We assume that $\psi^{-1}(0)$ and $\phi^{-1}(0)$ have isolated singularities at 0 (note that here the maps ϕ and ψ do not need to have isolated critical points at the origin). Some authors investigated the following difference:

$$D_{\delta,\alpha} = \chi\big(\phi^{-1}(\delta) \cap \{f_k \geq \alpha\} \cap B_\varepsilon\big) - \chi\big(\phi^{-1}(\delta) \cap \{f_k \leq \alpha\} \cap B_\varepsilon\big),$$

where (δ, α) is a regular value of ψ such that $0 \leq |\alpha| \ll |\delta| \ll \varepsilon$.

In [**Du2**], the third author proved that:

$$D_{\delta,\alpha} \equiv \dim_{\mathbb{R}} \frac{\mathcal{O}_{\mathbb{R}^n,0}}{I} \bmod 2,$$

where $\mathcal{O}_{\mathbb{R}^n,0}$ is the ring of analytic function-germs at the origin and I is the ideal generated by $f_1, \dots, f_{k-1}$ and all the $k \times k$ minors $\frac{\partial(f_k, f_1, \dots, f_{k-1})}{\partial(x_{i_1}, \dots, x_{i_k})}$. This is only a mod 2 relation and we may ask if it is possible to get a more precise relation.

When $k = n$ and $f_k = x_1^2 + \cdots + x_n^2$, according to Aoki et al. ([**AFN1**], [**AFS**]), $D_{\delta,0} = \chi\big(\phi^{-1}(\delta) \cap B_\varepsilon\big) = 2\deg_0 H$ and $2\deg_0 H$ is the number of semi-branches of $\phi^{-1}(0)$, where

$$H = \Big(\frac{\partial(f_n, f_1, \dots, f_{n-1})}{\partial(x_1, \dots, x_n)}, f_1, \dots, f_{n-1}\Big).$$

They proved a similar formula in the case $f_k = x_n$ in [**AFN**] and Szafraniec generalized all these results to any f_k in [**Sz**].

When $k = 2$ and $f_2 = x_1$, Fukui [**Fu1**] stated that

$$D_{\delta,0} = -\text{sign}(-\delta)^n \deg_0 H,$$

where $H = (f_1, \frac{\partial f_1}{\partial x_2}, \dots, \frac{\partial f_1}{\partial x_n})$. Several generalizations of Fukui's formula are given in [**Fu1**], [**Du1**], [**FK**] and [**Du3**]. Note that in [**Du3**], the third author gave degree formulas for the Euler characteristic of map-germs from $(\mathbb{R}^n, 0)$ to $(\mathbb{R}^2, 0)$ called partially parallelizable.

In this paper we aim to state some general results about the Euler characteristic of the Milnor fiber and use them to do a topological/geometrical description of the fibers of the real Milnor fibration in low dimension (see Section 4). For this, we will use techniques and tools from singularity theory, Morse theory and differential geometry.

2. Tools from Morse Theory

We recall here results about Morse theory for manifolds with boundary. Our reference is [**HL**] where the results are given for a C^∞ manifold M with boundary ∂M. For simplicity, we will present the results for a manifold with boundary of the type $M \cap \{g \leq 0\}$, where M is a C^∞ manifold and $g : M \to \mathbb{R}$ is a C^∞ function such that 0 is a regular value of g. In fact this is the case that we will need in the following section.

Let M be a C^∞ manifold of dimension d. Let $g : M \to \mathbb{R}$ be a C^∞ function such that 0 is a regular value of g. We assume that $M \cap g^{-1}(0)$ is not empty. This implies that $M \cap \{g \leq 0\}$ is a smooth manifold with boundary. Let $f : M \to \mathbb{R}$ be a smooth function. A critical point of $f_{|M \cap \{g \leq 0\}}$ is a critical point of $f_{|M \cap \{g < 0\}}$ or a critical point of $f_{|M \cap g^{-1}(0)}$.

DEFINITION 2.1. Let $q \in M \cap g^{-1}(0)$. We say that q is a correct critical point of $f_{|M\cap\{g\leq 0\}}$ if q is a critical point of $f_{|M\cap\{g=0\}}$ and q is not a critical point of $f_{|M}$.

We say that q is a correct non-degenerate critical point of $f_{|M\cap\{g\leq 0\}}$ if q is a correct critical point of $f_{|M\cap\{g\leq 0\}}$ and q is a non-degenerate critical point of $f_{|M\cap g^{-1}(0)}$.

If q is a correct critical point of $f_{|M\cap\{g\leq 0\}}$ then $\nabla f(q) \neq 0$, $\nabla f(q)$ and $\nabla g(q)$ are colinear and there exists $\tau(q) \neq 0$ such that $\nabla f(q) = \tau(q)\nabla g(q)$.

DEFINITION 2.2. Let q be a correct critical point of $f_{|M\cap\{g\leq 0\}}$. We say that $\nabla f(q)$ points inwards (resp. outwards) if $\tau(q) < 0$ (resp. $\tau(q) > 0$).

DEFINITION 2.3. Let $f : M \to \mathbb{R}$ be a C^∞ function. We say that $f_{|M\cap\{g\leq 0\}}$ is correct if all the critical points of $f_{|M\cap g^{-1}(0)}$ are correct.

We say that $f_{|M\cap\{g\leq 0\}}$ is a Morse correct function if $f_{|M\cap\{g<0\}}$ admits only non-degenerate critical points and if $f_{|M\cap\{g=0\}}$ admits only non-degenerate correct critical points.

PROPOSITION 2.4. *Let M be a C^∞ manifold and let $g : M \to \mathbb{R}$ be a smooth function such that 0 is a regular value of g. The set of C^∞ functions $f : M \to \mathbb{R}$ such that $f_{|M\cap\{g\leq 0\}}$ is a Morse correct function is dense in $C^\infty(M,\mathbb{R})$.*

THEOREM 2.5. *Let M be a C^∞ manifold and let $g : M \to \mathbb{R}$ be a smooth function such that 0 is a regular value of g. We assume that $M \cap \{g \leq 0\}$ is compact. Let $f : M \to \mathbb{R}$ be such that $f_{|M\cap\{g\leq 0\}}$ is Morse correct. Let $\{p_i\}$ be the set of critical points of $f_{|M\cap\{g<0\}}$ and let $\{\lambda_i\}$ be the set of their respective indices. Let $\{q_j\}$ be the set of critical points of $f_{|M\cap\{g=0\}}$ and let $\{\mu_j\}$ be the set of their respective indices. Then we have:*

$$\chi(M\cap\{g\leq 0\}\cap\{f\geq 0\}) - \chi(M\cap\{g\leq 0\}\cap\{f=0\}) = \sum_{i:f(p_i)>0} (-1)^{\lambda_i} + \sum_{j:\substack{f(q_j)>0\\ \tau(q_j)<0}} (-1)^{\mu_j}$$

$$\chi(M\cap\{g\leq 0\}\cap\{f\leq 0\}) - \chi(M\cap\{g\leq 0\}\cap\{f=0\}) = (-1)^d \sum_{i:f(p_i)<0} (-1)^{\lambda_i} + (-1)^{d-1} \sum_{j:\substack{f(q_j)<0\\ \tau(q_j)>0}} (-1)^{\mu_j}.$$

3. Euler Characteristic for Real Mappings Fibers

Let $\psi = (f_1,\ldots,f_k) : (\mathbb{R}^n,0) \to (\mathbb{R}^k,0)$, $2 \leq k \leq n$, be a real analytic map-germ with an isolated critical point at the origin. This means that 0 is isolated in

$$\Sigma(\psi) = \{x \in \mathbb{R}^n \mid \mathrm{rank}(\nabla f_1(x),\ldots,\nabla f_k(x)) < k\}.$$

This implies that for any $l \in \{1,\ldots,k\}$ and for any l-tuple $(i_1,\ldots,i_l)$ of pairwise distinct elements of $\{1,\ldots,k\}$, the mapping $(f_{i_1},\ldots,f_{i_l}) : (\mathbb{R}^n,0) \to (\mathbb{R}^l,0)$ has an isolated critical point at the origin as well, for otherwise 0 would not be isolated in $\Sigma(\psi)$.

In what follows we are going to relate the Euler characteristic of the Milnor fiber of ψ to the topological degrees of the ∇f_i's at the origin. For this, we apply the tools from Section 2 about Morse theory for manifolds with boundary.

Let $\phi = (f_1, \ldots, f_{k-1}) : (\mathbb{R}^n, 0) \to (\mathbb{R}^{k-1}, 0)$ and $V = \phi^{-1}(0)$. Since ϕ has an isolated critical point at the origin, V is an analytic set with an isolated singularity at 0. Furthermore, $f_{k|V\setminus\{0\}}$ has no critical point in a neighborhood of the origin.

Let $\omega : \mathbb{R}^n \to \mathbb{R}$ be defined by $\omega(x) = x_1^2 + \cdots + x_n^2$. It is known that for $\varepsilon > 0$ small enough, the sphere $S_\varepsilon = \omega^{-1}(\varepsilon^2)$ intersects V transversally. Let p^ε be a critical point of $f_{k|V\cap S_\varepsilon}$. This means that there exists $\tau(p^\varepsilon)$ such that:

$$\nabla f_{k|V}(p^\varepsilon) = \tau(p^\varepsilon)\nabla\omega_{|V}(p^\varepsilon).$$

Note that $\tau(p^\varepsilon) \neq 0$ since f_k has no critical point on $V \setminus \{0\}$.

LEMMA 3.1. *If ε is small enough then $f_k(p^\varepsilon) \neq 0$. Furthermore $\nabla f_{k|V}(p^\varepsilon)$ points outwards (resp. inwards) for $V \cap B_\varepsilon$ if and only if $f_k(p^\varepsilon) > 0$ (resp. $f_k(p^\varepsilon) < 0$).*

Proof. If for any ε small enough, there is a critical point p^ε of $f_{k|V\cap S_\varepsilon}$ such that $f_k(p^\varepsilon) = 0$ then by the Curve Selection Lemma, there is a smooth subanalytic curve $p : [0, \nu] \to V$, $p(0) = 0$, such that $p(t)$ is a critical point of $f_{k|V\cap S_{\|p(t)\|}}$ and $f_k(p(t)) = 0$. Hence, we have:

$$0 = (f_k \circ p)'(t) = \langle \nabla f_k(p(t)), p'(t)\rangle = \tau(p(t))\langle \nabla\omega_{|V}(p(t)), p'(t)\rangle,$$

because $p'(t)$ lies in $T_{p(t)}V$. Therefore $(\omega \circ p)'(t) = 0$ and $\omega \circ p$ is constant. But $\omega(p(t))$ tends to 0 as t tends to 0, so $\omega\circ p$ is zero everywhere, which is a contradiction.

Now let us assume that $\tau(p^\varepsilon) > 0$. By the Curve Selection Lemma, there exists a smooth subanalytic curve $p : [0, \nu] \to V$ passing through p^ε such that $p(0) = 0$ and for $t \neq 0$, $p(t)$ is a critical point of $f_{k|V\cap S_{\|p(t)\|}}$ with $\tau(p(t)) > 0$. Therefore we have:

$$(f_k \circ p)'(t) = \tau(p(t))\langle \nabla\omega_{|V}(p(t)), p'(t)\rangle = \tau(p(t))(\omega \circ p)'(t).$$

But $(\omega\circ p)' > 0$ for otherwise $(\omega\circ p)' \leq 0$ and $\omega\circ p$ would be decreasing. Since $\omega(p(t))$ tends to 0 as t tends to 0, this would imply that $\omega \circ p(t) \leq 0$, which is impossible. Hence we can conclude that $(f_k \circ p)' > 0$ and $f_k \circ p$ is strictly increasing. Since $f_k \circ p(t)$ tends to 0 as t tends to 0, we see that $f_k \circ p(t) > 0$. Similarly if $\tau(p^\varepsilon) < 0$ then $f_k(p^\varepsilon) < 0$. □

Let δ be a small regular value of ϕ. The following lemma deals with critical points of $f_{k|\phi^{-1}(\delta)\cap B_\varepsilon}$ lying in $\phi^{-1}(\delta) \cap S_\varepsilon$.

LEMMA 3.2. *For all δ and ε such that $0 < |\delta| \ll \varepsilon \ll 1$, $f_{k|\phi^{-1}(\delta)\cap B_\varepsilon}$ is a correct function. Furthermore, we have:*

- *at all correct critical points of $f_{k|\phi^{-1}(\delta)\cap S_\varepsilon}$ with $f_k > 0$, $\nabla f_{k|\phi^{-1}(\delta)}$ points outwards,*
- *at all correct critical points of $f_{k|\phi^{-1}(\delta)\cap S_\varepsilon}$ with $f_k < 0$, $\nabla f_{k|\phi^{-1}(\delta)}$ points inwards,*
- *there are no correct critical points of $f_{k|\phi^{-1}(\delta)\cap S_\varepsilon}$ in ${f_k}^{-1}(0)$.*

PROOF. Since $f_{k|V\cap B_\varepsilon}$ is correct, $f_{k|\phi^{-1}(\delta)\cap B_\varepsilon}$ is also correct if δ is small enough. We prove the other items similarly, using the previous lemma and the fact that δ is close to 0. □

Next we prove the main result on this section, where we get the main relationship between the Euler characteristic of a generic fiber and the topological degree of the coordinate functions.

PROPOSITION 3.3. *Given ψ and ϕ as above, the following holds:*

(i) *If n is even, we have:*

$$\chi\big(\phi^{-1}(\delta) \cap {f_k}^{-1}(0) \cap B_\varepsilon\big) = 1 - \deg_0 \nabla f_1;$$

(ii) *If n is odd, we have* $\deg_0 \nabla f_1 = 0$ *and:*

$$\chi\big(\phi^{-1}(\delta) \cap {f_k}^{-1}(0) \cap B_\varepsilon\big) = 1.$$

PROOF. Applying Morse theory for manifolds with boundary to $f_{k|\phi^{-1}(\delta)\cap B_\varepsilon}$, we have:

$$\chi\big(\phi^{-1}(\delta) \cap \{f_k \geq 0\} \cap B_\varepsilon\big) - \chi\big(\phi^{-1}(\delta) \cap f_k^{-1}(0) \cap B_\varepsilon\big) = 0,$$

because $f_{k|\phi^{-1}(\delta)}$ has no critical points for $\Sigma(\psi) = \{0\}$ and because, by the previous lemma, $\nabla f_{k|\phi^{-1}(\delta)\cap B_\varepsilon}$ points outwards at the critical points of $f_{k|\phi^{-1}(\delta)\cap S_\varepsilon}$ lying in $\{f_k > 0\}$. Similarly, we have:

$$\chi\big(\phi^{-1}(\delta) \cap \{f_k \leq 0\} \cap B_\varepsilon\big) - \chi\big(\phi^{-1}(\delta) \cap {f_k}^{-1}(0) \cap B_\varepsilon\big) = 0.$$

Summing these two equalities up and using the Mayer-Vietoris sequence, we obtain that:

$$\chi\big(\phi^{-1}(\delta) \cap B_\varepsilon\big) = \chi\big(\phi^{-1}(\delta) \cap {f_k}^{-1}(0) \cap B_\varepsilon\big).$$

Applying this procedure $k-1$ times, we obtain that:

$$\chi\big(\phi^{-1}(\delta) \cap {f_k}^{-1}(0) \cap B_\varepsilon\big) = \chi\big({f_1}^{-1}(\alpha_1) \cap B_\varepsilon\big),$$

where α_1 is a small regular value of f_1.

By Khimshiashvili's formula, we know that:

$$\chi\big({f_1}^{-1}(\alpha_1) \cap B_\varepsilon\big) = 1 - \text{sign}(-\alpha_1)^n \deg_0 \nabla f_1.$$

Hence, if n is even, we find that:

$$\chi\big(\phi^{-1}(\delta) \cap {f_k}^{-1}(0) \cap B_\varepsilon\big) = 1 - \deg_0 \nabla f_1.$$

If n is odd, just changing α_1 to $-\alpha_1$, we get that $\deg_0 \nabla f_1 = 0$ and $\chi\big(\phi^{-1}(\delta) \cap {f_k}^{-1}(0) \cap B_\varepsilon\big) = 1$. □

COROLLARY 3.4. *Let γ be a small regular value of ψ. If n is even, we have:*

$$\chi\big(\psi^{-1}(\gamma) \cap B_\varepsilon\big) = 1 - \deg_0 \nabla f_1 = \cdots = 1 - \deg_0 \nabla f_k,$$

and $\deg_0 \nabla f_1 = \cdots = \deg_0 \nabla f_k$. *If n is odd, we have* $\deg_0 \nabla f_1 = \cdots = \deg_0 \nabla f_k = 0$ *and* $\chi\big(\psi^{-1}(\gamma) \cap B_\varepsilon\big) = 1$.

4. Topological-Geometrical Description - Application to Low Dimensions

Given a map germ $\psi : (\mathbb{R}^n, 0) \to (\mathbb{R}^p, 0)$ with an isolated singularity at the origin, in [**Mi**], Milnor proposed to call it trivial if the fiber of the Milnor fibration is diffeomorphic to an $(n-p)-$dimensional closed disc, and he asked ([**Mi**], p. 100): "For what dimensions $n \geq p \geq 2$ do non-trivial examples exist?"

In [**CL**] P. Church and K. Lamotke used a construction of E. Looijenga (see [**Lo**] for details) and proved for what pairs (n, p) non-trivial examples exist. Their main result states that for $0 \leq n - p \leq 2$ non-trivial examples occur precisely for the pairs $(2, 2)$, $(4, 3)$ and $(4, 2)$. For $n - p = 3$, all examples are trivial except for the pairs $(5, 2)$, $(8, 5)$ and possibly $(6, 3)$, and if the Poincaré conjecture is true, for $n - p \geq 4$ non-trivial examples occur for all pairs (n, p).

In the same paper the authors proved the following interesting result which says for what pairs of dimensions (n,p), the point $0 \in \mathbb{R}^n$ is isolated in $\psi^{-1}(0)$. This is equivalent to saying that the link $K = \psi^{-1}(0) \cap S_\varepsilon$ is empty.

LEMMA 4.1. *If 0 is an isolated point of $\psi^{-1}(0)$, then $n = p$ or $(n,p) = (4,3),(8,5),(16,9)$. If $n = p$ the singularity is trivial unless $p = 2$. The other cases are never trivial.*

PROOF. See [**CL**], page 151.

EXAMPLE 4.2. All map-germs $\psi : (\mathbb{R}^3, 0) \to (\mathbb{R}^2, 0)$ with an isolated singularity at the origin are trivial.

EXAMPLE 4.3. For any map-germ $\psi : (\mathbb{R}^4, 0) \to (\mathbb{R}^2, 0)$ with an isolated singularity at origin the link K is not empty.

As a consequence of Lemma 4.4 below (see Corollary 4.5) we get an easy characterization of trivial fibration for the pair $(4,2)$.

4.1. Topology of the fiber for the case $n = 4$. The tools developed in Section 3 help us to do a geometric-topological description of the Milnor fibers in low dimensions. In this direction, let us consider the case of a real polynomial map-germ $\psi = (f,g) : (\mathbb{R}^4, 0) \to (\mathbb{R}^2, 0)$ with an isolated singularity at the origin. Recall that, since $p = 2$ by [**Mi**], Lemma 11.4, we know that the Milnor fiber is connected. Hence, if we denote its closure by F_ψ, we have that it is an orientable connected compact surface with boundary. Moreover, in [**Ru**] Lee Rudolph showed a relation between the Euler characteristic of F_ψ and $\beta_1(F_\psi)$, the first Betti number of the fiber F_ψ, which he defined as the Milnor number. It is

$$\chi(F_\psi) = 1 - \text{rank}(H_1(F_\psi)),$$

where, as usual, $H_1(F_\psi)$ stands for the first homology group of F_ψ. So, by Proposition 3.3 we have that

$$\beta_1(F_\psi) = \deg_0 \nabla f.$$

Therefore, as in the case of isolated singularities of complex curves, the topological degree of a coordinate function is the first Betti number of the Milnor fiber.

From the classification of orientable closed (compact and without boundary) connected surfaces S, it is known that $\chi(S)$ is an even integer and less than or equal to 2 and its Betti numbers are $\beta_0 = 1, \beta_1 = 2g(S), \beta_2 = 1$, where $g(S)$ is the genus of S.

On the other hand, consider an orientable compact surface with boundary, let us say S_B. Then the boundaries are closed curves diffeomorphic to circles and the Euler characteristic is

$$\chi(S_B) = \chi(S) - k,$$

where now S is the closed surface obtained from S_B after gluing k open discs to the boundaries.

We can gather all the information above in the following result:

LEMMA 4.4. *Let F be the closed surface obtained from F_ψ by gluing the open discs to the boundary and $k \geq 1$ the number of closed curves components of the boundaries. Then:*

(i.) $\chi(F) = k + 1 - \deg_0 \nabla f,$
(ii.) $2g(F) = \deg_0 \nabla f + 1 - k.$

PROOF. Item (i) is easy. To see item (ii), just use the fact that $\beta_1(F_\psi) = \beta_1(F) + k - 1$. □

COROLLARY 4.5. *Let* $\psi = (f, g) : (\mathbb{R}^4, 0) \to (\mathbb{R}^2, 0)$ *have an isolated singularity at the origin. Then,* ψ *is trivial if, and only if,* $\deg_0 \nabla f = 0$.

PROOF. If F_ψ is a closed disc, we have that $k = 1$ and gluing an open disc to the boundary we obtain F homeomorphic to the two-sphere. Applying, for instance, Lemma 4.4, item (ii.), we get that $\deg_0 \nabla f = 0$, since $g(F) = 0$. Conversely, if we assume that $\deg_0 \nabla f = 0$, by Lemma 4.4, item (i.), we have that $\chi(F) = k + 1$ and since $\chi(F) \leq 2$ and $k \geq 1$, we get $k = 1$. So, $\chi(F) = 2$ and F_ψ is a closed disc. □

EXAMPLE 4.6. Let $\psi(x, y, u, v) = (x, 3x^2y + y^3 + u^2 + v^2)$. It is easy to see that $\sum(f) = \{(0, 0, 0, 0)\}$. Since $\nabla f = (1, 0, 0, 0)$, then $\deg_0 \nabla f = 0$. So, ψ is a trivial fibration.

4.2. Geometrical/Topological description. Since we assume that the map $\psi = (f, g) : (\mathbb{R}^4, 0) \to (\mathbb{R}^2, 0)$ has an isolated singularity at the origin, then we have that $\deg_0 \nabla f$ is well defined in a small enough sphere centered at origin. Using Lemma 4.4, item (ii.), we can construct (see above) a table of possible topological types.

The notations D, S^2, T, T^2 stand for: the closed disc, the two-sphere, the torus and bitorus, respectively. For $S^2 - 2D$, for instance, we mean that the fiber is diffeomeorphic to *the two-sphere minus two open discs removed*, i.e. the cylinder, and so on.

To illustrate the situation, in the column under the pair (f, g) appear an example of a map-germ with the properties described in the row.

	$\deg_0 \nabla f$	k	Topology	$(f, g) : (\mathbb{R}^4, 0) \to (\mathbb{R}^2, 0)$
a)	0	1	D	$(x, 3x^2y + y^3 + u^2 + v^2)$
b)	1	2	$S^2 - 2D$	$z^2 + w^2$
c)	2	1	$T - D$	$z^2 + w^3$
d)	2	3	$S^2 - 3D$	$zw(\overline{z} + \overline{w})$
e)	2	3	$S^2 - 3D$	$z^3 + w^3 + zw(\overline{z} - z + \overline{w} - w)$
f)	3	2	$T - 2D$	$z^2 + w^2\overline{z}$
g)	3	4	$S^2 - 4D$	$\overline{zw}(z^2 + w^4)$
h)	4	1	$T^2 - D$	$z^2 + w^5$
i)	4	3	$T - 3D$	$w^3 + z^3$
j)	4	5	$S^2 - 5D$	$z^2\overline{w} + w^2\overline{z}$

REMARK 4.7. Examples b), c), h) and i) are Pham-Brieskorn singularities ([**Mi**], 71), and are complex. Example j) is a twisted Pham-Brieskorn singularity ([**Se, RSV**]), while example g) is based on a family of singularities studied in ([**PS**]). Example d) is from A'Campo's paper ([**AC**]) and we call it the *A'Campo example*, while example e) is apparently new. While the fibers of examples d) and e) are topologically equivalent, their Milnor fibrations are all distinct. A'Campo's example has a link consisting of three mutually linked circles, while the link for example e) has two of the three circles unlinked.

References

[AC] N. A'Campo, *Le Nombre de Lefschetz d'une Monodromie*, Indagationes Math **35** (1973), 113–118. MR0320364 (47:8903)

[AFN] K. Aoki, T. Fukuda, T. Nishimura, *On the number of branches of the zero locus of a map germ* $(\mathbf{R}^n, 0) \to (\mathbf{R}^{n-1}, 0)$, Topology and Computer Science: Proceedings of the Symposium held in honor of S. Kinoshita, H. Noguchi and T. Homma on the occasion of their sixtieth birhtdays, (1987), 347–363. MR1112604 (92e:58024)

[AFN1] K. Aoki, T. Fukuda, T. Nishimura, *An algebraic formula for the topological types of one parameter bifurcation diagrams*, Archive for Rational Mechanics and Analysis **108** (1989), 247–265. MR1012176 (90k:58024)

[AFS] K. Aoki, T.Fukuda, W.Z. Sun, *On the number of branches of a plane curve germ*, Kodai Math. J. **9** (1986), 179–187. MR842865 (87k:58034)

[AT] R. Araújo dos Santos, M. Tibar, *Real map germs and higher open book structures*, Geom. Dedicata **147** (2010), 177-185. MR2660575 (2011g:58062)

[CA] J. L. Cisneros-Molina, R. N. Araújo dos Santos, *About the Existence of Milnor Fibrations*, London Mathematical Society, LNS **380** (2010), 82–103. MR2759082

[CL] P. T. Church, K. Lamotke, *Non-trivial polynomial isolated singularities.* Nederl. Akad. Wetensch. Ser. A 78=Indag. Math. **37** (1975), 149–154. MR0365592 (51:1844)

[Du1] N. Dutertre, *Degree formulas for a topological invariant of bifurcations of function germs*, Kodai Math. J. **23**, no. 3 (2000), 442–461. MR1787676 (2001i:58087)

[Du2] N. Dutertre, *On the Milnor fibre of a real map-germ*, Hokkaido Mathematical Journal **31** (2002), 301–319. MR1914962 (2003d:58060)

[Du3] N. Dutertre, *On the Euler characteristics of real Milnor fibres of partially parallelizable maps of* $(\mathbb{R}^n, 0) \to (\mathbb{R}^2, 0)$, Kodai Math. J. **32**, no. 2 (2009), 324–351. MR2549547 (2011c:58071)

[FK] T. Fukui, A. Khovanskii, *Mapping degree and Euler characteristic*, Kodai Math. J. **29**, no. 1 (2006), 144–162. MR2222171 (2007g:58047)

[Fu] T. Fukui, *An algebraic formula for a topological invariant of bifurcation of 1-parameter family of function-germs*, Stratifications, singularities, and differential equations, II (Marseille, 1990; Honolulu, HI, 1990), Travaux en cours **55** (1997), 45–54. MR1473240 (98h:58020)

[Fu1] T. Fukui, *Mapping degree formula for 2-parameter bifurcation of function-germs*, Topology **32** (1993), 567–571. MR1231962 (94g:58024)

[GA] T. Gaffney, R. N. Araújo dos Santos *Real Integral Closure and Milnor Fibrations*, London Mathematical Society, LNS **380** (2010), 146–157. MR2759077

[HL] H. Hamm, L. D. Trãng, *Un théorème de Zariski du type de Lefschetz*, Ann. Sci. Ecol. Norm. Sup. (3) **6** (1973), 317–355. MR0401755 (53:5582)

[Kh] G.M. Khimshiashvili, *On the local degree of a smooth map*, Soobshch. Akad. Nauk Gruz. SSR **85** (1977), 309–311.

[Ki] H. King, *Topological type of isolated critical points*, Ann. Math. (2) **107**, no. 2 (1978), 385–397. MR0494153 (58:13081)

[Lo] E. Looijenga, *A note on polynomial isolated singularities*, Nederl. Akad. Wetensch. Proc. Ser. A 74=Indag. Math. **33** (1971), 418–421. MR0303557 (46:2694)

[Mi] J. Milnor, *Singular points of complex hypersurfaces*, Ann. of Math. Studies, No. 61 (1968). MR0239612 (39:969)

[PS] A. Pichon, J. Seade, *Real singularities and open-book decompositions of the 3-sphere*, Ann. Fac. Sci. Toulouse **12** (2003), 245–265. MR2123256 (2005m:57031)

[TYA] M. Tibar, C. Ying, R. Araújo dos Santos, *Singular open book structures from real mappings*, arXiv:1011.1757v2.

[RS] M. A. Ruas, R. N. Araújo dos Santos, *Real Milnor fibrations and (C)-regularity*, Manuscripta Math. **117** no. 2, (2005), 207–218. MR2150482 (2006g:32047)

[RSV] M. A. Ruas, J. Seade, A. Verjovsky, *On Real Singularities with a Milnor Fibration*, Trends in Mathematics: Trends in Singularities (2002), 191–213. MR1900787 (2003c:32031)

[Ru] L. Rudolph, *Isolated critical points of mappings from* $\mathbf{R}^4 \to \mathbf{R}^2$ *and a natural splitting of the Milnor number of a classical fibered link*, I. Basic Theory; examples. Comment. Math. Helv. **62**, no. 4 (1987), 630–645. MR920062 (88k:57009)

[Se] J. Seade, *On the Topology of Isolated Singularities in Analytic Spaces*, monograph. Series "Progress in Mathematics" **241** 2005. MR2186327 (2007a:32031)

[Sz] Z. Szafraniec, *On the number of branches of a 1-dimensional semi-analytic set*, Kodai Math. J. **11** 1988, 78–85. MR930423 (89i:32012)

Universidade de São Paulo - ICMC - Av. Trabalhador São-Carlense, 400 - Centro, Postal Box 668, 13560-970, São Carlos - São Paulo, Brazil
E-mail address: rnonato@icmc.usp.br

Department of Mathematics and Statistics, University of North Florida, 1 UNF Dr., Jacksonville, Florida 32224
E-mail address: ddreibel@unf.edu

Université de Provence, Centre de Mathématiques et Informatique, 39 rue Joliot-Curie, 13453 Marseille Cedex 13, France
E-mail address: dutertre@cmi.univ-mrs.fr

Contemporary Mathematics
Volume **569**, 2012
http://dx.doi.org/10.1090/conm/569/11238

Compact 3-manifolds supporting some $\mathbb{R}^2$-actions

C. Maquera and W. T. Huaraca

ABSTRACT. In this paper we consider $\mathbb{R}^2$-actions of class C^2 on compact 3-manifolds whose singular orbits (orbits of dimension less than 2) are a finite number of circle orbits. Using some results obtained by Chatelet-Rosenberg-Weil [**C-R-W**], we give a description of closed 3-manifolds supporting certain $\mathbb{R}^2$-actions having as singular set a union of a finite number of circle orbits. Also is given a description of the orbit structure of this actions.

1. Introduction

Singular foliations can be defined in different ways and have been studied by several authors (see for instance [**C**], [**St**], [**Su**]). For a recent account of the theory we refer the reader to [**C-S**] and [**M**]. Singular foliations defined by orbits of an action of a Lie group are in the category of foliations introduced by Stefan [**St**] and Sussmann [**Su**] and appear in control theory.

The geometric description and characterization of locally free C^2-actions ϕ of $\mathbb{R}^2$ on a compact orientable 3-dimensional manifold N (in this case, the orbits of ϕ define a codimension one foliation in N) are given in [**A-C**], [**C-R-W**] and [**R-R-W**]. To our knowledge, the case when ϕ is not locally free (that is, the foliation defined by all the orbits ϕ has singularities) has been little studied.

In [**B-M**] Brito-Maquera have initiated this study in the particular case that the manifold is the compact solid torus and that the action has only one singular orbit which is a circle in the interior of the solid torus.

For foliations, Álvarez López-Arraut-Biasi [**A-A-B**] proved that if the open manifold $N \setminus \{p_1, \dots, p_k\}$ is foliated by closed planes, then $N = \mathbb{S}^n$ (the n-sphere) and $k = 1$. For recent results on foliations with singularities see also Camacho–Scárdua ([**C-S**]) and Seade–Scárdua [**S-S**].

The aim of this paper is to continue the study iniciated in [**B-M**]. More precisely, we study a family of actions of $\mathbb{R}^2$ on 3-manifolds, that among other conditions, have a finite number of singular orbits, which are all diffeomorphic to circles. Our main result is the following:

THEOREM 1.1. *Let N be a C^∞ compact connected and orientable 3-manifold. Let ϕ be a C^2 action of $\mathbb{R}^2$ on N whose singular set, $Sing(\phi)$, is the union of a finite number of circle orbits $S_1, \dots S_k$ such that every separatrix of every S_i is*

2010 *Mathematics Subject Classification.* Primary 37C85.
Key words and phrases. Foliation, action of $\mathbb{R}^2$, circle orbit.
The first author was supported by CNPq and FAPESP Grant # 2009/17493-4.

an isolated homoclinic orbits (see Definition 2.5). Then, for every $i \in \{1, 2, ..., k\}$ there exist Σ_i, a compact surface with boundary, of genus zero and transverse to the action ϕ, and a neighborhood W_i of S_i which is a compact submanifold of N with boundary. Furthermore:

(i) *W_i is homeomorphic to the suspension of Σ_i.*
(ii) *each connected component of the boundary of W_i is a 2-torus T that is either transverse or tangent to the action ϕ. Moreover, if T is transverse to ϕ, then the induced foliation by ϕ on T is topologically equivalent to the foliation defined by the linear flow on $\mathbb{T}^2$.*
(iii) *each connected component of $N \setminus (\cup_i^k W_i)$ is diffeomorphic to $\mathbb{T}^2 \times (0, 1)$.*

Rosenberg-Roussarie-Weil showed in [**R-R-W**] that every closed 3-manifold that admits a locally free action of $\mathbb{R}^2$ is a bundle over $\mathbb{S}^1$ with fibre $\mathbb{T}^2$.

The orbit structure in each connected component of $N \setminus (\cup_i^k W_i)$ is given by the theory developed by Chatelet-Rosenberg-Weil [**C-R-W**].

This work is organized as follows. In Section 2 we study the orbit structure in a neighborhood of an isolated circle orbit, Lemma 2.3. Next, in this section we define the family of actions that are considered in this paper. In case that the circle orbit has separatrices, in Section 3, we obtain the orbit structure in a neighborhood of the circle orbits and all their separatrices, Proposition 3.3. Finally, in Section 4 is given proof of Theorem 1.1.

2. Preliminares

Let $\phi : \mathbb{R}^2 \times N \to N$ be a C^2 action on a C^∞ compact 3-manifold N. Recall that $\mathcal{O}_p = \{\phi(p, r) : r \in \mathbb{R}^2\}$ is called the *orbit* of $p \in N$ and $G_p = \{r \in \mathbb{R}^2 : \phi(p, r) = p\}$ is called the *isotropy subgroup* of p. Moreover, it is well known that an orbit is homeomorphic to a: point, line, circle, cylinder, plane or torus. The action ϕ induces a C^2 foliation (eventually, with singularities) $\mathcal{F}_\phi$ on N whose leaves are orbits of ϕ. The *singular set*, denoted by $\mathrm{Sing}(\phi)$, is the set of points in N whose orbits have dimension at most one.

For each basis $\{w_1, w_2\}$ of $\mathbb{R}^2$ we have two commutative flows $\phi^{tw_1} = \phi(tw_1, \cdot)$ and $\phi^{tw_2} = \phi(tw_2, \cdot)$ that generate the action ϕ. The vector fields X_{w_1} and X_{w_2} associated to these flows are called *infinitesimal generators* of ϕ.

Let K be a subset of N. We denote by $\mathrm{cl}(K)$ the closure of K in N. The asymptotic behavior of an orbit is given by the following notion.

DEFINITION 2.1. *The* limit set *of an orbit $\mathcal{O}_p$ is a ϕ-invariant compact set $\lim \mathcal{O}_p = \cap_{i=1}^\infty \mathrm{cl}(\mathcal{O}_p \setminus K_i)$, where K_i is a compact subset of $\mathcal{O}_p$ such that $K_i \subset K_{i+1}$ and $\mathcal{O}_p = \cup_{i=1}^\infty K_i$. It is not difficult to show that $\mathrm{cl}(\mathcal{O}_p) = \mathcal{O}_p \cup \lim \mathcal{O}_p$.*

The notions of minimal and exceptional minimal sets that we use here are the standard ones.

2.1. The orbit structure in a neighborhood of a circle orbit. Here we assume that ϕ has a circle orbit S which is an isolated singular orbit. We shall study the orbit structure of ϕ in a neighborhood of S.

A classical result in foliation theory states that the leaves structure of a foliation in a neighborhood of a compact leaf is determined by the holonomy of this leaf (see [**C-LN**]). We shall determine the holonomy of S and thus obtain information about the ϕ-orbits in a neighborhood of S.

Adapted infinitesimal generators. We denote by G the isotropy subgroup of S. Note that G is isomorphic to $\mathbb{R} \times \mathbb{Z}$. Let H be a 1-dimensional subspace of $\mathbb{R}^2$ such that $\mathbb{R}^2 = H \oplus G^0$, where G^0 is the connected component of G that contains 0. We consider a basis $\{w_1, w_2\}$ of $\mathbb{R}^2$ such that w_1 and w_2 generate the subgroups $G \cap H$ and G^0, respectively. Note that $\{X_i = X_{w_i}\,; i = 1, 2\}$ is a set of infinitesimal generators of ϕ such that S is a periodic orbit of X_1 of period one, and $X_2(q) = 0$ for every $q \in S$. We say that $\{X_1, X_2\}$ is a set of *infinitesimal generators adapted to* ϕ *at* S.

Adapted chart. For $\varepsilon > 0$, define $I^3(\varepsilon) = \{(\theta, x) \in \mathbb{R} \times \mathbb{R}^2; |\theta| < \varepsilon \text{ and } |x| < \varepsilon\}$. Take ε small enough and let $h : V \to I^3(\varepsilon)$ be a chart of N at $q_0 \in S$ (here V is a neighborhood of q_0 in N) such that $h(q_0) = 0$ and if $(\theta, x_1, x_2) \in I^3(\varepsilon)$, then the vector fields X_1 and X_2 in this chart can be written as

$$\begin{aligned} X_1(\theta, x_1, x_2) &= \frac{\partial}{\partial \theta}, \\ X_2(\theta, x_1, x_2) &= a(x_1, x_2)\frac{\partial}{\partial \theta} + b(x_1, x_2)\frac{\partial}{\partial x_1} + c(x_1, x_2)\frac{\partial}{\partial x_2}. \end{aligned} \tag{2.1}$$

The above chart is called *adapted to* S *at* q_0. The vector field

$$\widehat{Y}_2(x_1, x_2) = b(x_1, x_2)\frac{\partial}{\partial x_1} + c(x_1, x_2)\frac{\partial}{\partial x_2} \tag{2.2}$$

defined on $A_0(\varepsilon) = h^{-1}(\{(0, x) \in I^3; |x| < \varepsilon\})$ has only q_0 as its singularity.

REMARK 2.2. *Note that* $\{X_1, \widehat{Y}_2\}$ *defines a local* $\mathbb{R}^2$*-action* $\widehat{\phi}$ *on* $I^3(\varepsilon)$ *and* $\mathcal{O}_{(\theta,x)}(\widehat{\phi}) = \mathcal{O}_{(\theta,x)}(h \circ \phi \circ h^{-1})$ *for each* $(\theta, x) \in I^3(\varepsilon)$.

Let U_0 be a neighborhood of q_0 in $A_0(\varepsilon)$ such that the Poincaré diffeomorphism of X_1 at q_0, $f_0 : A_0(\varepsilon) \to U_0$, is well defined. Note that f_0 is of class C^2. For $\varepsilon > 0$ sufficiently small, let $\tau : A_0(\varepsilon) \to [0, 1 + \varepsilon)$ be the time of the first return map. Let W_0 denote the interior of $\cup_{q \in \mathrm{cl}(A_0(\varepsilon))}\{X_1^t(q); 0 \le t \le \tau(q)\}$.

As an immediate consequence we obtain the following result.

LEMMA 2.3. *The diffeomorphism* f_0 *preserves the orbits of* $\widehat{Y}_2$ *and* $\mathcal{F}_\phi|_{W_0}$ *is topologically equivalent to the suspension of* $\widehat{Y}_2$ *by* f_0.

Since the vector field $\widehat{Y}_2$ has only a finite number of sectors and f_0 preserves their orbits, we have:

REMARK 2.4. *Any orbit of* ϕ *containing a separatrix of* $\widehat{Y}_2$ *is diffeomorphic to the cylinder* $\mathbb{S}^1 \times \mathbb{R}$.

DEFINITION 2.5. *Let* ℓ *be a separatrix of the vector field* $\widehat{Y}_2$, *and* $\mathcal{O}_\ell$ *the orbit of* ϕ *containing* ℓ.

(1) *The orbit* $\mathcal{O}_\ell$ *is called a separatrix of* ϕ.
(2) *An orbit* $\mathcal{O}$ *of* ϕ *is called homoclinic if its limit set* $\lim(\mathcal{O})$ *is exactly a circle orbit.*

The family. We consider the family $\mathcal{A}$ of C^2 actions ϕ of $\mathbb{R}^2$ on a closed orientable and connected 3-manifold N of class C^∞ such that:

(1) The singular set of ϕ is a union of a finite number of circle orbits;
(2) All separatrices of the singular orbits are the only homoclinic orbits.

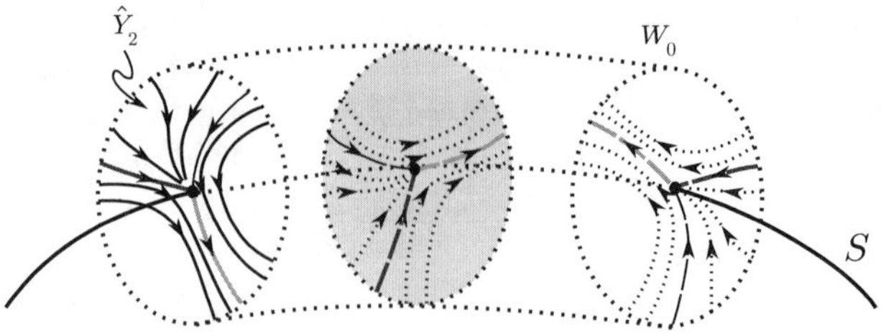

FIGURE 1. The suspension of Y_2 by f_0.

3. Semi-global orbit structure in a neighborhood of a circle orbit

From now, we will fix an action ϕ in $\mathcal{A}$. Let $S_1, \ldots, S_k$ be the singular orbits of ϕ. Also, for simplicity we denote the orbit S_i by S and by Σ the section desired. Fix a point q_0 in S and let $A_0(\varepsilon)$, X_1, X_2 and Y_2 be as in Subsection 2.1.

The local orbit structure of ϕ in a neighborhood of S is given by Lemma 2.3. In case that Y_2 has separatrices, in Proposition 3.3, we obtain the orbit structure in a neighborhood of the union of S and all their separatrices.

PROPOSITION 3.1. *There is a subset K of N of the form $\cup_{i=1}^{s}\gamma_i$ satisfying the following properties.*

(1) *γ_i is a simple and closed curve containing q_0;*
(2) *$\gamma_i \setminus \{q_0\}$ is smooth and is contained in a homoclinic orbit $\mathcal{O}_i$;*
(3) *$\mathcal{O}_i \neq \mathcal{O}_j$ for $i \neq j$;*
(4) *the separatrices of S are exactly the orbits $\mathcal{O}_1, \mathcal{O}_2, ..., \mathcal{O}_s$;*
(5) *the vector field X_1 is transverse to K in the following sense: for every $i = 1, \ldots, s$, the vector field X_1 is never tangent to γ_i.*

PROOF. Since the vector field $\widehat{Y}_2$ has only a finite number of sectors, by Lemma 2.3, we have that S admits only a finite number of separatrices. Let $\mathcal{O}_1, \mathcal{O}_2, ..., \mathcal{O}_s$ be the separatrices of S. Since these orbits are homoclinic, it follows that each orbit $\mathcal{O}_i$ contains an even number of separatrices of the vector field Y_2.

For every $i = 1, \ldots, s$, let $u_i \in \mathbb{R}^2 \setminus \{0\}$ be a generator of the isotropy subgroup G_i of the orbit $\mathcal{O}_i$. We choose two separatrices ℓ and $\ell^{'}$ of the vector field Y_2 that are contained in $\mathcal{O}_i$ and such that:

$$f_0^j(\ell) \neq \ell^{'} \quad \text{and} \quad f_0^j(\ell^{'}) \neq \ell, \quad \text{for all} \quad i = 1, \ldots, k,$$

where $k \in \mathbb{N}$ is minimal satisfying that $f_0^k(\ell) = \ell$.

Since $\phi^{u_i}(\ell) = \ell$, $\phi^{u_i}(\ell^{'}) = \ell^{'}$ and, $\phi^{tu_i}(\ell) \neq \ell$ and $\phi^{tu_i}(\ell^{'}) \neq \ell^{'}$ for all t in $(0,1)$, we can take two points p in ℓ and q in $\ell^{'}$ such that $q \neq \phi^{tu_i}(p)$ for all $t \in \mathbb{R}$. As p and q are in $\mathcal{O}_i$, there exists an unique $v_i \in \mathbb{R}^2$ such that $\phi(v_i, p) = q$. Moreover, the fact that $q \neq \phi^{tu_i}(p)$ for all $t \in \mathbb{R}$ implies that $\{v_i, u_i\}$ are linearly independent, consequently the vector fields X_{u_i} and X_{v_i} are transverse in $\mathcal{O}_i$.

Finally, we consider the closed interval ℓ_0 in ℓ with endpoints q_0 and p, and the closed interval $\ell_0^{'}$ in $\ell^{'}$ with endpoints q_0 and q. Thus, we can take γ_i as:

$$\gamma_i = \ell_0 \cup \ell_0^{'} \cup \{\phi(tv_i, p); t \in [0,1]\}.$$

By construction, the set $K = \bigcup_{i=1}^{s} \gamma_i$ satisfies items (1) to (5) above. □

REMARK 3.2. As X_1 is transverse to K then

(1) *there exists an open C^2 surface $V(K)$ containing K in its interior such that X_1 is also transverse to $V(K)$.*

Furthermore, if $D(s)$ is an open subset of $\mathbb{R}^2$ obtained from an open disc by removal of mutually disjoint closed discs from it, then as $K = \cup_{i=1}^{s}\gamma_i$,

(2) *we have that $V(K)$ is a C^2 embedding of $D(s)$.*

Let G be the isotropy subgroup of S. Since $\lim(\mathcal{O}_i) = S$ we have that

(3) *$G_i \subset G$ and $G_i \cap G^0 = \{0\}$, for all $i = 1, 2, ..., k$, consequently, there exists an unique $t_0 > 0$ such that $\phi^{t_0 w_1}(K) = K$ and $\phi^{t w_1}(K) \neq K$ for $0 < t < t_0$.*

Hence, there exists a neighborhood $V_0(K)$ of K in $V(K)$ such that

(4) *the Poincaré first return map $P : V_0(K) \to V(K)$ is well defined and it is a C^2 diffeomorphism onto its image.*

From (2), there exists a C^2 embedding $g : D(s) \to V(K)$. For $\delta > 0$, we consider the C^2 map $h : D(s) \times (-\delta, \delta) \to N$ which is defined by

$$h(x_1, x_2, t) = X_1^t(g(x_1, x_2)),$$

where $(x_1, x_2) \in D(s)$ and $t \in (-\delta, \delta)$. For δ small enough, h is a diffeomorphism onto its image W. Hence, in the coordinates (h^{-1}, W) the infinitesimal generators adapted to S are of the form:

$$\begin{aligned} X_1(x_1, x_2, t) &= \frac{\partial}{\partial t}, \\ X_2(x_1, x_2, t) &= a_2(x_1, x_2)\frac{\partial}{\partial x_1} + b_2(x_1, x_2)\frac{\partial}{\partial x_2} + c_2(x_1, x_2)\tfrac{\partial}{\partial t}. \end{aligned} \tag{3.1}$$

The vector field Z_2:

$$Z_2(x_1, x_2) = a_2(x_1, x_2)\frac{\partial}{\partial x_1} + b_2(x_1, x_2)\frac{\partial}{\partial x_2}$$

defined on $V(K)$ has q_0 as the only singularity, and

(5) *Z_2 is an extension of the vector field Y_2 defined on $A_0(\varepsilon)$ as in Subsection 3.1;*

(6) *P preserves the orbits of Z_2;*

(7) *the vector fields X_1, Z_2 define a local $\mathbb{R}^2$-action $\widehat{\phi}$ on $D(s) \times (-\delta, \delta)$ such that*

$$\mathcal{O}_{(x_1,x_2,t)}(\widehat{\phi}) = \mathcal{O}_{(x_1,x_2,t)}(h \circ \phi \circ h^{-1}), \qquad (x_1, x_2, t) \in D(s) \times (-\delta, \delta).$$

Let $\eta > 0$ be such that the first return time map $\tau : V_0(K) \to (1 - \eta, 1 + \eta)$ is well defined. Considering the open set W_K, which is the interior of

$$\bigcup_{q \in \mathrm{cl}(V_0(K))} \{X_1^t(q); 0 \leq t \leq \tau(q)\},$$

we obtain the following proposition.

PROPOSITION 3.3. *The foliation $\mathcal{F}_\phi|_{W_K}$ is topologically equivalent to the suspension of Z_2 by P.*

4. Proof of main Theorem

The proof of Theorem 1.1 is a consequence of Theorem 4.5 and Theorem 4.6.

4.1. Building the section Σ. The construction of Σ depends on the existence of separatrices for the vector field Y_2. We will build the section Σ by considering the two cases: in 4.1.1 we assume that Y_2 has no separatrix (that is, S admits no separatrix), and in 4.1.2 we assume that Y_2 has separatrices.

First of all, by a well known result for vector fields with isolated singularities (see [**Ha**, Section 8 of Chapter VII]), we have:

PROPOSITION 4.1. *The vector field Y_2 satisfies only one of the following statements:*

(a) *for every δ in $(0, \varepsilon)$, the open 2-disc $A_0(\delta)$ contains a closed orbit of Y_2;*
(b) *Y_2 has finite number of sectors (elliptic, hyperbolic and/or parabolic).*

DEFINITION 4.2. *Let R be an open connected subset of N such that its closure $\mathrm{cl}(R)$ is compact. We say that $\mathrm{cl}(R)$ is a singular Reeb component of $\mathcal{F}|_\phi$ if:*

(1) *R is saturated by orbits diffeomorphic to $\mathbb{R}^2$, which are closed in R;*
(2) *the boundary of R contains a circle orbit.*

EXAMPLE 4.3. *Let us fix a point p_0 in the 2-sphere $\mathbb{S}^2$ and consider the circle $S = \{p_0\} \times \mathbb{S}^1$ in the 3-manifold $N = \mathbb{S}^2 \times \mathbb{S}^1$. Since $N \setminus S$ is an open solid torus, we consider the foliation $\mathcal{F}$ in N such that: S is a leaf and $\mathcal{F}|_{N \setminus S}$ is the classical Reeb component in the open solid torus. In this situation, N is a singular Reeb component for $\mathcal{F}$.*

Let D_0 be an open 2-disc which is transverse to S at $q_0 = (p_0, \theta_0)$. Then the foliation induced by $\mathcal{F}$ in D_0 has q_0 as the only singularity and all the other leaves are diffeomorphic to $\mathbb{S}^1$ (see Figure 2).

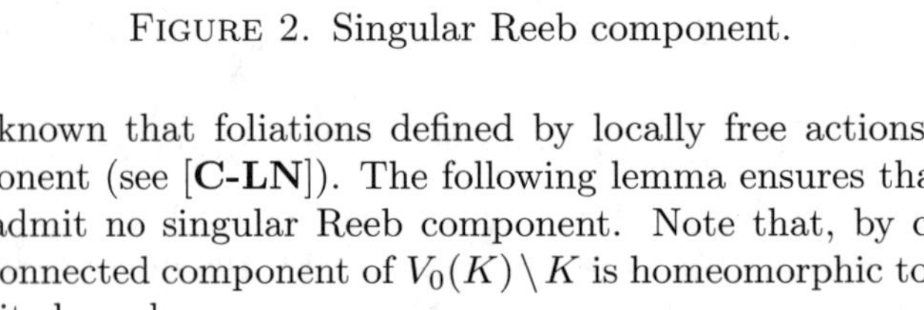

FIGURE 2. Singular Reeb component.

It is well known that foliations defined by locally free actions of $\mathbb{R}^2$ admits no Reeb component (see [**C-LN**]). The following lemma ensures that the foliation defined by ϕ admit no singular Reeb component. Note that, by construction of $V_0(K)$, every connected component of $V_0(K) \setminus K$ is homeomorphic to $\mathbb{S}^1 \times [0, 1]$ and contains q_0 in its boundary.

LEMMA 4.4. *Consider the vector fields Y_2 and Z_2 as in previous sections. Let A be a connected component of $V_0(K) \setminus K$. If one of the following conditions is satisfied*

(i) *the orbit of Y_2 through $p \in A_0(\varepsilon)$ is periodic and contained in $A_0(\varepsilon)$, or*
(ii) *the orbit of Z_2 through $p \in A$ is periodic and contained in A,*

then the orbit $\mathcal{O}_p$ is a torus $\mathbb{T}^2$ or a cylinder $\mathbb{S}^1 \times \mathbb{R}$. Consequently, $\mathcal{F}|_\phi$ admits no singular Reeb component.

PROOF. Firstly we assume that there exists $p \in A_0(\varepsilon)$ whose orbit $\mathcal{O}_p(Y_2)$ under the vector field Y_2 is periodic and contained in $A_0(\varepsilon)$. By contradiction, we suppose that the orbit $\mathcal{O}_p$ is homeomorphic to $\mathbb{R}^2$. Then $\mathcal{O}_p(Y_2)$ bounds a closed 2-disc D contained in $\mathcal{O}_p$. But the fact that X_1 is transverse to Y_2 implies that X_1 has at least a singularity contained in D. This contradiction proves (i).

Finally, the proof of the other case is analogous by considering the vector field Z_2. □

4.1.1. *First case: ϕ has no separatrices.* In this case we will show the following result:

THEOREM 4.5. *Suppose that ϕ has no separatrices. There exists a torus T that bounds a solid torus $W \subset W_0$ containing S such that if $\Sigma = W \cap A_0(\varepsilon)$, then $\mathcal{F}_\phi|_W$ is topologically equivalent to the suspension of $Y_2|_\Sigma$ by $f_0|_\Sigma$. Furthermore, only and only one of the following possibilities is valid:*

(i) *T is the ϕ-orbit of a point $p \in A_0(\varepsilon')$ whose Y_2-orbit bounds the closed disc Σ.*
(ii) *The foliation $\mathcal{F}_\phi$ induces on T a foliation having all leaves circles which are nullhomotopics in W. Moreover, all the ϕ-orbits (resp. Y_2-orbits) of points from $\Sigma \setminus \{q_0\}$ are cylindrical (resp. periodic).*
(iii) *The foliation $\mathcal{F}_\phi$ induces on T a foliation topologically equivalent to the foliation defined by a linear flow on $\mathbb{T}^2$. Consequently,* $\mathrm{Sat}(W \setminus S)$, *the saturated by ϕ of $W \setminus S$, is homeomorphic to $\mathbb{T}^2 \times (0,1)$ and either all the ϕ-orbits are cylindrical or all the orbits are dense planes.*

PROOF. Firstly we assume (a) of Proposition 4.1. Let C_δ be a periodic Y_2-orbit contained in $A_0(\delta)$. By Lemma 4.4 we have that the ϕ-orbit $\mathcal{O}_\delta$ containing C_δ is either a torus or is a cylinder. In the first case $T := \mathcal{O}_\delta$ is the boundary of a solid torus $W \subset W_0$ containing S, and thus (i) is verified.

Then, we can assume that ϕ has no torus orbit through points of $A_0(\varepsilon)$ (the orbits are cylindrical or planes). This implies that the Poincaré map f_0 of X_1 is a topological contraction (or a topological expansion). In this situation, we are going to show that (ii) is satisfied. Let D be the closed disc contained in $A_0(\varepsilon')$ whose boundary C is a closed Y_2-orbit.

Claim 1: all the ϕ-orbits (resp. Y_2-orbits) of points in $D \setminus \{q_0\}$ are cylindrical (resp. periodic). In fact, we consider a C^∞ curve $\gamma : [0,1) \to A_0(\varepsilon)$ such that $\gamma(0) = q_0$ and $\gamma((0,1))$ is transverse to Y_2. Let C be a periodic Y_2-orbit of period τ such that $\{\gamma(t_0)\} = C \cap \gamma((0,1))$ and let $\mathcal{O}_C$ be the ϕ-orbit containing C. Since ϕ has no torus orbits in W_0, we have that $S \subset \lim(\mathcal{O}_C)$. Thence, there exists a strictly decreasing sequence $\{t_n\}$, with $t_n \in (0,1)$, converging to 0 such that the sequence $\{\gamma(t_n)\} = f_0^n(C) \cup \gamma((0,1))$ converges to q_0.

Moreover, since the vector fields X_1 and Y_2 commute, we have that $Y_2^\tau(\gamma(t_n)) = \gamma(t_n)$ for all $n \geq 0$. Hence Y_2^τ and f_0 induce two C^2 diffeomorphisms g and f on $\gamma([0,t_0))$, respectively, such that $f(q_0) = q_0 = g(q_0)$, $g(q_n) = q_n$ and f is a topological contraction. Consequently, by a result of N. Kopell [**K**], we have that $g = id$ and our claim is proved.

To complete the proof of (ii) we need construct the 2-torus T. Let C_1 be the closed orbit of Y_2 such that $C_1 = f_0(C)$. We consider the 2-torus which is obtained by gluing the closed annulus R contained in $A_0(\varepsilon)$ and bounded by C and C_1 with the cylinder

$$\{X_1^t(p); p \in C \text{ and } t \in [0, \tau(p)]\}$$

where τ is the first return time map for the vector field X_1. By deforming this torus we obtain a torus T satisfying (ii).

Finally we are going to show (iii). Since Y_2 has no separatrices, it follows from (b) of Proposition 4.1 that q_0 is an attractor or a repeller. Let T be a 2-torus that bounds W, a solid torus which is a neighborhood of S. Since Y_2 is attractor, we can take W such that T is transverse to the orbits of ϕ. Hence, as the action is of class C^2, we have that the foliation on T induced by $\mathcal{F}_\phi$ is topologically equivalent to the foliation defined by the linear flow on $\mathbb{T}^2$. This finishes the proof of (iii). □

Conclusion: In this case we take Σ as the connected component of $A_0(\varepsilon) \setminus T$ that contains q_0.

4.1.2. **Second case:** *ϕ has separatrices.* Let K, $V(K), W_K$, Z_2 and P be as in Section 3.

THEOREM 4.6. *Suppose that ϕ has separatrices. There exists a closed submanifold W contained in W_K that contains S and all their separatrices such that each connected component T of ∂W is a 2-torus and if $\Sigma = W \cap V(K)$, then $\mathcal{F}_\phi|_W$ is topologically equivalent to the suspension of $Z_2|_\Sigma$ by $P|_\Sigma$. Furthermore, only and only one of the following possibilities is valid:*

(i) *T is the ϕ-orbit by a point $p \in A_0(\varepsilon')$ whose Y_2-orbit is a connected component of the boundary of Σ.*
(ii) *The foliation $\mathcal{F}_\phi$ induces on T a foliation topologically equivalent to the foliation defined by the linear flow on $\mathbb{T}^2$. Consequently,* Sat(T), *the saturated by ϕ of T, is homeomorphic to $\mathbb{T}^2 \times (0,1)$ and either all the ϕ-orbits are cylindrical or all the orbits are dense planes.*

PROOF. We can write $V(K) \setminus K = \cup_{i=1}^r A_i$, where each A_i is homeomorphic to a annulus. For i fixed, we have two possibilities:

(a) Z_2 has a periodic orbit C contained in A_i.
(b) Z_2 has no periodic orbits contained in A_i.

Firstly we assume (a). By Lemma 4.4 we have that the ϕ-orbit $\mathcal{O}$ containing C is either a torus or a cylinder. In the first case we take $T_i := \mathcal{O}$, and thus (i) is verified. Otherwise, we claim that the limit set lim(O) contains a $\mathbb{T}^2$ orbit T_i whose intersection with A_i is a closed orbit of Z_2. Recall that $K = \cup_{i=1}^s \gamma_i$, then we may suppose that γ_i is a connected component of the boundary of A_i. Let $\mathcal{O}_i$ be the separatrix containing γ_i. By contradiction, we assume that ϕ has no torus orbit through points of A_i (the orbits are cylindrical or planes). This implies that the Poincaré map P of X_1 is a topological contraction (or a topological expansion) and that lim$(\mathcal{O})$ contains S. Furthermore, since Z_2^t and P commute, we have that for all n in $\mathbb{N}$ the set $P^n(C)$ is a closed orbit of Z_2 having period constant and converging to γ_i. This is a contradiction, since γ_i contains q_0 which is a singularity for Z_2.

Finally if we assume (b) above, we have that γ_i is an attractor in A_i for the vector field Z_2. Hence, we can construct a closed curve C_0 contained in A_i and transverse to Z_2. Let $C_1 = P(C_0)$ and consider the 2-torus obtained by gluing the closed annulus R contained in A_i and bounded by C_0 and C_1, with the cylinder

$$\{X_1^t(p); p \in C \text{ and } t \in [0, \tau(p)]\}$$

where τ is the first return time map for the vector field X_1. By deforming this torus we obtains a torus T satisfying (ii). □

References

[A-A-B] J. A. Alvarez; J. L. Arraut and C. Biasi, *Foliations by planes and Lie group actions.* Ann. Polon. Math. **82** (2003), 1, 61–69. MR2041398 (2004m:57071)

[A-C] J. L. Arraut and M. Craizer, *Foliations of M^3 defined by $\mathbb{R}^2$-actions.* Ann. Inst. Fourier, Grenoble **45**, 4 (1995), 1091–1118. MR1359841 (96j:57030)

[A-M] J. L. Arraut and C. A. Maquera, *On the orbit structure of $\mathbb{R}^n$-actions on n-manifolds.* Qual. Theory Dyn. Syst. **4**, no. 2 (2003), 169–180. MR2129717 (2005m:37055)

[B-M] Brito L. F.M. and Maquera C. *Orbit structure of certain $\mathbb{R}^2$-actions on the solid torus.* Ann. Fac. Sci. Toulouse Math. (6), **17** (2008), no. 3, 613–633. MR2488234 (2010d:37046)

[C-LN] C. Camacho and A. Lins Neto, *Geometric Theory of Foliations.* Birkhäuser, Boston, Massachusetts, 1985. MR824240 (87a:57029)

[C-S] C. Camacho and B. A. Scárdua, *On codimension one foliations with Morse singularities on three-manifolds.* Topology Appl. **154** (2007) 1032–1040. MR2298620 (2008b:57039)

[C] D. Cearveau, *Distributions involutives singulières.* Ann. Inst. Fourier **29**, no.3 (1979), 261-294. MR552968 (83f:58057)

[C-R-W] G. Chatelet, H. Rosenberg and D. Weil, *A classification of the topological types of $\mathbb{R}^2$-action on closed orientable 3-manifolds.* Publ. Math. IHES **43** (1974), 261-272. MR0346809 (49:11533)

[H] A. Haefliger, *Variétés feuilletées.* Ann. Scuola Norm. Sup. Pisa, Série 3, **16** (1962), 367-397. MR0189060 (32:6487)

[Ha] P. Hartman, *Ordinary differential equations.* Corrected reprint of the second (1982) edition [Birkhäuser, Boston, MA; MR0658490 (83e:34002)]. With a foreword by Peter Bates. Classics in Applied Mathematics, 38. Society for Industrial and Applied Mathematics (SIAM), Philadelphia, PA, 2002. MR1929104 (2003h:34001)

[K-H] A. Katok and B. Hasselblatt, *Introduction to the modern theory of dynamical systems.* Cambridge University Press, 1995. MR1326374 (96c:58055)

[K] N. Kopell, *Commuting diffeomorphisms.* Global Analysis (Proc. Sympos. Pure Math., Vol. XIV, Berkeley, Calif., 1968) pp. 165–184 MR0270396 (42:5285)

[M] A. S. de Medeiros, *Singular foliations and differential p-forms.* Ann. Fac. Sci. Toulouse Math., (6) **9**, no.3 (2000), 451–466. MR1842027 (2002f:32056)

[R-R-W] H. Rosenberg, R. Roussarie and D. Weil, *A classification of closed orientable manifolds of rank two.* Ann. of Math. **91** (1970), 449-464. MR0270391 (42:5280)

[S] R. Sacksteder, *Foliations and pseudogroups.* Amer. J. Math. **87** (1965), 79-102. MR0174061 (30:4268)

[S-S] B. A. Scárdua and J. Seade, *Codimension one foliations with Bott-Morse singularities I.* J. Differential Geom. **83** (2009), no. 1, 189-212. MR2545034 (2010i:57047)

[St] P. Stefan, *Acessible sets, orbits, and foliations with singularities.* Proc. London Math. Soc. **29** (1974), 699–713. MR0362395 (50:14837)

[Su] H. J. Sussmann, *Orbits of families of vector fields and integrability of distributions.* Trans. Amer. Math. Soc. **180** (1973), 171–188. MR0321133 (47:9666)

Departamento de Matemática, USP - Universidade de São Paulo, Instituto de Ciências Matemáticas e de Computação, Caixa Postal 668, CEP 13560-970, São Carlos, SP, Brazil.

E-mail address: cmaquera@icmc.usp.br

Instituto de Matemática, UFAL - Universidade Federal de Alagoas, CEP 57072-900, Maceió, AL, Brazil.

E-mail address: walterhv@gmail.com

Contemporary Mathematics
Volume **569**, 2012
http://dx.doi.org/10.1090/conm/569/11263

Timelike canal hypersurfaces of spacelike submanifolds in a de Sitter space

Masaki Kasedou

ABSTRACT. We introduce the notion of timelike canal hypersurfaces to investigate the differential geometry of spacelike submanifolds of codimension at least two in a de Sitter space. The timelike canal hypersurface is given by parallel normal vector fields of the spacelike submanifold, and their geometrical properties are related. We also consider their generic properties from the viewpoint of the theory of Legendrian singularities.

1. Introduction

Canal surfaces are classically known. They are constructed from the families of circles which are perpendicular to the velocity vectors of space curves. It is known that the geometrical properties of a canal surface are related to those of the original curve. This notion can be analogously introduced in other situations. Izumiya, Pei, Romero Fuster and Takahashi [**3**] investigated the differential geometry of submanifolds of codimension at least two in a hyperbolic space. Spacelike canal hypersurfaces are constructed from submanifolds, and the parabolic set on the spacelike canal hypersurface is related to the horospherical points of the submanifold in a hyperbolic space.

We can introduce the analogous notion of the canal hypersurfaces in a de Sitter space. It is known that a de Sitter space is a Lorentzian space form with positive curvature. In [**5**] we studied geometrical relations between spacelike canal hypersurfaces and spacelike submanifolds in a de Sitter space. In this paper, we also consider *timelike* canal hypersurfaces of spacelike submanifolds of codimension at least two in a de Sitter space, and find some new geometrical relations between spacelike submanifolds and timelike hypersurfaces.

In §2 we review the differential geometry of timelike hypersurfaces in a de Sitter space studied in [**4**]. The singularities of the de Sitter Gauss map are related to the de Sitter parabolic points, where the de Sitter Gauss-Kronecker curvature vanishes. In §3 we study the differential geometry of spacelike submanifolds in a de Sitter space, whose codimension is at least two. We consider the spacelike parallel unit normal vector field of the spacelike submanifolds. In §4 we construct a timelike canal hypersurface from a unit parallel orthonormal frame of a spacelike submanifold, and

---

2010 *Mathematics Subject Classification.* Primary 53B30, 57R45, 53B25.
Key words and phrases. Lorentz geometry, de Sitter space, singularity theory.
This work was supported by the JSPS International Training Program (ITP).

study a geometrical relation between the timelike canal hypersurfaces and spacelike submanifolds.

In §5, 6 we naturally interpret the de Sitter θ-maps of spacelike submanifolds as a wavefront set of the theory of Legendrian singularities, and review the theory of contact of submanifolds due to Montaldi [**6**]. In §7 we consider generic properties of spacelike submanifolds in a de Sitter space.

2. Timelike hypersurfaces in a de Sitter space

In this section we review the basic notions to study the differential geometry of timelike hypersurfaces and submanifolds in a de Sitter space considered in further sections. Let $\mathbb{R}^{n+1} = \{\mathbf{x} = (x_0, \ldots, x_n) \mid x_i \in \mathbb{R}\ (i = 0, \ldots, n)\}$ be an $(n+1)$-dimensional vector space. For two vectors $\mathbf{x} = (x_0, \ldots, x_n)$ and $\mathbf{y} = (y_0, \ldots, y_n)$ in $\mathbb{R}^{n+1}$, the *pseudo scalar product* of $\mathbf{x}$ and $\mathbf{y}$ is defined as $\langle \mathbf{x}, \mathbf{y} \rangle = -x_0 y_0 + \sum_{i=1}^{n} x_i y_i$. We call $(\mathbb{R}^{n+1}, \langle , \rangle)$ a *Minkowski* $(n+1)$*-space* and write $\mathbb{R}^{n+1}_1$ instead of $(\mathbb{R}^{n+1}, \langle , \rangle)$. We say that a vector $\mathbf{x} \in \mathbb{R}^{n+1}_1 \setminus \{\mathbf{0}\}$ is *spacelike, timelike* or *lightlike* if $\langle \mathbf{x}, \mathbf{x} \rangle > 0$, $\langle \mathbf{x}, \mathbf{x} \rangle < 0$ or $\langle \mathbf{x}, \mathbf{x} \rangle = 0$ respectively. The norm of a vector $\mathbf{x} \in \mathbb{R}^{n+1}_1$ is defined as $||\mathbf{x}|| = \sqrt{|\langle \mathbf{x}, \mathbf{x} \rangle|}$.

We respectively define the *hyperbolic* n*-space* and the *de Sitter* n*-space* by

$$\begin{aligned} H^n_\pm(-1) &= \{\mathbf{x} \in \mathbb{R}^{n+1}_1 \mid \langle \mathbf{x}, \mathbf{x} \rangle = -1, \operatorname{sgn}(x_0) = \pm 1\}, \\ S^n_1 &= \{\mathbf{x} \in \mathbb{R}^{n+1}_1 \mid \langle \mathbf{x}, \mathbf{x} \rangle = 1\}, \end{aligned}$$

and we write $H^n(-1) = H^n_+(-1) \cup H^n_-(-1)$. We also define the *future* (resp. *past*) *lightcone* at the origin by $LC^*_\pm = \{\mathbf{x} \in \mathbb{R}^{n+1}_1 \mid \langle \mathbf{x}, \mathbf{x} \rangle = 0,\ \operatorname{sgn}(x_0) = \pm 1\}$, and we write $LC^* = LC^*_+ \cup LC^*_-$. For any n vectors $\mathbf{x}_1, \mathbf{x}_2, \ldots, \mathbf{x}_n \in \mathbb{R}^{n+1}_1$, we define their pseudo-external product $\mathbf{x}_1 \wedge \mathbf{x}_2 \wedge \ldots \wedge \mathbf{x}_n$ as

$$\mathbf{x}_1 \wedge \mathbf{x}_2 \wedge \ldots \wedge \mathbf{x}_n = \det \begin{pmatrix} -\mathbf{e_0} & \mathbf{e_1} & \cdots & \mathbf{e_n} \\ x_0^1 & x_1^1 & \cdots & x_n^1 \\ \vdots & \vdots & \ddots & \vdots \\ x_0^n & x_1^n & \cdots & x_n^n \end{pmatrix},$$

where $\{\mathbf{e_0}, \ldots \mathbf{e_n}\}$ is the standard basis of $\mathbb{R}^{n+1}_1$ and $\mathbf{x_i} = (x_0^i, \ldots, x_n^i)$ for $i = 1, \ldots, n$. Since $\langle \mathbf{x}, \mathbf{x}_1 \wedge \ldots \wedge \mathbf{x}_n \rangle = \det(\mathbf{x}, \mathbf{x}_1, \ldots, \mathbf{x}_n)$, the product $\mathbf{x}_1 \wedge \ldots \wedge \mathbf{x}_n$ is pseudo-orthogonal to all $\mathbf{x}_i$ for $i = 1, \ldots, n$.

For a vector $\mathbf{v} \in \mathbb{R}^{n+1}_1 \setminus \{\mathbf{0}\}$ and a real number c, we define a *hyperplane with the pseudo normal* $\mathbf{v}$ in the Minkowski space by

$$\mathrm{HP}(\mathbf{v}, c) = \{\mathbf{x} \in \mathbb{R}^{n+1}_1 \mid \langle \mathbf{x}, \mathbf{v} \rangle = c\}.$$

We say that a hyperplane $HP(\mathbf{v}, c)$ is spacelike, timelike or lightlike if the vector $\mathbf{v}$ is timelike, spacelike or lightlike respectively. We now consider hyperquadrics in the de Sitter space. We say that $HP(\mathbf{v}, c) \cap S^n_1$ is an *elliptic hyperquadric* or a *hyperbolic hyperquadric* if $HP(\mathbf{v}, c)$ is spacelike or timelike. We say that $HP(\mathbf{v}, c) \cap S^n_1$ is a *de Sitter hyperhorosphere* if $\mathbf{v}$ is lightlike and $c \neq 0$.

Let $\mathbf{v}$ be spacelike. We say that the hyperbolic hyperquadric $HP(\mathbf{v}, c) \cap S^n_1$ is a *flat timelike hyperquadric* or a *non-flat timelike hyperquadric* if $c = 0$ or $0 < c < ||v||$ respectively. If $c > ||\mathbf{v}||$ then the hyperbolic hyperquadric is a spacelike submanifold. If we consider $c = ||\mathbf{v}||$ then the timelike hyperplane $HP(\mathbf{v}, c)$ is tangent to de Sitter space at the point $(1/c)\mathbf{v}$.

We now briefly review the differential geometry of timelike hypersurfaces in the de Sitter space studied in [**4**]. Let U be an open subset of the Euclidean $(n-1)$-space and $\mathbf{u} = (u_1, \cdots, u_{n-1}) \in U$. We say that an embedding map $\mathbf{X} : U \longrightarrow S_1^n$ is a *timelike hypersurface* if the tangent space of $\mathbf{X}$ is a timelike hyperplane at any point. We set $M = \mathbf{X}(U)$.

We define the *de Sitter Gauss map* of a timelike hypersurface by

$$\mathbf{x}^d : U \longrightarrow S_1^n,\ \mathbf{x}^d(\mathbf{u}) = \frac{\mathbf{X}(\mathbf{u}) \wedge \mathbf{X}_{u_1}(\mathbf{u}) \wedge \ldots \wedge \mathbf{X}_{u_{n-1}}(\mathbf{u})}{||\mathbf{X}(\mathbf{u}) \wedge \ldots\ldots \wedge \mathbf{X}_{u_{n-1}}(\mathbf{u})||}.$$

We also define the *de Sitter shape operator* by $A_p^d := -d_{\mathbf{u}}\mathbf{x}^d : T_pM \longrightarrow T_pM$, the *de Sitter Gauss-Kronecker curvature* $K_d(\mathbf{u})$ as the determinant of the de Sitter shape operator and *de Sitter principal curvatures* $\{\kappa_{d,i}(\mathbf{u})\}_i$ as eigenvalues of the de Sitter shape operator. We call a point p with $K_d(\mathbf{u}) = 0$ a *de Sitter parabolic point* and a point p with $A_p^d = \kappa_d \cdot \mathrm{id}_{T_pM}$ a *de Sitter umbilic point.*

PROPOSITION 2.1 ([**4**]). *Suppose that a timelike hypersurface M is totally umbilic, then the de Sitter principal curvatures $\kappa_{d,i}(\mathbf{u})$ are constant κ_d and M is a part of some hyperbolic hyperquadric $HP(\mathbf{v}, c) \cap S_1^n$.*

(1) *If $\kappa_d \neq 0$ then M is a part of some non-flat timelike hyperquadric with*

$$\mathbf{v} = \frac{1}{\sqrt{\kappa_d^2 + 1}}(\kappa_d\mathbf{X}(\mathbf{u}) + \mathbf{x}^d(\mathbf{u}))\ \textit{and}\ c = \frac{\kappa_d}{\sqrt{\kappa_d^2 + 1}} < ||\mathbf{v}||.$$

(2) *If $\kappa_d = 0$ then M is a part of some flat timelike hyperquadric. In this case, we have $\mathbf{v} = \mathbf{x}^d(\mathbf{u})$ and $c = 0$.*

We now consider a Weingarten-type formula for the timelike hypersurfaces. We define the *de Sitter first fundamental form* by $ds^2 = \sum_{i,j=1}^{n-1} g_{ij}du_idu_j$, where $g_{ij} = \langle \mathbf{X}_{u_i}, \mathbf{X}_{u_j} \rangle$. Since the tangent space of M is always timelike, so ds^2 is a Lorentzian metric of index one. We also define the coefficients of the *de Sitter second fundamental form* as $h_{ij} = -\langle \mathbf{X}_{u_i}, \mathbf{x}^d_{u_j} \rangle$. We have the following Weingarten-type formula.

PROPOSITION 2.2 ([**4**]). *Let $\mathbf{X}$ be a timelike hypersurface. The differential map of the de Sitter Gauss map $\mathbf{x}^d$ is given by*

$$\mathbf{x}^d_{u_i} = \sum_{j=1}^{n-1}(h_i^j)\mathbf{X}_{u_j},$$

where $(h_i^j) = (h_{ik})(g_{kj})^{-1}$.

We define the *de Sitter height function* $H^d : U \times S_1^n \longrightarrow \mathbb{R}$ by

$$H^d(\mathbf{u}, \mathbf{v}) = \langle \mathbf{X}(\mathbf{u}), \mathbf{v} \rangle,$$

and write $h^d_{\mathbf{v}_0}(\mathbf{u}) = H^d(\mathbf{u}, \mathbf{v}_0)$.

We now naturally interpret the image of the de Sitter Gauss map of a timelike hypersurface M as a wave front set in the framework of contact geometry discussed in [**2**]. Let $\pi : PT^*S_1^n \longrightarrow S_1^n$ be the projective tangent bundles with the canonical structures. Consider the tangent bundle $\tau : T(PT^*S_1^n) \longrightarrow PT^*S_1^n$ and the differential map $d\pi : T(PT^*S_1^n) \longrightarrow TS_1^n$ of π. For any $X \in T(PT^*S_1^n)$, there exists an element $\alpha \in T^*S_1^n$ such that $\tau(X) = [\alpha] \in PT^*S_1^n$. For any $x \in S_1^n$ and

$w \in T_x S_1^n$, the property $\alpha(w) = 0$ does not depend on the choice of a representative element α. Thus we can define a canonical contact structure on $PT^*S_1^n$ by $K = \{X \in T(PT^*S_1^n) \mid \tau(X)(d\pi(X)) = 0\}$.

On the other hand, for a local coordinates $(U, \mathbf{v})$ on S_1^n, where $\mathbf{v} = (v_1, \dots, v_n)$, we consider the local trivialization $PT^*(U) \equiv U \times P\mathbb{R}^{n-1}$ and call $(v_1, \dots, v_n, [\xi_1, \cdots, \xi_n])$ homogeneous coordinates, where $[\xi_1 : \cdots : \xi_n]$ are the homogeneous coordinates on the dual projective space $P\mathbb{R}^{n-1}$. It is easy to show that $X_\bullet \in K_\bullet^\pm$ if and only if $\sum_{i=1}^n \mu_i \xi_i = 0$, where $\bullet = (\mathbf{v}, [\xi])$ and $d\pi_\bullet^\pm(X_\bullet) = \sum_{i=1}^n \mu_i \partial/\partial v_i \in T_\bullet S_1^n$. An immersion $i : L \longrightarrow PT^*S_1^n$ is said to be a *Legendrian immersion* if $\dim L = n-1$ and $di_q(T_q L) \subset K_{i(q)}$ for any $q \in L$. The map $\pi \circ i$ is called a *Legendrian map* and the image $W(i) = \mathrm{Im}(\pi \circ i)$, the *wave front* of i. Moreover, i (or the image of i) is called the *Legendrian lift* of $W(i)$.

Let $F : (\mathbb{R}^{n-1} \times \mathbb{R}^n, (\mathbf{u}_0, \mathbf{v}_0)) \longrightarrow (\mathbb{R}, 0)$ be a function germ. We say that $F(\mathbf{u}, \mathbf{v}) = 0$ is a *Morse family* of hypersurfaces if the map germ $\Delta^* F : (\mathbb{R}^{n-1} \times \mathbb{R}^n, (\mathbf{u}_0, \mathbf{v}_0)) \longrightarrow (\mathbb{R}^n, \mathbf{0})$ defined as $\Delta^* F = (F, \partial F/\partial u_1, \cdots, \partial F/\partial u_{n-1})$ is non singular. In this case, we may regard $\{(\mathbf{u}, \mathbf{v}) \in (\mathbb{R}^{n-1} \times \mathbb{R}^n, (\mathbf{u}_0, \mathbf{v}_0)) \mid F(\mathbf{u}, \mathbf{v}) = 0\}$ as a family of hypersurfaces of $\mathbb{R}^{n-1}$ with a parameter $\mathbf{v}$. We have a smooth $(n-1)$-dimensional submanifold $\Sigma_*(F) = \{(\mathbf{u}, \mathbf{v}) \in (\mathbb{R}^{n-1} \times \mathbb{R}^n, (\mathbf{u}_0, \mathbf{v}_0)) \mid \Delta^* F(\mathbf{u}, \mathbf{v}) = 0\}$, and the Legendrian immersion germ $\mathcal{L}_F : (\Sigma_*(F), (\mathbf{u}_0, \mathbf{v}_0)) \longrightarrow PT^*\mathbb{R}^n$ is defined as

$$\mathcal{L}_F(\mathbf{u}, \mathbf{v}) = \left(\mathbf{v}, \left[\frac{\partial F}{\partial v_1}(\mathbf{u}, \mathbf{v}) : \cdots : \frac{\partial F}{\partial v_n}(\mathbf{u}, \mathbf{v})\right]\right). \tag{2.1}$$

We have the following fundamental theorem of [**8, 1**].

PROPOSITION 2.3 ([**8, 1**]). *All Legendrian submanifold germs in $PT^*\mathbb{R}^n$ are constructed by the above method.*

We call F a generating family of $\mathcal{L}_F(\Sigma_*(F))$. The wave front is

$$W(\mathcal{L}_F) = \left\{\mathbf{v} \in \mathbb{R}^n \,\middle|\, \exists \mathbf{u} \in \mathbb{R}^{n-1} \text{ such that } F(\mathbf{u}, \mathbf{v}) = \frac{\partial F}{\partial u_1}(\mathbf{u}, \mathbf{v}) = \cdots = \frac{\partial F}{\partial u_{n-1}}(\mathbf{u}, \mathbf{v}) = 0\right\}.$$

We call it the *discriminant set* of F.

PROPOSITION 2.4 ([**4**], Propositions 4.1 and 4.2). *Let $H^d : U \times S_1^n \longrightarrow \mathbb{R}$ be the de Sitter height function on the timelike hypersurface $\bar{\mathbf{X}}$. Then*

(1) *$H^d(\mathbf{u}, \mathbf{v}) = \partial H^d/\partial u_i(\mathbf{u}, \mathbf{v}) = 0$ for $i = 1, \dots, n-1$ if and only if $\mathbf{v} = \mathbf{x}^d(\mathbf{u})$.*

(2) *Let $\mathbf{v} = \mathbf{x}^d(\mathbf{u})$ then $p = \mathbf{X}(\mathbf{u})$ is a de Sitter parabolic point if and only if the rank of the Hessian matrix Hess$(h_\mathbf{v})(\mathbf{u})$ is zero.*

Therefore the discriminant set of the de Sitter height function is the image of the de Sitter Gauss map of the timelike hypersurfaces in the de Sitter space, and a singular point of the de Sitter Gauss map corresponds to a de Sitter parabolic point on the timelike hypersurface.

PROPOSITION 2.5 ([**4**]). *Let $H^d : (U \times S_1^n, (\mathbf{u}_0, \mathbf{v}_0)) \longrightarrow (\mathbb{R}, 0)$ be the de Sitter height function germ. $H^d(\mathbf{u}, \mathbf{v}) = 0$ is a Morse family of hypersurfaces.*

Therefore a Legendrian immersion germ $\mathcal{L}_{H^d} : (\Sigma_*(H^d), (\mathbf{u}_0, \mathbf{v}_0)) \longrightarrow PT^*S_1^n$ is defined by the above formula (2.1). The wave front set of the Legendrian immersion germ $\mathcal{L}_{H^d}$ is the image of the de Sitter Gauss map germ $\mathbf{x}^d$.

3. Spacelike submanifolds in the de Sitter space

In this section we study the differential geometry of spacelike submanifolds of codimension at least two in the de Sitter space. Let $r \geq 2$ be an integer and $\mathbf{X} : U \longrightarrow S_1^n$ an embedding. We say that the image $M = \mathbf{X}(U)$ is a *spacelike submanifold* in the de Sitter space if every tangent vector $\mathbf{v} \in T_pM \setminus \{\mathbf{0}\}$ is spacelike for every point p on M. We take a *spacelike unit normal vector field* $\mathbf{n} : U \longrightarrow S_1^n$ with $\mathbf{n}(\mathbf{u}) \perp \mathbf{X}(\mathbf{u}), \mathbf{X}_{u_i}(\mathbf{u})$ for $i = 1, \dots, n-r$ and $\mathbf{u} \in U$. Let $\pi_p^t : T_p\mathbb{R}_1^{n+1} \longrightarrow T_pM$ be the orthonormal projection to the tangent space. We define a *spacelike shape operator* with respect to $\mathbf{n}$ by $S_p(\mathbf{n}) = -\pi_p^t(d_p\mathbf{n})$. We also define *spacelike principal curvatures* $\kappa_i(\mathbf{n})(\mathbf{u})$ for $i = 1, \dots, n-r$ as eigenvalues of $S_p(\mathbf{n})$, and the *spacelike Gauss-Kronecker curvature* $K(\mathbf{n})(\mathbf{u})$ with respect to $\mathbf{n}$ as the determinant of $S_p(\mathbf{n})$.

We now consider the totally umbilic case of spacelike submanifolds. We say that the spacelike unit normal vector field $\mathbf{n}$ is *parallel* if $\mathbf{n}_{u_i}(\mathbf{u}) \in T_{\mathbf{X}(\mathbf{u})}M$ for any $\mathbf{u} \in U$ and $i = 1, \dots, n-r$. We have following proposition.

PROPOSITION 3.1. *Suppose that the spacelike unit normal vector field* $\mathbf{n}$ *is parallel and the spacelike submanifold* M *is totally umbilic with respect to* $\mathbf{n}$. *Then the principal curvatures* $\kappa_i(\mathbf{n})(\mathbf{u})$ *are of constant value* $\kappa(\mathbf{n})$ *and* M *is a part of some hyperbolic hyperquadric* $HP(\mathbf{v}, c) \cap S_1^n = \{\mathbf{x} \in S_1^n \mid \langle \mathbf{x}, \mathbf{v} \rangle = c\}$. *Moreover, we have the following cases:*

(1) *If* $\kappa(\mathbf{n}) \neq 0$ *then* M *is a part of some non-flat timelike hyperquadric with*

$$c = \frac{\kappa(\mathbf{n})}{\sqrt{\kappa^2(\mathbf{n}) + 1}} \quad \textit{and} \quad \mathbf{v} = \frac{1}{\sqrt{\kappa^2(\mathbf{n}) + 1}}(\kappa(\mathbf{n})\mathbf{X}(\mathbf{u}) + \mathbf{n}(\mathbf{u})).$$

(2) *If* $\kappa(\mathbf{n}) = 0$ *then* M *is a part of some flat timelike hyperbolic hyperquadric with* $\mathbf{v} = \mathbf{n}(\mathbf{u})$ *and* $c = 0$.

PROOF. We will show that the principal curvatures are constant. By assumption, we have $d_p\mathbf{n} = -\kappa(\mathbf{n})(\mathbf{u})\,\mathrm{id}_{T_pM}$. This means that $\mathbf{n}_{u_i}(\mathbf{u}) = -\kappa(\mathbf{n})(\mathbf{u})\mathbf{X}_{u_i}(\mathbf{u})$ for $i = 1, \dots, n-r$, so that we have

$$\mathbf{n}_{u_i u_j}(\mathbf{u}) = -\kappa(\mathbf{n})_{u_j}(\mathbf{u})\mathbf{X}_{u_i}(\mathbf{u}) - \kappa(\mathbf{n})(\mathbf{u})\mathbf{X}_{u_i u_j}(\mathbf{u}).$$

It follows that $\kappa(\mathbf{n})_{u_j}(\mathbf{u})\mathbf{X}_{u_i}(\mathbf{u}) = \kappa(\mathbf{n})_{u_i}(\mathbf{u})\mathbf{X}_{u_j}(\mathbf{u})$. Since $\mathbf{X}_{u_i}(\mathbf{u})$ and $\mathbf{X}_{u_j}(\mathbf{u})$ are linearly independent, $\kappa(\mathbf{n})_{u_i}(\mathbf{u}) \equiv 0$ for all i, j and $\mathbf{u} \in U$. Therefore $\kappa(\mathbf{n})(\mathbf{u}) \equiv \kappa(\mathbf{n})$ for some real constant value $\kappa(\mathbf{n})$.

We now assume that $\kappa(\mathbf{n}) \neq 0$. We have $\mathbf{n}_{u_i}(\mathbf{u}) = -\kappa(\mathbf{n})\mathbf{X}_{u_i}(\mathbf{u})$, so that there exists a constant vector $\mathbf{v}$ such that $\mathbf{X}(\mathbf{u}) = \mathbf{v} - (1/\kappa(\mathbf{n}))\mathbf{n}(\mathbf{u})$. Then $\langle \mathbf{v}, \mathbf{v} \rangle = 1 + (1/\kappa(\mathbf{n}))^2$ and $\langle \mathbf{X}(\mathbf{u}) - \mathbf{v}, \mathbf{X}(\mathbf{u}) - \mathbf{v} \rangle = (1/\kappa(\mathbf{n}))^2$. So $\langle \mathbf{X}(\mathbf{u}), \mathbf{v} \rangle = 1$. Therefore the spacelike submanifold M is a part of a hyperbolic hyperquadric $HP(\mathbf{v}, +1) \cap S_1^n$ and $\mathbf{v}$ is a spacelike vector.

If $\kappa(\mathbf{n}) = 0$, then we have $\mathbf{n}_{u_i}(\mathbf{u}) \equiv 0$. There exists a constant vector $\mathbf{v}$ such that $\mathbf{n}(\mathbf{u}) \equiv \mathbf{v}$. So $\langle \mathbf{X}(\mathbf{u}), \mathbf{v} \rangle = 0$ for any $\mathbf{u} \in U$. This means that M is a part of a hyperbolic hyperquadric $HP(\mathbf{v}, 0) \cap S_1^n$ and $\mathbf{v}$ is spacelike. This completes the proof. □

Let $g_{ij} = \langle \mathbf{X}_{u_i}, \mathbf{X}_{u_j} \rangle$ be a coefficient of the first fundamental form and $h_{ij}(\mathbf{n}) = -\langle \mathbf{n}_{u_i}, \mathbf{X}_{u_j} \rangle$ be that of the second. We have a Weingarten-type formula with respect to the spacelike normal vector field $\mathbf{n}$ of spacelike submanifolds:

$$\pi_p^t \circ \mathbf{n}_{u_i}(\mathbf{u}) = -\sum_{k=1}^{n-r}(h_{ij}(\mathbf{n})(\mathbf{u}))(g^{jk}(\mathbf{u}))\mathbf{X}_{u_k}(\mathbf{u}),$$

where π_p^t is the orthogonal projection $\pi_p^t : T_p\mathbb{R}_1^{n+1} \longrightarrow T_pM$. Therefore we have the following expression of the spacelike Gauss-Kronecker curvature:

$$K(\mathbf{n})(\mathbf{u}) = \det(h_{ij}(\mathbf{n})(\mathbf{u}))/\det(g_{ij}(\mathbf{u})).$$

If $\mathbf{n}$ and $\mathbf{n}'$ are parallel spacelike unit normal vector fields with $\mathbf{n}(\mathbf{u}_0) = \mathbf{n}'(\mathbf{u}_0) = \mathbf{n}_0$, then the corresponding second fundamental forms coincide. This means that the spacelike Gauss-Kronecker curvature and spacelike principal curvatures are determined by the spacelike normal direction $\mathbf{n}_0$ at each point on the spacelike submanifolds M, so we may write $K_{p_0}(\mathbf{n}_0)$ and $\kappa_i(\mathbf{n}_0)$.

4. Timelike canal hypersurfaces

In this section we study a geometrical relation between timelike canal hypersurfaces and spacelike submanifolds. We assume that there exists a unit parallel orthonormal frame of the spacelike submanifold M denoted by

$$\{\mathbf{X}(\mathbf{u}), \mathbf{n}_0(\mathbf{u}), \mathbf{n}_1(\mathbf{u}), \dots, \mathbf{n}_{r-1}(\mathbf{u})\},$$

where $\mathbf{n}_0(\mathbf{u})$ is a timelike unit normal section and the others are spacelike unit normal sections. Let $\theta \notin (\pi/2)\mathbb{Z}$ be a fixed real number. We define a map $\bar{\mathbf{X}}_\theta^t : U \times S_1^{r-1} \longrightarrow S_1^n$ by

$$\bar{\mathbf{X}}_\theta^t(\mathbf{u}, \bar{\mu}) = \cos\theta\mathbf{X}(\mathbf{u}) + \sin\theta\mathbf{e}(\mathbf{u}, \bar{\mu}),$$

where $\mathbf{e}(\mathbf{u}, \bar{\mu}) = \sum_{i=0}^{r-1} \mu_i \mathbf{n}_i(\mathbf{u})$ is a spacelike unit normal. We call $\bar{\mathbf{X}}_\theta^t$ a *timelike canal hypersurface* of the spacelike submanifold M and its image is denoted by CM_θ^t. We remark that $\bar{\mathbf{X}}_\theta^t$ is not always an embedding map.

PROPOSITION 4.1. *An element $(\mathbf{u}, \bar{\mu}) \in U \times S_1^{r-1}$ is a singular point of the map $\bar{\mathbf{X}}_\theta^t$ if and only if the value $\cot\theta$ is one of the corresponding spacelike principal curvatures of the spacelike submanifold M with respect to the spacelike unit normal direction $\mathbf{e}(\mathbf{u}, \bar{\mu})$.*

PROOF. Let $(\mathbf{u}, \bar{\mu}) = (u_1, \dots, u_{n-r}, \mu_0, \dots, \mu_{r-1}) \in U \times S_1^{r-1}$. Without loss of generality, we assume that $\mu_{r-1} > 0$ and write $\mu_{r-1} = \sqrt{1 + \mu_0^2 - \sum_{k=1}^{r-2} \mu_k^2}$, then $(u_1, \dots, u_{n-r}, \mu_0, \dots, \mu_{r-2})$ is a local coordinate on $U \times S_1^{r-1}$. Partial derivatives of $\bar{\mathbf{X}}_\theta^t$ are

$$\begin{aligned}
(\bar{\mathbf{X}}_\theta^t)_{u_i}(\mathbf{u}, \bar{\mu}) &= \cos\theta\mathbf{X}_{u_i}(\mathbf{u}) + \sin\theta\mathbf{e}_{u_i}(\mathbf{u}, \bar{\mu}) \in T_pM, \\
(\bar{\mathbf{X}}_\theta^t)_{\mu_j}(\mathbf{u}, \bar{\mu}) &= \sin\theta\mathbf{e}_{\mu_j}(\mathbf{u}, \bar{\mu}) \in N_pM,
\end{aligned}$$

where N_pM is an orthonormal space of M at $p = \mathbf{X}(\mathbf{u})$. Since $\mathbf{n}_0, \cdots, \mathbf{n}_{r-1}$ are linearly independent and $\sin\theta \neq 0$, the vectors

$$\begin{aligned}
\mathbf{e}_{\mu_0}(\mathbf{u}, \bar{\mu}) &= \mathbf{n}_0(\mathbf{u}) + \frac{\mu_0}{\mu_{r-1}}\mathbf{n}_{r-1}(\mathbf{u}), \\
\mathbf{e}_{\mu_j}(\mathbf{u}, \bar{\mu}) &= \mathbf{n}_j(\mathbf{u}) - \frac{\mu_j}{\mu_{r-1}}\mathbf{n}_{r-1}(\mathbf{u}) \quad \text{for } j = 1, \dots, r-2
\end{aligned}$$

are also linearly independent for any $(\mathbf{u}, \bar{\mu})$. By applying the Weingarten-type formula for the spacelike submanifolds of codimension r in the de Sitter space, we have

$$\begin{pmatrix} (\bar{\mathbf{X}}_\theta^t)_{u_1} \\ \vdots \\ (\bar{\mathbf{X}}_\theta^t)_{u_{n-r}} \end{pmatrix} = \left(\cos\theta E_n - \sin\theta I\!I_{\mathbf{e}(\mathbf{u},\bar{\mu})} I^{-1}\right) \begin{pmatrix} \mathbf{X}_{u_1} \\ \vdots \\ \mathbf{X}_{u_{n-r}} \end{pmatrix},$$

where E_n is an identity matrix and $I, I\!I_{\mathbf{e}(\mathbf{u},\bar{\mu})}$ are the first and second fundamental form matrices $(g_{ij})(\mathbf{u})$ and $(h_{ij}(\mathbf{e}(\mathbf{u},\bar{\mu})))(\mathbf{u})$ of the spacelike submanifold M. Since $\mathbf{X}$ is an embedding map, the partial derivatives $(\bar{\mathbf{X}}_\theta^t)_{u_i}$ for $i = 1, \dots, n-r$ are linearly dependent if and only if $\cot\theta \neq 0$ is one of the eigenvalues of the matrix $I\!I_{\mathbf{e}(\mathbf{u},\bar{\mu})} I^{-1}$, this means that the value $\cot\theta$ is a principal curvature of the spacelike submanifold M with respect to the spacelike unit normal direction $\mathbf{e}(\mathbf{u}, \bar{\mu})$. This completes the proof. □

We use the notions of the de Sitter height functions and de Sitter Gauss maps introduced in [4]. Let $V \subset U \times S_1^{r-1}$ be a regular part of the timelike canal hypersurface $\bar{\mathbf{X}}_\theta^t$ and $CM_\theta^t := \bar{\mathbf{X}}_\theta^t(V)$. The de Sitter height function $\bar{H}_\theta^d : V \times S_1^n \longrightarrow \mathbb{R}$ of the timelike canal hypersurface is given by

$$\bar{H}_\theta^d((\mathbf{u}, \bar{\mu}), \mathbf{v}) = \langle \bar{\mathbf{X}}_\theta^t(\mathbf{u}, \bar{\mu}), \mathbf{v} \rangle.$$

For any $\mathbf{v}_0 \in S_1^n$, we write $\bar{h}_{\mathbf{v}_0}^d(\mathbf{u}, \bar{\mu}) = \bar{H}_\theta^d((\mathbf{u}, \bar{\mu}), \mathbf{v}_0)$.

By calculation, the de Sitter Gauss map $\bar{\mathbf{x}}_{CM_\theta^t}^d : V \longrightarrow S_1^n$ of the timelike canal hypersurface is given by

$$\bar{\mathbf{x}}_{CM_\theta^t}^d(\mathbf{u}, \bar{\mu}) = \sin\theta \mathbf{X}(\mathbf{u}) - \cos\theta \mathbf{e}(\mathbf{u}, \bar{\mu}).$$

Let $\bar{K}_d(\mathbf{u}, \bar{\mu})$ and $\bar{\kappa}_{d,i}(\mathbf{u}, \bar{\mu})$ (for $i = 1, \dots, n-1$) be the de Sitter Gauss-Kronecker curvature and de Sitter principal curvatures.

We also observe the shape of the spacelike canal surface. The following proposition shows that $\bar{\mathbf{x}}_{CM_\theta^t}^d$ is analogous to the notion of the canal surfaces in a Euclidean space.

PROPOSITION 4.2. *The timelike canal hypersurface $\bar{\mathbf{x}}_{CM_\theta^t}^d$ has principal directions $\partial/\partial\mu_j$ (for $j = 0, \dots, r-2$) of the same principal curvature* $\cot\theta$ *at each point* $\bar{p}$.

PROOF. Let $(u_1, \dots, u_{n-r}, \mu_0, \dots, \mu_{r-2})$ be the local coordinates introduced in the proof of Proposition 4.1. The partial derivatives of $\bar{\mathbf{x}}_{CM_\theta^t}^d$ are

$$\begin{aligned} (\bar{\mathbf{x}}_{CM_\theta^t}^d)_{u_i}(\mathbf{u}, \bar{\mu}) &= \sin\theta \mathbf{X}_{u_i}(\mathbf{u}) - \cos\theta \mathbf{e}_{u_i}(\mathbf{u}, \bar{\mu}), \\ (\bar{\mathbf{x}}_{CM_\theta^t}^d)_{\mu_j}(\mathbf{u}, \bar{\mu}) &= -\cos\theta \mathbf{e}_{\mu_j}(\mathbf{u}, \bar{\mu}). \end{aligned}$$

The differential of de Sitter Gauss map $\bar{\mathbf{x}}_{CM_\theta^t}^d$ is the de Sitter shape operator A_p^d of the spacelike canal hypersurface $\bar{\mathbf{X}}_\theta^t$. By Proposition 2.2, the matrix of A_p^d with respect to the basis $\{\partial/\partial u_i\}$ is given by $(h_{ij})(g_{ij})^{-1}$, where (g_{ij}) and (h_{ij}) are the de Sitter first and second fundamental forms. Since $(\bar{\mathbf{x}}_{CM_\theta^t}^d)_{u_i}, (\bar{\mathbf{X}}_\theta^t)_{u_i}, (\bar{\mathbf{X}}_\theta^t)_{\mu_j} \in T_pM$

and $(\bar{\mathbf{x}}^d_{CM^t_\theta})_{\mu_j} \in N_pM$ for any $i = 1, \ldots, n-r$ and $j = 0, \ldots, r-2$, we have

$$\begin{aligned} \mathrm{I} = (g_{ij}) &= \left(\begin{array}{c|c} \langle (\bar{\mathbf{X}}^t_\theta)_{u_i}, (\bar{\mathbf{X}}^t_\theta)_{u_j} \rangle & \langle (\bar{\mathbf{X}}^t_\theta)_{u_i}, (\bar{\mathbf{X}}^t_\theta)_{\mu_j} \rangle \\ \hline \langle (\bar{\mathbf{X}}^t_\theta)_{\mu_i}, (\bar{\mathbf{X}}^t_\theta)_{u_j} \rangle & \langle (\bar{\mathbf{X}}^t_\theta)_{\mu_i}, (\bar{\mathbf{X}}^t_\theta)_{\mu_j} \rangle \end{array} \right) \\ &= \left(\begin{array}{c|c} * & O \\ \hline O & (\sin^2\theta \langle \mathbf{e}_{\mu_i}, \mathbf{e}_{\mu_j} \rangle)_{ij} \end{array} \right), \\ \mathrm{II}_{\mathbf{e}(\mathbf{u},\bar{\mu})} = (h_{ij}) &= \left(\begin{array}{c|c} * & O \\ \hline O & (\sin\theta\cos\theta \langle \mathbf{e}_{\mu_i}, \mathbf{e}_{\mu_j} \rangle)_{ij} \end{array} \right). \end{aligned}$$

Therefore we have

$$A_p = \left(\begin{array}{c|c} * & O \\ \hline O & \cot\theta E_{r-1} \end{array} \right),$$

where E_{r-1} is the identity matrix of order $r-1$. Therefore the de Sitter shape operator A_p has an eigenvalue $\cot\theta$ of multiplicity at least $r-1$, and each vector $\partial/\partial\mu_j$ for $j = 0, \ldots, r-2$ is the principal direction. This completes the proof. □

On the other hand, the other $n-r$ principal curvatures are not always constant. We also have the following proposition which implies the relations between spacelike submanifolds and timelike canal hypersurfaces.

PROPOSITION 4.3. *Let $\bar{\mathbf{x}}^d_{CM^t_\theta}$ be the de Sitter Gauss map of the timelike canal hypersurface $\bar{\mathbf{X}}^t_\theta$ and $\kappa_i(\mathbf{v})$ be the principal curvatures of a spacelike submanifold M with respect to the spacelike normal direction $\mathbf{v}$ at $\mathbf{X}(\mathbf{u})$. Then we have*

$$rank\, d_{(\mathbf{u},\bar{\mu})}\bar{\mathbf{x}}^d_{CM^t_\theta} = (n-1) - \sharp\{i \mid \kappa_i(\mathbf{v})(\mathbf{u}) = \tan\theta\} \geq r-1.$$

PROOF. Let $(u_1, \ldots, u_{n-r}, \mu_0, \ldots, \mu_{r-2})$ be the local coordinates on the regular part $V \subset U \times S^{r-1}_1$. Since $(\bar{\mathbf{x}}^d_{CM^t_\theta})_{\mu_j}(\mathbf{u}, \bar{\mu})$ are linearly independent and orthogonal to $(\bar{\mathbf{x}}^d_{CM^t_\theta})_{u_i}$ for all $i = 1, \ldots, n-r$ and $j = 0, \ldots, r-2$, so the de Sitter Gauss map is degenerate if and only if $(\bar{\mathbf{x}}^d_{CM^t_\theta})_{u_i}$ for $i = 1, \ldots, n-r$ are linearly dependent. Due to an argument similar to that in Proposition 4.1, this condition is equivalent to that the value $\tan\theta$ is a principal curvature of the spacelike submanifold M with respect to the spacelike unit normal direction $\mathbf{e}(\mathbf{u}, \bar{\mu})$. This completes the proof. □

COROLLARY 4.4. *Let $\bar{p} = \bar{\mathbf{X}}^t_\theta(\mathbf{u}, \bar{\mu})$ and $p = \mathbf{X}(\mathbf{u})$, then the following conditions are equivalent:*

(1) *$\bar{p}$ is a singular point of the de Sitter Gauss map $\bar{\mathbf{x}}^d_{CM^t_\theta}$.*
(2) *$\bar{p}$ is a de Sitter parabolic point of $\bar{\mathbf{x}}^d_{CM^t_\theta}(V)$, that is $\bar{K}_d(\mathbf{u}, \bar{\mu}) = 0$.*
(3) *$\tan\theta \neq 0$ is one of the spacelike principal curvatures of the spacelike submanifold M at p with respect to the spacelike normal direction $\bar{\mathbf{x}}^d_{CM^t_\theta}(\mathbf{u}, \bar{\mu})$.*

Therefore each singularity of the de Sitter Gauss map is related to a timelike principal curvature of the spacelike submanifold, and it corresponds to a de Sitter parabolic point of the timelike canal hypersurface.

We now define de Sitter θ-height functions and θ-maps of spacelike submanifolds. Let $\theta \notin (\pi/2)\mathbb{Z}$ be a real number. We define a map $H_\theta : U \times S^n_1 \longrightarrow \mathbb{R}$ by

$$H_\theta(\mathbf{u}, \mathbf{v}) = \langle \mathbf{X}(\mathbf{u}), \mathbf{v} \rangle - \sin\theta$$

and $h_{\theta,\mathbf{v}}(\mathbf{u}) = H_\theta(\mathbf{u},\mathbf{v})$. We call H_θ a *de Sitter θ-height function* of M. We also define a *de Sitter θ-map* $\mathbf{e}^d_\theta : U \times S_1^{r-1} \longrightarrow S_1^n$ of a spacelike submanifold by

$$\mathbf{e}^d_\theta(\mathbf{u},\bar{\mu}) = \sin\theta\mathbf{X}(\mathbf{u}) - \cos\theta\mathbf{e}(\mathbf{u},\bar{\mu}).$$

We remark that $\bar{\mathbf{x}}^d_{CM^t_\theta}$ is a restriction of $\mathbf{e}^d_\theta$ to V.

PROPOSITION 4.5. *Let $\mathbf{X}$ be a spacelike submanifold in the de Sitter space. The discriminant set of the de Sitter height function H_θ is the image of de Sitter θ-map $\mathbf{e}^d_\theta(\mathbf{u},\bar{\mu})$.*

PROOF. The discriminant set is $D_{H_\theta} = \{\mathbf{v} \in S_1^n \mid H_\theta(\mathbf{u},\mathbf{v}) = H_{\theta,u_i}(\mathbf{u},\mathbf{v}) = 0 \text{ for all } i\}$. Let $\mathbf{v}$ is an element of D_{H_θ}. Since $\{\mathbf{X}(\mathbf{u}), \mathbf{X}_{u_1}(\mathbf{u}), \dots, \mathbf{X}_{u_{n-r}}(\mathbf{u}), \mathbf{n}_0(\mathbf{u}), \dots, \mathbf{n}_{r-1}(\mathbf{u})\}$ is a basis of $\mathbb{R}^{n+1}_1$,

$$\mathbf{v} = \alpha\mathbf{X}(\mathbf{u}) + \sum_{i=0}^{n-r} \beta_i \mathbf{X}_i(\mathbf{u}) + \sum_{j=0}^{r-1} \gamma_j \mathbf{n}_j(\mathbf{u}),$$

for some $\alpha, \beta_i, \gamma_j \in \mathbb{R}$. By calculation, $\mathbf{v}$ is an element of D_{H_θ} if and only if $\alpha = \sin\theta$ and $\beta_i = 0$ for all i. The condition $\langle \mathbf{v},\mathbf{v}\rangle = 1$ implies $-\gamma_0^2 + \sum_{j=1}^{r-1}\gamma_j^2 = \cos^2\theta$. Let $\bar{\mu} = (\mu_0, \dots, \mu_{r-1})$ have $\mu_j = \gamma_j/\sin\theta$ for $j = 0, \dots, r-1$, then we have $\bar{\mu} \in S_1^{r-1}$ and $\mathbf{v} = \sin\theta\mathbf{X}(\mathbf{u}) + \cos\theta\mathbf{e}(\mathbf{u},\bar{\mu})$. Conversely, let $\mathbf{v} = \sin\theta\mathbf{X}(\mathbf{u}) + \cos\theta\mathbf{e}(\mathbf{u},\bar{\mu})$ for some μ, then we can check that $\mathbf{v}$ is an element of the discriminant set of the de Sitter θ-height function. This completes the proof. □

Let $(\mathbf{u}_0,\bar{\mu}_0) \in V$ and $\mathbf{v}_0 = \mathbf{e}^d_\theta(\mathbf{u}_0,\bar{\mu}_0)$, then

$$\operatorname{Hess} h_{\theta,\mathbf{v}_0}(\mathbf{u}_0) = \cos\theta I(\mathbf{u}_0)(\tan\theta\,\mathrm{id}_{T_pM} - I\!I_{\mathbf{v}_0}(\mathbf{u}_0)I^{-1}(\mathbf{u}_0)),$$

where $I(\mathbf{u}_0)$ is the first fundamental form matrix of M at $\mathbf{u}_0 \in U$ and $I\!I_{\mathbf{v}_0}(\mathbf{u}_0)$ is the second fundamental form matrix of M with respect to the spacelike normal direction $\mathbf{v}_0$.

5. De Sitter θ-maps as wavefronts

In this section we naturally interpret de Sitter θ-maps of spacelike submanifolds as a wavefront set of the theory of Legendrian singularities as discussed in Section 2.

PROPOSITION 5.1. *Let $\mathbf{X}$ be a spacelike submanifold in the de Sitter space and $H_\theta : (U \times S_1^n, (\mathbf{u},\mathbf{v})) \longrightarrow (\mathbb{R}, 0)$ be the de Sitter θ-height function germ. $H_\theta(\mathbf{u},\mathbf{v}) = 0$ is a Morse family of hypersurfaces.*

PROOF. Let $\mathbf{v}$ be an element of the discriminant set of H_θ. It is sufficient to show that $\Delta^* H_\theta = (H_\theta, H_{\theta u_1}, \dots, H_{\theta u_{n-r}})$ is regular at $(\mathbf{u},\mathbf{v})$. Let $\mathbf{X}(\mathbf{u}) = (X_0(\mathbf{u}), \dots, X_n(\mathbf{u}))$ and $(x_0, \dots, x_n) \in S_1^n$. Without loss of generality, we may assume that $v_n > 0$, then $v_n = \sqrt{1 + v_0^2 - \sum_{k=1}^{n-1} v_k^2}$. The Jacobi matrix of $\Delta^* H_\theta$ is

$$J\Delta^* H_\theta = \left(\begin{array}{c|c} * & \dfrac{\partial H_\theta}{\partial v_j}(\mathbf{u},\mathbf{v})_{\,j=0,\dots,n-1} \\ \hline * & \left(\dfrac{\partial^2 H_\theta}{\partial u_i \partial v_j}(\mathbf{u},\mathbf{v})\right)_{\substack{j=0,\dots,n-1 \\ i=1,\dots,n-r}} \end{array}\right).$$

By calculation, we have

$$\frac{\partial H_\theta}{\partial v_j}(\mathbf{u},\mathbf{v}) = \begin{cases} -X_0(\mathbf{u}) + X_n(\mathbf{u})\dfrac{v_0}{v_n} & \text{if } j=0 \\ X_j(\mathbf{u}) - X_n(\mathbf{u})\dfrac{v_j}{v_n} & \text{if } j=1,\dots,n-1 \end{cases}$$

$$\frac{\partial H_\theta}{\partial u_i \partial v_j}(\mathbf{u},\mathbf{v}) = \begin{cases} -X_{0,u_i}(\mathbf{u}) + X_{n,u_i}(\mathbf{u})\dfrac{v_0}{v_n} & \text{if } j=0 \\ X_{j,u_i}(\mathbf{u}) - X_{n,u_i}(\mathbf{u})\dfrac{v_j}{v_n} & \text{if } j=1,\dots,n-1. \end{cases}$$

The $(n-r+1)\times n$ submatrix of $J\Delta^* H_\theta$ with these entries will be denoted by $B : J\Delta^* H_\theta = (* \mid B)$.

On the other hand, $\mathbf{v}, \mathbf{X}(\mathbf{u}), \mathbf{X}_{u_1}(\mathbf{u}), \dots, \mathbf{X}_{u_{n-r}}(\mathbf{u})$ are linearly independent and a transformation

$$T : \mathbb{R}^{n+1} \longrightarrow \mathbb{R}^{n+1}, \qquad T(x_0, x_1, \dots, x_n) \longmapsto T(-x_0, x_1, \dots, x_n)$$

is regular, so that $T\mathbf{v}, T\mathbf{X}(\mathbf{u}), T\mathbf{X}_{u_1}(\mathbf{u}), \dots, T\mathbf{X}_{u_{n-r}}(\mathbf{u})$ are also linearly independent. Therefore the rank of the matrix

$$C = \begin{pmatrix} -v_0 & v_1 & \cdots & v_n \\ -X_0 & X_1 & \cdots & X_n \\ -X_{0,u_1} & X_{1,u_1} & \cdots & X_{n,u_1} \\ \vdots & \vdots & \ddots & \vdots \\ -X_{0,u_{n-r}} & X_{1,u_{n-r}} & \cdots & X_{n,u_{n-r}} \end{pmatrix}$$

is equal to $n-r+2$. We now show that we obtain the matrix B from C by row operations. We subtract the first row multiplied by $\mathbf{X}_n(\mathbf{u})/\lambda_n$ from the second row, and then subtract the first row multiplied by $\mathbf{X}_{n,u_k}(\mathbf{u})/\lambda_n$ from the $(2+k)$-th row for $k = 1, \cdots, n-2$. This results in the matrix

$$\left(\begin{array}{ccc|c} -v_0 \; v_1 & \cdots & v_{n-1} & v_n \\ \hline & & & 0 \\ & B & & \vdots \end{array}\right).$$

Since the rank of this matrix is $n-r+2$, the rank of the matrix B is $n-r+1$. Therefore rank $J\Delta^* H_\theta$ is $n-r+1$ at $(\mathbf{u},\mathbf{v})$. This completes the proof. □

Therefore we may construct a Legendrian immersion germ

$$\mathcal{L}_{H_\theta}(\mathbf{u},\bar{\mu},\mathbf{v}) = \left(\mathbf{v}, \left[\frac{\partial H_\theta}{\partial v_1}(\mathbf{u},\bar{\mu},\mathbf{v}) : \dots : \frac{\partial H_\theta}{\partial v_n}(\mathbf{u},\bar{\mu},\mathbf{v})\right]\right),$$

where $\Sigma_*(H_\theta) = \{(\mathbf{u},\bar{\mu},\mathbf{v}) \mid \mathbf{v} = \mathbf{e}^d_\theta(\mathbf{u},\bar{\mu}) \text{ for some } \bar{\mu} \in S^{r-1}_1\}$. A generating family of $\mathcal{L}_{H_\theta}$ is H_θ and the wave front of the Legendrian immersion germ $\mathcal{L}_{H_\theta}$ is the de Sitter θ-map germ $\mathbf{e}^d_\theta$.

On the other hand, we have already considered another Legendrian immersion germ $\mathcal{L}_H$ and related it to the timelike canal hypersurface in §2. A generating family of $\mathcal{L}_{\bar{H}^d}$ is the de Sitter height function $\bar{H}^d$ of the timelike canal hypersurface. Since the de Sitter Gauss map of CM_θ and de Sitter θ-map coincide as map germs, we have the following relation of the Legendrian immersion germs.

PROPOSITION 5.2. *Let $V \subset U \times S^{r-1}_1$ be a set of regular points of $\bar{\mathbf{X}}_\theta$. For every point $(\mathbf{u},\bar{\mu}) \in V$ and $\mathbf{v} = \bar{\mathbf{x}}^d_{CM_\theta}(\mathbf{u},\bar{\mu}) = \mathbf{e}^d_\theta(\mathbf{u},\bar{\mu})$, the Legendrian immersion germs coincide: $\mathcal{L}_{\bar{H}^d} = \mathcal{L}_{H_\theta}$ as map germs at $(\mathbf{u},\bar{\mu},\mathbf{v}) \in \Sigma_*(\bar{H}^d)$.*

PROOF. Let $\mathbf{X} = (\mathbf{X}_0, \ldots, \mathbf{X}_n)$ and $\mathbf{e} = (\mathbf{e}_0, \ldots, \mathbf{e}_n)$. Without loss of generality, we assume that $v_n > 0$ for $\mathbf{v} = (v_0, \ldots, v_n) \in S^n_1$. Since $(\mathbf{u}, \bar{\mu})$ is a parameter of the timelike canal hypersurface, $\mathcal{L}_{\bar{H}^d}$ is given by

$$\mathcal{L}_{\bar{H}^d}(\mathbf{u}, \bar{\mu}, \mathbf{v}) = \left(\mathbf{v}, \left[\frac{\partial \bar{H}^d}{\partial v_0}(\mathbf{u}, \bar{\mu}, \mathbf{v}) : \ldots : \frac{\partial \bar{H}^d}{\partial v_{n-1}}(\mathbf{u}, \bar{\mu}, \mathbf{v})\right]\right),$$

where $\mathbf{v} = \bar{\mathbf{x}}^d_{CM_\theta}(\mathbf{u}, \bar{\mu})$ coincides with $\mathbf{e}^d_\theta(\mathbf{u}, \bar{\mu})$. By calculation, we have

$$\begin{aligned}
\frac{\partial \bar{H}^d}{\partial v_0}(\mathbf{u}, \bar{\mu}, \mathbf{v}) &= -\mathbf{X}_n(\mathbf{u})\mathbf{e}_0(\mathbf{u}, \bar{\mu}) + \mathbf{X}_0(\mathbf{u})\mathbf{e}_n(\mathbf{u}, \bar{\mu}), \\
\frac{\partial \bar{H}^d}{\partial v_j}(\mathbf{u}, \bar{\mu}, \mathbf{v}) &= \mathbf{X}_n(\mathbf{u})\mathbf{e}_j(\mathbf{u}, \bar{\mu}) - \mathbf{X}_j(\mathbf{u})\mathbf{e}_n(\mathbf{u}, \bar{\mu}), \qquad j = 1, \ldots, n-1.
\end{aligned}$$

On the other hand, the fiber components of $\mathcal{L}_{H_\theta}$ at $\mathbf{v} \in S^n_1$ are

$$\begin{aligned}
\frac{\partial H_\theta}{\partial v_0}(\mathbf{u}, \bar{\mu}, \mathbf{v}) &= \cos\theta(-\mathbf{X}_n(\mathbf{u})\mathbf{e}_0(\mathbf{u}, \bar{\mu}) + \mathbf{X}_0(\mathbf{u})\mathbf{e}_n(\mathbf{u}, \bar{\mu})), \\
\frac{\partial H_\theta}{\partial v_j}(\mathbf{u}, \bar{\mu}, \mathbf{v}) &= \cos\theta(\mathbf{X}_n(\mathbf{u})\mathbf{e}_j(\mathbf{u}, \bar{\mu}) - \mathbf{X}_j(\mathbf{u})\mathbf{e}_n(\mathbf{u}, \bar{\mu})), \qquad j = 1, \ldots, n-1,
\end{aligned}$$

so that $[\partial \bar{H}^d/\partial v_0, \ldots, \partial \bar{H}^d/\partial v_{n-1}] = [\partial H_\theta/\partial v_0, \ldots, \partial H_\theta/\partial v_{n-1}]$. Therefore we have $\mathcal{L}_{\bar{H}^d}(\mathbf{u}, \bar{\mu}, \mathbf{v}) = \mathcal{L}_{H_\theta}(\mathbf{u}, \bar{\mu}, \mathbf{v})$. This completes the proof. □

6. Contact with hyperbolic hyperquadrics in the de Sitter space

In this section we review the theory of contact of submanifolds due to Montaldi [**6**]. Let X_i and Y_i (for $i = 1, 2$) be submanifolds of $\mathbb{R}^n$ with $\dim X_1 = \dim X_2$ and $\dim Y_1 = \dim Y_2$. We say that the contact of X_1 and Y_1 at $\mathbf{y}_1$ is the same type as the contact of X_2 and Y_2 at $\mathbf{y}_2$ if there is a diffeomorphism germ $\Phi : (\mathbb{R}^n, \mathbf{y}_1) \longrightarrow (\mathbb{R}^n, \mathbf{y}_2)$ such that $\Phi(X_1) = X_2$ and $\Phi(Y_1) = Y_2$. In this case we write $K(X_1, Y_1; \mathbf{y}_1) = K(X_2, Y_2; \mathbf{y}_2)$. Function germs $g_i : (\mathbb{R}^n, \mathbf{a}_i) \longrightarrow (\mathbb{R}, 0)$, $i = 1, 2$ are $\mathcal{K}$ *equivalent* if there are a diffeomorphism germ $\Phi : (\mathbb{R}^n, \mathbf{a}_1) \longrightarrow (\mathbb{R}^n, \mathbf{a}_2)$ and a function germ $\lambda : (\mathbb{R}^n, \mathbf{a}_1) \longrightarrow \mathbb{R}$ with $\lambda(\mathbf{a}_1) \neq 0$ such that $g_1 = \lambda \cdot (g_2 \circ \Phi)$. In [**6**] Montaldi proved the following theorem.

THEOREM 6.1 (Montaldi [**6**]). *Let X_i and Y_i (for $i = 1, 2$) be submanifolds of $\mathbb{R}^n$ with $\dim X_1 = \dim X_2$ and $\dim Y_1 = \dim Y_2$. Let $g_i : (X_i, \mathbf{x}_i) \longrightarrow (\mathbb{R}^n, \mathbf{y}_i)$ be immersion germs and $f_i : (\mathbb{R}^n, \mathbf{y}_i) \longrightarrow (\mathbb{R}^p, \mathbf{0})$ be submersion germs with $(Y_i, \mathbf{y}_i) = (f_i^{-1}(\mathbf{0}), \mathbf{y}_i)$. Then $K(X_1, Y_1; \mathbf{y}_1) = K(X_2, Y_2; \mathbf{y}_2)$ if and only if $f_1 \circ g_1$ and $f_2 \circ g_2$ are $\mathcal{K}$-equivalent.*

We now consider the case of $p = 1$ and apply this theorem to the contact between spacelike submanifolds and non-flat timelike hyperquadrics. Let $\mathbf{v}_0 = \mathbf{e}^d_\theta(\mathbf{u}_0, \bar{\mu}_0)$ be a spacelike vector. We consider an immersion germ $g : (U, \mathbf{u}_0) \longrightarrow (S^n_1, p_0)$ and a submersion germ $f_{\theta, \mathbf{v}_0} : (S^n_1, p_0) \longrightarrow (\mathbb{R}, 0)$ defined as $g(\mathbf{u}) = \mathbf{X}(\mathbf{u})$ and $f_{\theta, \mathbf{v}_0}(\mathbf{x}) = \langle \mathbf{x}, \mathbf{v}_0 \rangle - \sin\theta$. We have $f_{\theta, \mathbf{v}_0} \circ g(\mathbf{u}) = \langle \mathbf{X}(\mathbf{u}), \mathbf{v}_0 \rangle - \sin\theta = h_{\theta, \mathbf{v}_0}(\mathbf{u})$, and $h^{-1}_{\theta, \mathbf{v}_0}(0) = HP(\mathbf{v}_0, \sin\theta) \cap S^n_1$ is a non-flat timelike hyperquadric. Moreover,

$$h_{\theta, \mathbf{v}_0}(\mathbf{u}_0) = \frac{\partial h_{\theta, \mathbf{v}_0}}{\partial u_i}(\mathbf{u}_0) = 0 \qquad \text{for } i = 1, \ldots, n-r,$$

so that the non-flat timelike hyperquadric $HP(\mathbf{v}_0, \sin\theta) \cap S^n_1$ is tangent to the spacelike submanifold M at $p_0 = \mathbf{X}(\mathbf{u}_0)$. Therefore we call $HP(\mathbf{v}_0, \sin\theta) \cap S^n_1$ a *tangent non-flat timelike hyperquadric* of M at p_0.

By the theory of Montaldi, the contact type between the spacelike submanifold M and the tangent non-flat timelike hyperquadric $HP(\mathbf{v}_0, \sin\theta) \cap S^n_1$ corresponds to the $\mathcal{K}$-equivalence class of the de Sitter θ-height function germs $h_{\theta,\mathbf{v}_0}$. Moreover, the $\mathcal{K}$-equivalence preserves the diffeomorphism type class of the zero level set germs $h^{-1}_{\theta,\mathbf{v}_0}(0)$. We call $h^{-1}_{\theta,\mathbf{v}_0}(0)$ a *tangent non-flat timelike hyperquadric indicatrix germ.*

On the other hand, the contact of timelike hypersurfaces and flat hyperbolic hyperquadrics was discussed in [**4**]. Let $\bar{\mathbf{X}} : V \longrightarrow S^n_1$ be a timelike hypersurface in the de Sitter space and $\bar{\mathbf{x}}^d$ be the de Sitter Gauss map of $\bar{\mathbf{X}}$. For $\bar{\mathbf{u}}_0 \in V$ and $\bar{\mathbf{v}}_0 = \bar{\mathbf{x}}^d(\bar{\mathbf{u}}_0)$, $\bar{\mathbf{X}}(\bar{\mathbf{u}})$ is an immersion and $\bar{h}_{\bar{\mathbf{v}}_0}(\mathbf{x}) = \langle \mathbf{x}, \bar{\mathbf{v}}_0\rangle$ is a submersion, so that the flat timelike hyperquadric $\bar{h}^{-1}_{\bar{\mathbf{v}}_0}(0) = HP(\bar{\mathbf{v}}_0, 0) \cap S^n_1$ is tangent to the timelike hypersurface $\bar{M} = \bar{\mathbf{X}}(U)$ at $\bar{p}_0 = \bar{\mathbf{X}}(\bar{\mathbf{u}}_0)$. By the above argument, the contact type of the submanifolds $K(\bar{M}, HP(\bar{\mathbf{v}}_0, 0) \cap S^n_1; \bar{p}_0)$ corresponds to the $\mathcal{K}$-equivalence class of the de Sitter height function germs $\bar{h}^d_{\bar{\mathbf{v}}_0}$ at $\bar{p}_0 = \bar{\mathbf{X}}(\bar{\mathbf{u}}_0)$. We call the germ $((\bar{h}^d_{\bar{\mathbf{v}}_0})^{-1}(0), \bar{\mathbf{u}}_0)$ a *tangent flat timelike hyperquadric indicatrix germ.*

We now review some notions of theory of Legendrian singularities which we need to study the contact between submanifolds and hyperbolic hyperquadrics. We say that Legendrian immersion germs $\mathcal{L}_i : (U_i, \mathbf{u}_i) \longrightarrow (PT^*\mathbb{R}^k, p_i)$ $(i = 1, 2)$ are *Legendrian equivalent* if there exists a contact diffeomorphism germ $H : (PT^*\mathbb{R}^k, p_1) \longrightarrow (PT^*\mathbb{R}^k, p_2)$ such that H preserves fibers of π and $H(U_1) = U_2$. A Legendrian immersion germ at a point is said to be *Legendrian stable* if for every map with the given germ there are a neighborhood in the space of Legendrian immersions with the Whitney C^∞-topology and a neighborhood of the base point such that each Legendrian map belonging to the first neighborhood has a point in the second neighborhood, at which its germ is Legendrian equivalent to the original germ.

PROPOSITION 6.2 (Zakalyukin [**9**]). *Let $\mathcal{L}_1, \mathcal{L}_2$ be Legendrian immersion germs such that the sets of regular points of $\pi \circ \mathcal{L}_1$ and $\pi \circ \mathcal{L}_2$ are dense. Then $\mathcal{L}_1, \mathcal{L}_2$ are Legendrian equivalent if and only if the corresponding wave front $W(\mathcal{L}_1)$ and $W(\mathcal{L}_2)$ are diffeomorphic as set germs.*

Let $F_i : (\mathbb{R}^n \times \mathbb{R}^k, (\mathbf{a}_i, \mathbf{b}_i)) \longrightarrow (\mathbb{R}, 0)$ $(i = 1, 2)$ be k-parameter unfoldings of function germs $f_i : (\mathbb{R}^n, \mathbf{a}_i) \longrightarrow (\mathbb{R}, 0)$. We say that F_1 and F_2 are $\mathcal{P}$-$\mathcal{K}$-equivalent if there exists a diffeomorphism germ $\Phi : (\mathbb{R}^n \times \mathbb{R}^k, (\mathbf{a}_1, \mathbf{b}_1)) \longrightarrow (\mathbb{R}^n \times \mathbb{R}^k, (\mathbf{a}_2, \mathbf{b}_2))$ of the form $\Phi(\mathbf{a}, \mathbf{b}) = (\phi_1(\mathbf{a}, \mathbf{b}), \phi_2(\mathbf{b}))$ for $(\mathbf{a}, \mathbf{b}) \in \mathbb{R}^n \times \mathbb{R}^k$ and a function germ $\lambda : (\mathbb{R}^n \times \mathbb{R}^k, (\mathbf{a}_1, \mathbf{b}_1)) \longrightarrow \mathbb{R}$ such that $\lambda(\mathbf{a}_1, \mathbf{b}_1) \neq 0$ and $F_1(\mathbf{a}, \mathbf{b}) = \lambda(\mathbf{a}, \mathbf{b}) \cdot (F_2 \circ \Phi)(\mathbf{a}, \mathbf{b})$.

THEOREM 6.3 ([**8, 1**]). *Let $F, G : (\mathbb{R}^n \times \mathbb{R}^k, \mathbf{0}) \longrightarrow (\mathbb{R}, \mathbf{0})$. Suppose that $F(\mathbf{a}, \mathbf{b}) = 0$, $G(\mathbf{a}, \mathbf{b}) = 0$ are Morse families of hypersurfaces and their Legendrian immersions are denoted by $\mathcal{L}_F, \mathcal{L}_G$, then*

(1) *$\mathcal{L}_F$ and $\mathcal{L}_G$ are Legendrian equivalent if and only if F and G are $\mathcal{P}$-$\mathcal{K}$-equivalent.*
(2) *$\mathcal{L}_F$ is Legendrian stable if and only if F is a $\mathcal{K}$-versal deformation of $f = F\,|_{\mathbb{R}^n \times \{\mathbf{0}\}}$.*

Let $g_i : (\mathbb{R}^n, \mathbf{a}_i) \longrightarrow (\mathbb{R}^k, \mathbf{b}_i)$ (for $i = 1, 2$) be map germs. We say that g_1 and g_2 are $\mathcal{A}$-equivalent if and only if there exist diffeomorphism germs $\phi : (\mathbb{R}^n, \mathbf{a}_1) \longrightarrow (\mathbb{R}^n, \mathbf{a}_2)$ and $\Phi : (\mathbb{R}^k, \mathbf{b}_1) \longrightarrow (\mathbb{R}^k, \mathbf{b}_2)$ such that $\Phi \circ g_1 = g_2 \circ \phi$. Let $Q(f)$ be the local ring of the function germ $f : (\mathbb{R}^n, \mathbf{u}_0) \longrightarrow (\mathbb{R}, 0)$ defined as

$$Q(f) = C^\infty_{\mathbf{u}_0}(\mathbb{R}^n)/\langle f\rangle_{C^\infty_{\mathbf{u}_0}(\mathbb{R}^n)},$$

where $C^\infty_{\mathbf{u}_0}(\mathbb{R}^n)$ is the local ring of C^∞ function germs at $\mathbf{u}_0 \in \mathbb{R}^n$.

PROPOSITION 6.4 ([**2**], Proposition A.4). *Let $F, G : (\mathbb{R}^n \times \mathbb{R}^k, \mathbf{0}) \longrightarrow (\mathbb{R}, \mathbf{0})$. Suppose that $F(\mathbf{a},\mathbf{b}) = 0$, $G(\mathbf{a},\mathbf{b}) = 0$ are Morse families of hypersurfaces and the Legendrian immersion germs $\mathcal{L}_F$ and $\mathcal{L}_G$ are Legendrian stable, then the following conditions are equivalent:*

(1) *$(W(\mathcal{L}_F), \lambda)$ and $(W(\mathcal{L}_G), \lambda')$ are diffeomorphic as set germs;*
(2) *$\mathcal{L}_F$ and $\mathcal{L}_G$ are Legendrian equivalent;*
(3) *$Q(f)$ and $Q(g)$ are isomorphic as $\mathbb{R}$-algebras, where $f = F\,|_{\mathbb{R}^n \times \{\mathbf{0}\}}$ and $g = G\,|_{\mathbb{R}^n \times \{\mathbf{0}\}}$.*

7. Generic properties of spacelike submanifolds

We consider the genericity condition for spacelike submanifolds. Let $U \subset \mathbb{R}^{n-r}$ be open. We define Sp-Emb(U, S^n_1) as the space of spacelike embeddings with the Whitney C^∞-topology. We also define a function $\mathcal{H} : S^n_1 \times S^n_1 \longrightarrow \mathbb{R}$ by $\mathcal{H}(\mathbf{x},\mathbf{v}) = \langle \mathbf{x},\mathbf{v}\rangle$ and write $\mathfrak{h}_{\mathbf{v}}(\mathbf{x}) = \mathcal{H}(\mathbf{x},\mathbf{v})$. Then $\mathfrak{h}_{\mathbf{v}}$ is a submersion for any $\mathbf{v} \in LC^*$. For any spacelike submanifold $\mathbf{X} \in$ Sp-Emb(U, S^n_1), we have $H = \mathcal{H}\circ(\mathbf{X}\times \mathrm{id}_{S^n_1})$. We also have the ℓ-jet extension $j^\ell_1 H : U \times S^n_1 \longrightarrow J^\ell(U,\mathbb{R})$ defined as $j^\ell_1 H(\mathbf{x},\mathbf{v}) = j^\ell \mathfrak{h}_{\mathbf{v}}(\mathbf{u})$. We consider the trivialization $J^\ell(U,\mathbb{R}) \equiv U \times \mathbb{R} \times J^\ell(n-r,1)$. For any submanifold $Q \subset J^\ell(n-r,1)$, we write $\widetilde{Q} = U \times \{0\} \times Q$. Then we have the following proposition as a corollary of Lemma 6 of Wassermann.

PROPOSITION 7.1 ([**7**]). *Let Q be a submanifold of $J^\ell(n-1,1)$. Then the set*

$$T_Q = \{\mathbf{X} \in \textit{Sp-Emb}(U, S^n_1) \mid j^\ell_1 H \textit{ is transversal to } \widetilde{Q}\}$$

is a residual subset of Sp-Emb(U, S^n_1). If Q is a closed subset, then T_Q is open.

We remark that if the function $F_{\mathbf{v}_0}$ is ℓ-determined relative to $\mathcal{K}$, then F is a $\mathcal{K}$-versal deformation if and only if $j^\ell_1 F$ is transversal to $\widetilde{\mathcal{K}}^\ell_{f,\mathbf{v}_0}$, where $\widetilde{\mathcal{K}}^\ell_{f,\mathbf{v}_0}$ is the $\mathcal{K}$-orbit of $j^\ell f_{\mathbf{v}_0}(\mathbf{0}) \in J^\ell(n-1,1)$. Due to Theorem 6.3, this condition is equivalent to the condition that the corresponding Legendrian immersion germ is Legendrian stable. Moreover, we have the following theorem, (see [**1**] and Appendix of [**2**]).

THEOREM 7.2. *If $n \leq 6$, there exists an open subset $\mathcal{O} \subset$ Sp-Emb(U, S^n_1) such that for any $\mathbf{X} \in \mathcal{O}$, the corresponding Legendrian immersion germ $\mathcal{L}_F$ is Legendrian stable.*

Finally we have the following theorem.

THEOREM 7.3. *Let $\theta_i \in \mathbb{R} \setminus (\pi/2)\mathbb{Z}$, and $(\mathbf{X}_i, \mathbf{u}_i)$, $(\bar{\mathbf{X}}^t_{\theta_i}, (\mathbf{u}_i, \bar{\mu}_i))$ be spacelike submanifold germs and corresponding regular timelike canal hypersurface germs for $i = 1, 2$. Suppose that the corresponding Legendrian immersion germs $\mathcal{L}_{H_{\theta_i}}$ are Legendrian stable, then the following conditions are equivalent.*

(1) *The Legendrian immersion germs $\mathcal{L}_{H_{\theta_i}}$ are Legendrian equivalent.*
(2) *The de Sitter θ_i-maps $\mathbf{e}^d_{\theta,i}$ are $\mathcal{A}$-equivalent.*
(3) *The contact types $K(\mathbf{X}_i(U), DH(v_i, sin\theta_i); p_i)$ are equivalent.*
(4) *The de Sitter θ_i-height function germs H_{i,θ_i} are $\mathcal{P}$-$\mathcal{K}$-equivalent.*
(5) *The function germs $h_{i,\theta_i,\mathbf{v}_i}$ are $\mathcal{K}$-equivalent.*
(6) *The local rings $Q(h_{i,\theta_i,\mathbf{v}_i})$ are isomorphic as $\mathbb{R}$-algebras.*
(7) *The Legendrian immersion germs $\mathcal{L}_{\bar{H}^d_i}$ are Legendrian equivalent.*

(8) *The de Sitter Gauss map germs* $\bar{\mathbf{x}}^d_{CM_{\theta_i}}$ *are* $\mathcal{A}$*-equivalent.*
(9) *The contact types* $K(\bar{\mathbf{X}}^t_{\theta_i,i}(V), HP(\bar{v}_i, 0) \cap S^n_1; \bar{p}_i)$ *are equivalent.*
(10) *The de Sitter height function germs* $\bar{H}^d_i$ *are* $\mathcal{P}$-$\mathcal{K}$*-equivalent.*
(11) *The function germs* $\bar{h}^d_{i,\bar{\mathbf{v}}_i}$ *are* $\mathcal{K}$*-equivalent.*
(12) *The local rings* $Q(\bar{h}^d_{i,\bar{\mathbf{v}}_i})$ *are isomorphic as* $\mathbb{R}$*-algebras.*

Here $p_i = \mathbf{u}_i$, $\mathbf{v}_i = \mathbf{e}^d_{\theta,i}(\mathbf{u}_i)$ *and* $\bar{\mathbf{v}}_i = \bar{\mathbf{x}}^d_{CM_{\theta_i}}(\mathbf{u}_i, \bar{\mu}_i)$. *In this case, the corresponding tangent flat timelike hyperquadric indicatrix germs* $((h^d_{i,\mathbf{v}_i})^{-1}(0), \mathbf{u}_i)$ *are diffeomorphic to each other, and the corresponding tangent non-flat timelike hyperquadric indicatrix germs* $((h_{i,\theta_i,\mathbf{v}_i})^{-1}(0), \mathbf{u}_i)$ *are also diffeomorphic to each other.*

We remark that the statement of Theorem 6.5 in [**4**] includes the equivalence of the conditions (2), (3), (4) and (6) of the above theorem.

PROOF. Suppose that $\mathcal{L}_{H_{\theta_i}}$ are Legendrian stable, then the sets of regular points $HS_{\mathbf{X}_1}$ and $HS_{\mathbf{X}_2}$ are dense. By Proposition 6.4, (1), (2) and (6) are equivalent. By applying Theorem 6.3, conditions (1) and (4) are equivalent and the corresponding θ_i-height functions H_{i,θ_i} are $\mathcal{K}$-versal deformations of the $h_{i,\theta_i,\mathbf{v}_i}$, so that (4) and (5) are also equivalent. By the arguments in Theorem 6.1, conditions (3) and (5) are equivalent.

On the other hand, by Proposition 5.2, the corresponding Legendrian immersion germs coincide $\mathcal{L}_{\bar{H}^d_i} = \mathcal{L}_{H_{\theta_i}}$ for $i = 1, 2$ and the $\mathcal{L}_{\bar{H}^d_i}$ are also Legendrian stable, so that (1) and (7) are equivalent. We may similarly show the equivalence of the conditions from (7) to (12). Therefore the conditions from (1) to (12) are equivalent. Since the $\mathcal{K}$-equivalence preserves the zero level sets, each pair of the indicatrix germs are diffeomorphic as set germs. This completes the proof. □

Acknowledgement. The author wishes to express his thanks to Professor Shyuichi Izumiya for valuable advice and discussions.

References

1. V.I. Arnold, S.M. Gusein-Zade and A.N. Varchenko, Singularities of Differentiable Maps, Volume I, Birkhäuser, Basel, 1986. MR777682 (86f:58018)
2. S. Izumiya, D. Pei and T. Sano, Singularities of hyperbolic Gauss maps, Proc. London Math Soc. **86** (2003), 485–512. MR1971160 (2004h:58056)
3. S. Izumiya, D. Pei, M.C. Romero Fuster and M. Takahashi, The horospherical geometry of submanifolds in hyperbolic space, J. London Math. Soc. (2) **71** (2005), 779–800. MR2132383 (2006f:53074)
4. S. Izumiya, Time-like hypersurfaces in de Sitter space and Legendrian singularities, J. Math. Sci. (1) **144** (2007), 3789–3803. MR2464746 (2009m:53190)
5. M. Kasedou, Spacelike submanifolds in de Sitter space, Demonstratio Mathematica (2) **43** (2010), 401–418. MR2668484 (2011h:53063)
6. J.A. Montaldi, On contact between submanifolds, Michigan Math. J. **33** (1986), 195–199. MR837577 (87i:58024)
7. G. Wassermann, Stability of Caustics, Math. Ann. **216** (1975), 43–50. MR0372912 (51:9116)
8. V.M. Zakalyukin, Lagrangian and Legendrian singularities, Funct. Anal. Appl. **10** (1976), 26–36.
9. V.M. Zakalyukin, Reconstructions of fronts and caustics depending on one parameter and versality of mappings, J. Soviet. Math. **27** (1984), 2713–2735.

DEPARTMENT OF MATHEMATICS, HOKKAIDO UNIVERSITY, SAPPORO 060-0810, JAPAN
E-mail address: kasedou@math.sci.hokudai.ac.jp

Contemporary Mathematics
Volume **569**, 2012
http://dx.doi.org/10.1090/conm/569/11228

Residues in K-theory

Daniel Lehmann

Dedicated to the memory of Vincent Cavalier

1. Introduction

In this paper, we present a synthesis of various examples of situations giving rise to residues of characteristic classes in K-theory. Most of these examples are already well known, and are summarized here without proof. More details can be found in [**AL**] (chapter 6); see also [**L1**] for the use of the Mayer-Vietoris complex, and [**S**] for synthesis on residues which are especially related to holomorphic foliations.

In general, residues of characteristic classes appear when some "vanishing theorem", due to the existence of some "geometrical structure" or "object" on a space V, does not apply because of the presence of a non-empty "singular set" Σ in V for this object : then, the concerned characteristic class may not vanish, but it will however "localize" near Σ (the prototype result in this direction is the Poincaré-Hopf theorem : the Euler-Poincaré invariant of a compact manifold, which vanishes when there exists on the manifold a vector field without singularity, may be more generally written as the sum of the indices of a singular vector field around its singular points, each index depending only on the local behavior of the vector field near the point).

There are essentially two kinds of situations giving rise to residues in K-theory:

- The first one happens when the vanishing theorem concerns characteristic classes of a bundle which is defined only off Σ, but whose stable class in K-theory extends naturally to all of V; in this category are for instance the Baum-Bott residues for singular holomorphic foliations, the residues for holomorphic maps which are submersions almost everywhere (including the Milnor numbers for hypersurfaces), etc.

- The second one occurs when the vanishing theorem arises from the exactness off Σ of a sequence of vector bundles, all defined above the whole space V : such a sequence defines an element in the relative K-theory $K(V, V \setminus \Sigma))$ (see for instance [**A**] or [**B**]); here we find for instance the genus formula for singular algebraic curves, the Mexican index (Gómez-Mont, Seade, Verjovski) for foliations preserving some singular subvariety (including applications to Poincaré inequality for algebraic solutions of an algebraic differential system), the index defined in cooperation with Camacho for foliations which are transverse almost everywhere to some submanifold (including in particular the index for dicritical singularities of vector fields), etc.

2010 *Mathematics Subject Classification.* Primary: 53C99, 57R20; Secondary: 19M05, 32A27, 32L20.

1.1. Background on characteristic classes in K-theory.

Let $E \to V$ be a complex vector bundle, and $c(E) = 1 + c_1(E) + c_2(E) + \cdots$ its total Chern class (depending only on its isomorphy class). Recall the formula

$$c(E \oplus F) = c(E).c(F),$$

which gives a morphism of semi-groups

$$c : \mathcal{E}_{\mathbb{C}}(V) \to H^+(V),$$

- where $\mathcal{E}_{\mathbb{C}}(V)$ denotes the set of isomorphy classes of complex vector bundles with base V (abelian semi-group for the Whitney sum),

- and $H^+(V)$ the set of formal series $1 + \sum_{k\geq 1} h_k$, with $h_k \in H^{2k}(V)$ (abelian group for the multiplication in cohomology).

Consequently, this morphism factorizes through a morphism of groups (still denoted by c) defined on the group $K_{\mathbb{C}}(V)$, *symmetrized* of the semi-group $\mathcal{E}_{\mathbb{C}}(V)$.

$$\begin{array}{ccc} \mathcal{E}_{\mathbb{C}}(V) & \stackrel{c}{\to} & H^+(V) \\ \downarrow & & \| \\ K_{\mathbb{C}}(V) & \stackrel{c}{\to} & H^+(V) \end{array}$$

We get : $c([E - F] = c(E).\big(c(F)\big)^{-1}$.

Thus, if $c(E) = 1 + c_1 + c_2 + \cdots$, and $c(F) = 1 + c'_1 + c'_2 + \cdots$, then:

$\big(c(F)\big)^{-1} = 1 - c'_1 + \big((c'_1)^2 - c'_2\big) + \cdots$, and

$c_1([E - F]) = c_1 - c'_1$,

$c_2([E - F]) = c_2 - c_1 c'_1 + (c'_1)^2 - c'_2$,

etc

The same can be done with real vector bundles, Pontrjagin classes (or Stiefel-Whitney classes[1]), and the group $K_{\mathbb{R}}(V)$.

The same can be done also for the relative K-theory $K(V, V \setminus \Sigma)$, Σ denoting some closed set in V (see [**A**] or [**B**]).

1.2. Background on the localization of characteristic classes.

A vanishing theorem concerns the existence, on a manifold V, of some "geometrical stucture" (generically denoted by $\mathcal{O}$ in this introduction), and may be more or less written in the following way.

[1]In fact, we shall restrict ourselves to characteristic classes with real or complex coefficients, as arising in differential geometry by Chern-Weil theory, so that Stiefel-Whitney classes are here irrelevant.

Pseudo-vanishing theorem : *If $\mathcal{O}$ exists on V, then "some" characteristic class(es) $\varphi(\xi) \in H^*(V)$ of "some" bundle ξ with base V or of "some" element $\xi \in K_{\mathbb{K}}(V)$ must vanish, φ denoting a polynomial belonging*

- to the algebra $I^*(G)$ of ad(G)-invariant polynomials on the Lie algebra of G if ξ is a bundle with structural group a Lie group G,

- to $\mathbb{R}[p_1, p_2, \cdots, p_j, \cdots]$ if $\xi \in K_{\mathbb{R}}(V)$ (polynomials with infinitely many variables),

- to $\mathbb{R}[c_1, c_2, \cdots, c_j, \cdots]$ if $\xi \in K_{\mathbb{C}}(V)$.

The existence of such a structure $\mathcal{O}$ off a closed subset Σ of V implies that the restriction of $\varphi(\xi)$ to $V \backslash \Sigma$ will vanish, and consequently $\varphi(\xi)$ will lift in $H^*(V, V \backslash \Sigma)$ by the map j of the exact sequence

$$(*) \qquad \cdots \to H^*(V, V \setminus \Sigma) \xrightarrow{j} H^*(V) \xrightarrow{\pi} H^*(V \setminus \Sigma) \to \cdots .$$

Moreover, among all possible lifts of $\varphi(\xi)$ in $H^*(V, V \setminus \Sigma)$, the data of $\mathcal{O}$ on $V \setminus \Sigma$ provides often the possibility to choose naturally one of them, $\varphi(\xi, \mathcal{O})$, which will be called the *residual class* of $\varphi(\xi)$ on Σ (associated to $\mathcal{O}$).

Notice that, for any open neighborhood U of Σ in V, the restriction

$$H^*(V, V \setminus \Sigma) \overset{\cong}{\to} H^*(U, U \setminus \Sigma)$$

is an isomorphism (excision theorem), so that $\varphi(\xi, \mathcal{O})$ depends only on the local behaviour of $\mathcal{O}$ near Σ.

If we assume the manifold V to be compact, connected and orientable, the cap-product by the fundamental class $[V] \in H_N(V)$ defines the Poincaré and Alexander duality isomorphisms

$$[\mathcal{P}] : H^*(V) \overset{\cong}{\to} H_{N-*}(V) \qquad \text{and} \qquad [\mathcal{A}] : H^*(V, V \setminus \Sigma) \overset{\cong}{\to} H_{N-*}(\Sigma)$$

making commutative the diagram

$$(**) \qquad \begin{array}{ccccc}
\varphi(\xi, \mathcal{O}) & \to & \varphi(\xi) & \to & 0 \\
 & & & & \\
H^{2k}(V, V \setminus \Sigma) & \xrightarrow{j} & H^{2k}(V) & \to & H^{2k}(V \setminus \Sigma) \\
| & & | & & \\
\mathcal{A}lexander & & \mathcal{P}oincar\acute{e} & & \downarrow \\
\downarrow & & \downarrow & & \\
\bigoplus_\alpha H_{N-2k}(\Sigma_\alpha) & \xrightarrow{\iota_*} & H_{N-2k}(V) & \to & H_{N-2k}(V, \Sigma) \\
 & & & & \\
Res(\varphi, \xi, \mathcal{O}) & \to & \varphi(\xi) \frown [V] & &
\end{array}$$

Denote by $(\Sigma_\alpha)_\alpha$ the family of the connected components of Σ (that we may separate by disjoint neighborhoods, since Σ is closed in V which is compact). Denote also by $\iota = (\iota_\alpha)_\alpha$ the inclusions $\iota : \Sigma \subset V$ and $\iota_\alpha : \Sigma_\alpha \subset V$. The homology $H_{N-*}(\Sigma)$ is then equal to $\oplus_\alpha H_{N-*}(\Sigma_\alpha)$, and what we shall call

residue $Res_\alpha(\varphi, \mathcal{O})$ of $\varphi(\xi)$ at Σ_α associated to the data of $\mathcal{O}$

will be the component of $[\mathcal{A}](\varphi_0(\xi, \mathcal{O}))$ in $H_{N-*}(\Sigma_\alpha)$. We then get the

Pseudo-formula of residues :

$$\varphi(\xi) \frown [V] = \sum_\alpha (\iota_\alpha)_* \big(Res_\alpha(\varphi, \mathcal{O})\big).$$

2. Globally defined stable classes of bundles defined only off Σ

2.1. Baum-Bott residues ([BB2]).

Let V be a compact complex manifold, $\mathcal{F}$ be a singular holomorphic foliation on V of complex codimension q. Let Σ be the closed subset of V equal to the singular set of $\mathcal{F}$: taking for $\mathcal{O}$ the regular part of $\mathcal{F}$, and denoting by $N\mathcal{F}$ the normal bundle to $\mathcal{F}$ above $V \setminus \Sigma$, then Bott's vanishing theorem is

$$c_I(N\mathcal{F}) = 0 \text{ for } |I| > q,$$

where

$I = (i_1, \cdots i_q)$ is a multi-index made of integers $i_u \geq 0$,

$|I| = i_1 + 2i_2 + \cdots + qi_q$ denotes its height,

and c_I the Chern monomial $(c_1)^{i_1}.(c_2)^{i_2}.\cdots(c_q)^{i_q}$.

The bundle $N\mathcal{F}$ doesn't extend above the whole manifold V, but its stable class does, as proved by Baum and Bott, using the existence of a locally free resolution for any real analytic coherent sheaf. Taking for ξ this extension $[N\mathcal{F}] \in K_{\mathbb{C}}(V)$, and for φ any Chern monomial c_I such that $|I| > q$, we get the Baum-Bott residues $BB_\alpha(\mathcal{F}, I) \in H_*(\Sigma_\alpha)$, such that

$$c_I\big([N\mathcal{F}]\big) \frown [V] = \sum_\alpha BB_\alpha(\mathcal{F}, I) \text{ for } |I| > q.$$

Case of a one dimensional foliation
The foliation $\mathcal{F}$ is then given by a morphism $\ell : \mathcal{L} \to TV$, where $\mathcal{L}$ denotes a holomorphic line bundle, and TV the complex tangent bundle. In this case, the extension ξ of the stable class of $N\mathcal{F}$ is equal to $[TV - \mathcal{L}]$, in such a way that

$$c_I\big([TV - \mathcal{L}]\big) \frown [V] = \sum_\alpha BB_\alpha(\mathcal{F}, I) \text{ for } |I| = n,$$

n denoting the complex dimension of V.

When Σ_α is an isolated point, Baum and Bott proved that $BB_\alpha(\mathcal{F}, I)$ is given by a Grothendieck residue. They first proved it in [**BB1**] when the foliation is defined by a vector field, i.e. when $\mathcal{L}$ is trivial (another proof of that, using a total Čech-de Rham complex associated to the vector field on a small sphere around Σ_α, is given in [**L2**]). In [**BB2**], they proved that $BB_\alpha(\mathcal{F}, I)$ was the same for a general bundle $\mathcal{L}$, as the residue for a holomorphic vector field defining the foliation near the point Σ_α.

2.2. Residues of holomorphic maps (Diop [D]).

Denote by V a complex manifold (complex dimension n), W a compact complex manifold (complex dimension q, $q < n$), and $f : V \to W$ a holomorphic map, of rank q off some closed subset Σ of V.

The kernel of the differential $df : TV \to f^{-1}(TW)$ is then a bundle E of rank $n-q$ above $V \setminus \Sigma$, so that we get the obvious vanishing "theorem" :

$$c_j(E) = 0 \text{ for } j > n - q.$$

Since the stable class of E has a natural extension $\xi = [TV - f^{-1}(TW)] \in K_{\mathbb{C}}(V)$, we get residues $D_\alpha(f,j)$ for $j > n-q$, such that

$$c_j\big([TV - f^{-1}(TW)]\big) \frown [V] = \sum_\alpha D_\alpha(f,j),$$

(see the computation of $D_\alpha(f,j)$ in [**D**]).

Remark: This result generalizes in particular the Riemann-Hurwitz formula for ramified covering; see also [**FP**].

Case $q = 1$ (Iversen ([**Iv**]); see also Fulton ([**F**]) and Harvey-Lawson ([**HL**])).

If W is a compact complex curve, and Σ is made of isolated points m_α, the differential df may be seen as a section of the bundle $T^*V \otimes f^{-1}(TW)$ whose rank is equal to the dimension of the base V. Since

$$c_n\big(T^*V \otimes f^{-1}(TW)\big) = (-1)^n \; c_n\big([TV - f^{-1}(TW)]\big)$$

in this case, Diop's formula is then gotten by using the Hopf's theorem for this section and becomes :

$$\Big(c_n(V) - c_{n-1}(V).f^*\big(c_1(W)\big)\Big) \frown [V] = (-1)^n \sum_\alpha \mu_\alpha,$$

μ_α denoting the Milnor number of the hypersurface $f^{-1}\big(f(m_\alpha)\big)$ at m_α. Notice that the left-hand side of the last formula may also be written $\chi(V) - \chi(F).\chi(W)$, where χ denotes the Euler-Poincaré invariant, and F the generic fiber of f.

There are also various works in the case $q = n$ (ramified coverings) : see for instance Brasselet ([**Bra1**], [**Bra2**]), Izawa ([**Iz**]), Izawa-Suwa ([**IS**]).

3. Sequences of globally defined bundles, exact off Σ

Let $(E_i \to V)_{0 \le i \le r}$ be a family of (complex or real) vector bundles defined above V,

$\mathcal{O}$ be an exact sequence above $V \setminus \Sigma$

$$0 \to E_0|_{V\setminus\Sigma} \xrightarrow{u_0} E_1|_{V\setminus\Sigma} \cdots \xrightarrow{u_{r-1}} E_r|_{V\setminus\Sigma} \to 0,$$

and φ be any characteristic class of degree > 0.

Set $\xi = \sum_{i=0}^r (-1)^i [E_i]$ in $K_{\mathbb{C}}(V)$ or $K_{\mathbb{R}}(V)$.

The data of the exact sequence defines a lift $\xi_{\mathcal{O}} \in K(V, V \setminus \Sigma)$ of ξ in the relative K-theory, hence the following diagram in the complex case (there would be a similar one with Pontrjagin classes in the real case) :

$$\begin{array}{ccccc}
\xi_{\mathcal{O}} & \to & \xi & \to & 0 \\
K(V, V\setminus\Sigma) & \overset{j}{\to} & K(V) & \to & K(V\setminus\Sigma) \\
\Big| & & \Big| & & \\
Chern & & Chern & & \downarrow \\
\downarrow & & \downarrow & & \\
H^*(V, V\setminus\Sigma) & \overset{j}{\to} & H^*(V) & \to & H^*(V\setminus\Sigma) \\
\Big| & & \Big| & & \\
\mathcal{A}lexander & & \mathcal{P}oincar\acute{e} & & \\
\downarrow & & \downarrow & & \\
\bigoplus_\alpha H_{N-*}(\Sigma_\alpha) & \overset{\iota_*}{\to} & H_{N-*}(V) & &
\end{array}$$

Remarks :

1) The characteristic class $\varphi(\xi)$ might be defined without any K-theory, but (unless $r = 1$ or 2) it would seem very artificial. Moreover, unless W is a locally complete intersection, the "normal bundle" to some variety W can be defined only in K-theory, so that any exact sequence where such a "bundle" occurs makes sense only in K-theory (this is the case, for instance for the genus formula for an algebraic curve which is not a locally complete intersection).

2) The diagram above makes sense when using cohomology and homology with integral coefficients. However, the use of connections and Chern-Weil theory will often be useful in the computations, losing eventually information on the torsion.

3.1. Computation of residues when the base is a smooth complex curve ([**CLS**]).

Assume

V is a smooth complex compact curve Γ,

the bundles $E_i \to \Gamma$ are holomorphic bundles,

the sequence $(u_i : E_i \to E_{i+1})$ is holomorphic, exact maybe only off Σ, but defined everywhere on Γ.

Denote by

$v_i : Coker\ u_{i-1} \to Ker\ u_{i+1}$ the map induced by u_i,

$\Delta_{i,\alpha}$ a square matrix representing v_i relatively to local trivializations of the E_i's near a point $m_\alpha \in \Sigma$,

and $\nu_{i,\alpha}$ the order of $det\ \Delta_{i,\alpha}$ at m_α.

The following formula

$$c_1\Big(\sum_{i=0}^{r}(-1)^i[E_i]\Big) \frown [\Gamma] = \sum_\alpha\sum_i(-1)^i\nu_{i,\alpha}$$

is then proved in [**CLS**].

If the curve Γ has singular points, it is often efficient to use the pull back by the normalization $\pi : \widehat{\Gamma} \to \Gamma$ for being able to use this formula. It is what we shall do in the following examples.

3.2. The genus formula.

Let $\iota : \Gamma \subset \mathbb{P}_n$ be a singular curve of geometrical genus g, immersed in the n-dimensional complex projective space.

For simplification, we assume $\Gamma = \cap_{\lambda=1}^{n-1} S_\lambda$ to be the complete intersection of $n-1$ algebraic hypersurfaces S_λ of respective degrees d_λ. Let $d = \prod_\lambda d_\lambda$ be the degree of Γ, and $\pi : \widehat{\Gamma} \to \Gamma$ its normalization. The normal bundle to Γ is given by

$$N\Gamma = \iota^{-1}\big(\sum_\lambda \mathcal{O}(d_\lambda)\big).$$

The sequence

$$0 \to T\widehat{\Gamma} \to (\iota \circ \pi)^{-1}(T\mathbb{P}_n) \to \pi^{-1}(N\Gamma) \to 0$$

is then exact off $\Sigma = \pi^{-1}$(singular set of Γ), and we get the formula

$$g = 1 + \frac{1}{2}\, d\big(-(n+1) + \sum_\lambda d_\lambda\big) - \frac{\mathcal{E}(\Gamma)}{2}\,,$$

where $\mathcal{E}(\Gamma)$ is the sum of the residues of $c_1(\xi)$ for the sequence above.

In particular, for $n = 2$:

$$g = \frac{1}{2}\,(d-1)(d-2) - \frac{\mathcal{E}(\Gamma)}{2}\,.$$

Using then the CLS formula above, and assuming that the only singular points of Γ are ordinary r-uple points ($r \geq 2$), or elementary (p,q)-singularities, we get :

$$\mathcal{E}(\Gamma) = \sum_r a_r r(r-1) + \sum_{p,q} b_{pq}\big(pq - (p+q) + 1\big),$$

where a_r is the number of ordinary r-uple points, and b_{pq} is the number of elementary (p,q)-singularities (see the proof in [**AL**]). In particular, we recover the usual formula

$$\frac{\mathcal{E}(\Gamma)}{2} = a_2 + b_{23}$$

when the only singular points of Γ are ordinary double points or simple cusps of type $(2,3)$ (see [**GH**] for instance).

3.3. The Mexican index ([GSV]).

Let $\iota : \Gamma \subset \mathbb{P}_n$ be a singular curve of geometrical genus g, immersed in the n-dimensional complex projective space as above. Assume moreover Γ to be invariant by an algebraic foliation of dimension 1 and degree δ

$$\ell : \mathcal{O}(1-\delta) \to TP_n.$$

The sequence

$$0 \to \mathcal{O}(1-\delta)|_\Gamma \to TP_n|_\Gamma \to N\Gamma \to 0$$

is then exact off $\Sigma_1 = \mathrm{Sing}(\Gamma) \cup \big(\mathrm{Sing}(\ell) \cap \Gamma\big)$. The corresponding residues are the GSV-indices, hence the formula

$$(\delta + n - \sum_\lambda d_\lambda)d = \sum_\alpha GSV_\alpha.$$

In fact, in [**GSV**], is described the index of a (germ of) vector field v on a holomorphic manifold V at a (may be singular) point m of an analytic subset W of V which is preserved by v (see also [**Se**]); this index coincides with the Euler-Poincaré invariant of the restriction of v to W, when m is a regular point of W. Its description in words of K-theory and global interpretation is done in [**LSS**] for vector fields and in [**LS**] for one dimensional foliations ; it is done in words of homological algebra in [**G**].

Application to Poincaré type inequality

In [**Cvl**], is proved the following formula

$$\sum_\alpha GSV_\alpha \geq \mathcal{B}(\Gamma) - \mathcal{E}(\Gamma), \text{ where}$$

$$\begin{aligned} \mathcal{B}(\Gamma) &= \quad \text{number of points in } \pi^{-1}(\text{singular set of } \Gamma) \\ & \qquad\qquad \text{when } \Gamma \text{ is singular,} \\ &= 1 \qquad\qquad \text{when } \Gamma \text{ is non-singular.} \end{aligned}$$

In particular, for $n = 2$, using the value of $\mathcal{E}(\Gamma)$ written above, we get

$$(\delta + 2 - d)d \geq -\sum_r a_r r(r-2) - \sum_{p,q} b_{pq}\big(pq - (p+q)\big),$$

when Γ has only ordinary r-uple points or elementary (p,q)-singularities.
For instance, we recover the inequality

$$\delta + 2 - d \;\geq\; 0$$

when Γ has only nodal singularities (Cerveau-Lins Neto ([**CLi**]), Soares ([**So**])). The same inequality has been proved by Carnicer ([**Cr**]) when the foliation has no dicritical singularity; Brunella observed in [**Br**] that GSV is necessarily non-negative in this case).

3.4. Foliation generically transverse to a submanifold ([Cml]).

Let $\ell : \mathcal{L} \to TV$ be a holomorphic one dimensional foliation on a complex manifold V, and V_0 be the regular part of $\mathcal{F}$.

Let

$W \subset V$ be a submanifold of V,
$N_V(W)$ be its normal bundle,
$TV|_W \xrightarrow{\pi} N_V(W)$ be the normal projection,
$F \to W$ be a sub vector bundle of $N_V(W)$ such that

$$0 \to \mathcal{L}|_W \xrightarrow{\pi \circ \ell} F \to 0$$

is an isomorphism above $W \setminus \Sigma$. (When W is a hypersurface, $F = N_V(W)$).

We get then the corresponding residues for $c_j\big([F - \mathcal{L}|_W]\big)$.

Remark: This construction works also if W has singularities; but then, we need first to define the stable class $[N_V(W)]$ in K-theory.

Application to dicritical singularities (generalizing [**Cm**]):

Let

M be the total space of the tautological line bundle $\mathcal{O}(-1) \to \mathbb{P}^{n-1}$,

$\gamma = c_1(\mathcal{O}(1))$ be the Chern class in $H^2(\mathbb{P}^{n-1})$ of the bundle above,

$M \to \mathbb{C}^n$ be the blow-up of $\mathbb{C}^n$ at 0, the exceptional divisor $D = \mathbb{P}^{n-1}$ being identified with the zero section of the bundle.

Let

v be a germ of holomorphic vector field at 0 on $\mathbb{C}^n$,

$0 \in \mathbb{C}^n$ be a dicritical isolated singularity of v,

d be the order of v at 0,

$\mathcal{L} \to TM$ be the foliation $\mathcal{F}$ on M, which is the blow-up of v.

Assume $\mathcal{F}$ to be generically transverse to D.

Let $W_1 \subset D$ be the tangency locus of $\mathcal{F}$ to D (it is an algebraic hypersurface of degree k in D), and $\mathcal{F}|_{W_1}$ is generically transverse to W_1.

Let $W_2 \subset W_1$ be tangency locus of $\mathcal{F}|_{W_1}$ to W_1.

We can iterate the process, and get therefore a stratification of D.

On D, $\mathcal{L}|_D$ and $\mathcal{O}(-d)$ are isomorphic, as well as $N_M(D)$ and $\mathcal{O}(-1)$, hence:

$$c([N_M(D) - \mathcal{L}|_D]) = (1-\gamma)/(1-d\gamma),$$

and the corresponding residues on W_1.

On W_1, $N_D(W_1)$ and $\mathcal{O}(k)|_{W_1}$ are isomorphic, hence

$$c([N_D(W_1) - \mathcal{L}|_{W_1}]) = (1+k\gamma)/(1-d\gamma),$$

and the corresponding residues on W_2, etc.

4. Methods in differential Geometry

The principle is to use the Chern-Weil theory in K-theory on the relative Mayer-Vietoris complex.

4.1. The Mayer-Vietoris complex.

Let $\Sigma = \coprod_\alpha \Sigma_\alpha$ be a closed set in a manifold V, whose family of connected components will be denoted by $(\Sigma_\alpha)_\alpha$. We assume these components separated by open neighborhoods U_α such that $U_\alpha \cap U_\beta = \emptyset$ when $\alpha \neq \beta$. Set $U_0 = V \setminus \Sigma$, $U_1 = \coprod_\alpha U_\alpha$ and $U_{01} = \coprod_\alpha (U_\alpha \setminus \Sigma_\alpha)$ (or $= U_0 \cap U_1$).

We call Mayer-Vietoris complex $MV^*(\mathcal{U})$ the total Čech-de Rham complex associated to the open cover $\mathcal{U} = (U_0, U_1)$ of V:

$$MV^*(\mathcal{U}) = \Omega^*_{DR}(U_0) \oplus \Omega^*_{DR}(U_1) \oplus \Omega^{*-1}_{DR}(U_{01}),$$
$$= \Omega^*_{DR}(V \setminus \Sigma) \oplus \prod_\alpha \Big(\Omega^*_{DR}(U_\alpha)\Big) \oplus \prod_\alpha \Big(\Omega^{*-1}_{DR}(U_\alpha \setminus \Sigma_\lambda)\Big),$$

where $\Omega^*_{DR}(\cdot)$ denotes the graded differential algebra of the de Rham forms, with the differential

$$D\big(\omega_0, \omega_1, \omega_{01}\big) = \big(d\omega_0, d\omega_1, -d\omega_{01} + \omega_1 - \omega_0\big),$$

where d denotes the exterior differential of forms, and the (non-commutative, even in the graded sense) product

$$\big(\omega_0, \omega_1, \omega_{01}\big) \smile \big(\omega'_0, \omega'_1, \omega'_{01}\big)$$
$$= \big(\omega_0 \wedge \omega'_0, \omega_1 \wedge \omega'_1, (-1)^k \omega_0 \wedge \omega'_{01} + \omega_{01} \wedge \omega'_1\big),$$

(k being the degree of ω_0, with $\omega_1 = (\omega_\alpha)_\alpha$ and $\omega_{01} = (\omega_{0\alpha})_\alpha$).

[We shall omit the symbol of restriction of forms when they are defined on an open set bigger than necessary.]

Then D is a differential ($D \circ D = 0$), and a derivation

$$\big((D(\varpi \smile \varpi') = D\varpi \smile \varpi' + (-1)^k\, \varpi \smile D\varpi',\ (k = \text{degree of } \varpi)\big).$$

We define the inclusion $\iota : \Omega^*_{DR}(V) \to MV^*(\mathcal{U})$ by the map

$$\omega \mapsto (\omega|_{U_0}, \omega|_{U_1}, 0) \ \big(\text{or more shortly } \iota(\omega) = (\omega, \omega, 0) \text{ by abuse of notation}\big),$$

and the projection $\pi : MV^*(\mathcal{U}) \to \Omega^*_{DR}(V \setminus \Sigma)$ by the map $(\omega_0, \omega_1, \omega_{01}) \mapsto \omega_0$. These two maps are obviously morphisms of differential graded algebras.

Denote by $MV^*(\mathcal{U}, V \setminus \Sigma)$ the kernel of π, i.e. the set of elements $\big((0), \omega_1, \omega_{01}\big)$, still denoted by (ω_1, ω_{01}), where the parentheses (0) around zero means that this zero is forced: this is a differential ideal (i.e. D-invariant) of $MV^*(\mathcal{U})$. Denote by $j : \big((0), \omega_1, \omega_{01}\big) \to \big(0, \omega_1, \omega_{01}\big)$ the natural inclusion of $MV^*(\mathcal{U}, V \setminus \Sigma)$ into $MV^*(\mathcal{U})$.

Then

(i) The inclusion $\iota : \Omega^*_{DR}(V) \to MV^*(\mathcal{U})$ induces an isomorphism of graded algebras in cohomology.

(ii) The exact sequence

$$0 \to MV^*(\mathcal{U}, V \setminus \Sigma) \xrightarrow{j} MV^*(\mathcal{U}) \xrightarrow{\pi} \Omega^*(V \setminus \Sigma) \to 0$$

induces in cohomology the long exact sequence of the pair $(V, V \setminus \Sigma)$

$$(*) \qquad \cdots \longrightarrow H^*(V, V \setminus \Sigma) \xrightarrow{j} H^*(V) \xrightarrow{\pi} H^*(V \setminus \Sigma) \longrightarrow H^{*+1}(V) \to \cdots.$$

In particular, the cohomology of the complexes $MV^*(\mathcal{U})$ and $MV^*(\mathcal{U}, V \setminus \Sigma)$ doesn't depend on the choice of U_1. Hence, the cohomology $H^*(V, V \setminus \Sigma)$ of $MV^*(\mathcal{U}, V \setminus$

Σ) may be computed from differential forms defined only on an arbitrary small neighborhood U_1 of Σ. This property is called *the excision principle*, and $H^*(V, V \setminus \Sigma)$ is called the *local cohomology* of Σ (with coefficients in $\mathbb{K}$).

Assume V to be compact, connected, oriented, and N-dimensional. Let

$$R_{01} = \coprod_{\alpha} R_{0\alpha}$$

be a piecewise differentiable hypersurface of V without boundary, included into U_{01}, and separating V into two manifolds with boundary R_0 and $R_1 = \coprod_\alpha R_\alpha$ such that $R_0 \subset U_0$ and $R_\alpha \subset U_\alpha$. The set $\Sigma = \coprod_\alpha \Sigma_\alpha$ is therefore included into the interior $\overset{\circ}{R}_1$ of R_1, each manifold with boundary R_α being included into U_α and containing Σ_α in its interior $\overset{\circ}{R}_\alpha$. We suppose the orientation of R_{01} defined by the outbound normal to R_0, i.e. $\partial R_0 = R_{01}$, $\partial R_\alpha = -R_{0\alpha}$ (the sign "minus" means $R_{0\alpha}$ with the opposite orientation), and $\partial R_{0\alpha} = \emptyset$. Denote by R the cover (R_0, R_1, R_{01}) of V defined in this way by R_{01}.

Denote by $\int_V^R : MV^N(\mathcal{U}) \to \mathbb{K}$ and call *integration relative to* R the map

$$\int_V^R (\omega_0, \omega_1, \omega_{01}) = \int_{R_0} \omega_0 + \int_{R_1} \omega_1 + \int_{R_{01}} \omega_{01},$$

which extends the usual integration of the N-forms $(\int_V^R \circ \iota = \int_V)$.

(*i*) The integral $\int_V^R$ of a D-cocycle doesn't depend on the choice of R_{01}.
(*ii*) The integral $\int_V^R$ of a D-coboundary vanishes.

This implies

(*i*) The integration induces the natural isomorphism $H^N(V; \mathbb{K}) \cong \mathbb{K}$ in cohomology.

(*ii*) The Poincaré duality $[\mathcal{P}] : H^*(V; \mathbb{K}) \to H_{N-*}(V; \mathbb{K})$ is induced by the map

$$\mathcal{P} : MV^*(\mathcal{U}) \to Hom_{\mathbb{K}}(MV^{N-*}(\mathcal{U}), \mathbb{K})$$

defined by

$$\mathcal{P}(\omega_0, \omega_1, \omega_{01}) : (\omega'_0, \omega'_1, \omega'_{01}) \mapsto \int_V^R (\omega_0, \omega_1, \omega_{01}) \smile (\omega'_0, \omega'_1, \omega'_{01})),$$

where we identified $H_*(V; \mathbb{K})$ with the homology of the complex

$$\big(MV^*(\mathcal{U})\big)^* = Hom_{\mathbb{K}}\big(MV^*(\mathcal{U}), \mathbb{K}\big).$$

(*iii*) If we assume moreover the neighborhood U_1 of Σ to be *regular* (which means that the inclusion $\Sigma \subset U_1$ is a homotopy equivalence, and in particular the singular cohomology of Σ may be identified with the de Rham cohomology of U_1), the Alexander duality

$$[\mathcal{A}] : H^*(V, V \setminus \Sigma; \mathbb{K}) \to H_{N-*}(\Sigma; \mathbb{K})$$

is induced by the map $\mathcal{A} : MV^*(\mathcal{U}, V \setminus \Sigma) \to Hom_{\mathbb{K}}(\Omega_{DR}^{N-*}(U_1), \mathbb{K})$ defined by

$$\mathcal{A}((o), \omega_1, \omega_{01}) : \omega'_1 \mapsto \int_V^R ((0), \omega_1 \wedge \omega'_1, \omega_{01} \wedge \omega'_1) = \int_{R_1} \omega_1 \wedge \omega'_1 + \int_{R_{01}} \omega_{01} \wedge \omega'_1,$$

where we identified $H_{N-*}(\Sigma;\mathbb{K})$ with the homology of the complex

$$\left(\Omega^*_{DR}(U_1)\right)^* = Hom_{\mathbb{K}}\left(\Omega^*_{DR}(U_1),\mathbb{K}\right).$$

(iv) The diagram

$$\begin{array}{ccccccc} 0\to & MV^*(\mathcal{U},V\setminus\Sigma) & \overset{j}{\to} & MV^*(\mathcal{U}) & \overset{\pi}{\to} & \Omega^*_{DR}(V\setminus\Sigma) & \to 0 \\ & \downarrow \mathcal{A} & & \downarrow \mathcal{P} & & \downarrow & \\ 0\to & \left(\Omega^*_{DR}(U_1)\right)^* & \overset{\iota}{\to} & \left(MV^*(\mathcal{U})\right)^* & \to & \cdots & \to 0 \end{array}$$

induces in homology and cohomology the diagram $(**)$

$$\begin{array}{ccccccc} \to & H^*(V,V\setminus\Sigma) & \overset{j}{\longrightarrow} & H^*(V) & \overset{\pi}{\longrightarrow} & H^*(V\setminus\Sigma) & \to \\ & \downarrow [\mathcal{A}] & & \downarrow [\mathcal{P}] & & \downarrow & \\ \to & H_{N-*}(\Sigma) & \overset{\iota}{\longrightarrow} & H_{N-*}(V) & \longrightarrow & H_{N-*}(V,\Sigma) & \to \end{array}$$

For computing the homology class of $\mathcal{A}\big((0),\omega_1,\omega_{01}\big)$ when $\big((0),\omega_1,\omega_{01}\big)$ is a cocycle in $MV^*(\mathcal{U},V\setminus\Sigma)$, we may obviously assume ω_1' to be closed. In particular, if the dimension of $\big((0),\omega_1,\omega_{01}\big)$ is equal to the (real) dimension of the manifold V (assumed to be compact, connected, and oriented), we may assume ω_1' to be the constant function equal to 1 :

$$\mathcal{A}\big((0),\omega_1,\omega_{01}\big)\frown[V] = \int_{R_1}\omega_1 + \int_{R_{01}}\omega_{01}.$$

4.2. Chern-Weil theory on $K_{\mathbb{C}}$.

Let $(E_i)_i$, $0\le i\le r$ be a family of complex[2] vector bundles on a manifold V. A connection ∇ on $\xi=\sum(-1)^iE_i$ is given by a family $(\nabla_i)_i$ of connections, ∇_i being a connection on E_i. The total Chern class of ξ is then the (heterogeneous) cohomology class of the (heterogeneous) differental form

$$c(\nabla)=\prod_i\left(c(\nabla_i)\right)^{\varepsilon_i},$$

where $c(\nabla_i)=1+c_1(\nabla_i)+c_2(\nabla_i)+\cdots$ denotes the (heterogeneous) differential form associated to ∇_i whose cohomology class is the real total Chern class $c(E_i)$, and $\varepsilon_i=\pm1$ according to the parity of i. This includes of course the case of ξ being the stable class of a vector bundle E.

Now, if ∇ and ∇' are two connections on ξ, there is, for any Chern monomial c_I, a natural form $c_I(\nabla,\nabla')$ such that

$$dc_I(\nabla,\nabla')=c_I(\nabla')-c_I(\nabla).$$

If ∇_0 and ∇_1 denote connections on $\xi|_{U_0}$ and $\xi|_{U_1}$ respectively,

$$\left(c_I(\nabla_0),c_I(\nabla_1),c_I(\nabla_0,\nabla_1)\right)$$

is a cocycle in $MV^*(\mathcal{U})$ whose cohomology classes is equal to $c_I(\xi)$.

[2]The same could be done with real bundles and Pontrjagin classes.

Coming back to the notations of the background about the localization of characteristic classes, the existence of $\mathcal{O}$ above U_0 provides a set of connections $\mathcal{C}(\mathcal{O})$ on $\xi|_{U_0}$ such that, for some c_I,

$$c_I(\nabla) = 0 \text{ for all } \nabla \in \mathcal{C}(\mathcal{O}) \text{ (hence the vanishing theorem),}$$

$$c_I(\nabla, \nabla') = 0 \text{ for all be } \nabla \text{ and } \nabla' \text{ both in } \mathcal{C}(\mathcal{O}).$$

Therefore,

$$\big((0), c_I(\nabla_1), c_I(\nabla, \nabla_1)\big)$$

is a cocycle in the relative Mayer-Vietoris complex $MV^*(\mathcal{U}, V \setminus \Sigma)$ if ∇ belongs to $\mathcal{C}(\mathcal{O})$, whose cohomology class will be the wanted residual characteristic class, not depending on the choices of ∇ and ∇_1. See details in [**L1**], [**AL**], [**LS**], [**S**].

References

[AL] Abouqateb A., Lehmann D. *Classes caractéristiques en Géométrie différentielle*, Editions "Ellipses", 2010.

[A] Atiyah M.F., *K-theory*, Lectures by M.F. Atiyah, notes by D.W. Anderson, Harvard University, 1964. MR1043170 (90m:18011)

[BB1] Baum P., Bott R. *On the zeroes of holomorphic vector fields*, Essays on Topology and related topics, Mémoires dédiés à G. de Rham, Springer, 1970, 29-47. MR0261635 (41:6248)

[BB2] Baum P., Bott R. *Singularities of holomorphic foliations*, J. of Diff. Geom. 7, 1972, 279-342. MR0377923 (51:14092)

[B] Bott R., *Lectures on K-theory*, Ed. Benjamin, 1969. MR0258020 (41:2667)

[Bra1] Brasselet J.P., *Une généralisation de la formule de Riemann-Hurwitz*, Bulletin de la Soc.Math. de France., mémoire 38, 1974, 99-106. MR0361176 (50:13622)

[Bra2] Brasselet J.P., *Sur une formule de M.H. Schwartz relative aux revêtements ramifiés*, C.R. Ac. Sc. Paris 283 A, 1976, 41-44. MR0419818 (54:7836)

[Br] Brunella M., *Some remarks on indices of holomorphic vector fields*, Publ. Mat (Barcelona) 41, 1997, 527-544. MR1485502 (98m:32052)

[Cm] Camacho C., *Dicritical singularities of holomorphic vector fields*, Contemp. Math. 269, 2001, 39-46. MR1810535 (2002b:32049)

[Cml] Camacho C., Lehmann D., *Residues of holomorphic foliations relative to a general submanifold*, Bull. London Math.37 Soc., 2005, 435-445.. MR2131398 (2005k:32039)

[Cr] Carnicer M., *The Poincaré problem in the non-dicritical case*, Ann. of Math. (2) 140, 1994, 289-294. MR1298714 (95k:32031)

[Cvl] Cavalier V., Lehmann D.,*On the Poincaré inequality for one-dimensional foliations*, Compositio Math. 142, 2006, 529-540. MR2218909 (2006m:32032)

[CLS] Cavalier V., Lehmann D., Soares M., *Classes de Chern des ensembles analytiques*, K-theory 34, 2005, 35-69. MR2162900 (2006d:32041)

[CLi] Cerveau D., Lins Neto A., *Holomorphic foliations on $CP(2)$ having an invariant algebraic curve*, Annales de l'Institut Fourier 41, 1991, 883-903. MR1150571 (93b:32050)

[D] Diop E.C.M., *Résidus d'applications holomorphes entre variétés*, Hokkaido Mathematical J., vol. 29, 171-200, 2000. MR1745509 (2001g:32077)

[F] Fulton W., *Intersection theory*, Ergebnisse der Mathematik und ihrer Grenzgebiete 3, vol. 2, Springer Verlag, 1984, 245-246. MR732620 (85k:14004)

[FP] Fulton W. and Pragacz P., *Schubert varieties and Degeneracy loci*, Lecture Notes in Mathematics 1689, vol. 2, Springer Verlag, 1998. MR1639468 (99m:14092)

[G] Gómez-Mont X., *An algebraic formula for the index of a vector field on a hypersurface with an isolated singularity*, J. Alg. Geom. 7, no. 4, 1998, 731-752. MR1642757 (99f:32061)

[GSV] Gómez-Mont X., Seade J., Verjovski A., *The index of a holomorphic flow with an isolated singularity*, Math. Ann. 291, 1991, 737-751. MR1135541 (93d:32066)

[GH] Griffiths P., Harris J., *Principles of Algebraic Geometry*, John Wiley and Sons, 1978. MR507725 (80b:14001)

[HL] Harvey F.R., Lawson H. Blaine, *A theory of characteristic currents associated with a singular connection*, Astérisque 213, 1993, 262-263. MR1230025 (95b:58005)

[Iv] Iversen B., *Critical points of an algebraic function*, Invent. Math. 12, 1971, 210-224. MR0342512 (49:7258)

[Iz] Izawa T., *Chern number formula for ramified coverings*, J. Math. Soc. Japan 52, 2000, 1-15. MR1727204 (2001i:32028)

[IS] Izawa T., Suwa T., *Multiplicity of functions on singular varieties*, Intern. J. Math. 14, 2003, 541-558. MR1993796 (2004f:14009)

[L1] Lehmann D., *Classes caractéristiques résiduelles*, Differential Geometry and its Applications (Brno 1989), World Scientific, 1990, 85-108. MR1062009 (91e:57043)

[L2] Lehmann D., *Résidus des sous-variétés invariantes d'un feuilletage singulier*, Annales de l'Institut Fourier 41 (1), 1991, 211-258. MR1112198 (92k:57054)

[LSS] Lehmann D., Soares M., Suwa T., *On the index of a holomorphic vector field tangent to a singular variety*, Bol. Soc. Bras. Mat. 26, 1995, 183-199. MR1364267 (96k:32080)

[LS] Lehmann D., Suwa T., *Generalization of variations and Baum-Bott residues for holomorphic foliations on singular varieties*, International J. of Math. 10(3), 1999, 367-384. MR1688141 (2000b:32059)

[Se] Seade J., *The index of a vector field on a complex surface with singularities*, The Lefschetz Centennial Conf., ed. A. Verjovsky, Contemp. Math. 58 III, Am. Math. Soc., 1987, 225-232. MR893868 (88g:32021)

[So] Soares M.G., *On the geometry of Poincaré's problem for one dimensional projective foliations*, An. Acad. Bras. Cienc. 73, 2001, 475-482. MR1870799 (2002j:32032)

[S] Suwa T., *Indices of Vector Fields and Residues of Singular Holomorphic Foliations*, Actualités mathématiques, Hermann, 1998. MR1649358 (99h:32042)

Département de Mathématiques, Université de Montpellier 2, France

Current address: 4 rue Becagrun, 30980 Saint Dionisy, France.

E-mail address: `lehmann@math.univ-montp2.fr`

Contemporary Mathematics
Volume **569**, 2012
http://dx.doi.org/10.1090/conm/569/11239

Multicusps

Yusuke Mizota and Takashi Nishimura

ABSTRACT. For a given multicusp $f = c_{(\theta_0,\ldots,\theta_i)}$ $(1 \le i)$, we present a direct sum decomposition theorem of the source space of ${}_i\overline{\omega}f$, where ${}_i\overline{\omega}f$ is a higher version of the reduced Kodaira-Spencer-Mather map $\overline{\omega}f$. As a corollary of our direct sum decomposition theorem, we show that for any $i \in \mathbb{N}$ and any $f = c_{(\theta_0,\ldots,\theta_i)}$, ${}_i\overline{\omega}f$ is bijective. The corollary is an affirmative answer to the question raised by M. A. S. Ruas during the 11th International Workshop on Real and Complex Singularities at the University of São Paulo in São Carlos (2010).

1. Introduction

Throughout this paper, let $\mathbb{K}$ be $\mathbb{R}$ or $\mathbb{C}$, and let S be a finite subset of $\mathbb{K}^n$. Let C_S (resp., C_0) be the $\mathbb{K}$-algebra of C^∞ or holomorphic function germs $(\mathbb{K}^n, S) \to \mathbb{K}$ (resp., $(\mathbb{K}^p, 0) \to \mathbb{K}$), and let m_S (resp., m_0) be the subset of C_S (resp., C_0) consisting of C^∞ or holomorphic function germs $(\mathbb{K}^n, S) \to (\mathbb{K}, 0)$ (resp., $(\mathbb{K}^p, 0) \to (\mathbb{K}, 0)$). A map germ $f : (\mathbb{K}^n, S) \to (\mathbb{K}^p, 0)$ is called a *multigerm*. For a C^∞ or holomorphic multigerm $f : (\mathbb{K}^n, S) \to (\mathbb{K}^p, 0)$, we let $f^* : C_0 \to C_S$ be the $\mathbb{K}$-algebra homomorphism defined by $f^*(u) = u \circ f$.

For a C^∞ or holomorphic map germ $f : (\mathbb{K}^n, S) \to \mathbb{K}^p$, let $\theta_S(f)$ be the C_S-module consisting of germs of C^∞ or holomorphic vector fields along f. The C_S-module $\theta_S(f)$ is naturally identified with $\underbrace{C_S \oplus \cdots \oplus C_S}_{p \text{ tuples}}$. We define $\theta_S(n)$ (resp., $\theta_0(p)$) as the C_S-module (resp., C_0-module) $\theta_S(id._{(\mathbb{K}^n,S)})$ (resp., $\theta_{\{0\}}(id._{(\mathbb{K}^p,0)})$), where $id._{(\mathbb{K}^n,S)}$ (resp., $id._{(\mathbb{K}^p,0)}$) is the germ of the identity map of $(\mathbb{K}^n, S)$ (resp., $(\mathbb{K}^p, 0)$).

For a given C^∞ or holomorphic multigerm $f : (\mathbb{K}^n, S) \to (\mathbb{K}^p, 0)$, define $tf : \theta_S(n) \to \theta_S(f)$ and $\omega f : \theta_0(p) \to \theta_S(f)$ as $tf(\eta) = df \circ \eta$ and $\omega f(\xi) = \xi \circ f$, respectively, where *df* is the differential of f (for details on tf and ωf, see [**3**]). For the f, put $T\mathcal{R}_e(f) = tf(\theta_S(n))$, $T\mathcal{L}_e(f) = \omega f(\theta_0(p))$, and $T\mathcal{A}_e(f) = T\mathcal{R}_e(f) + T\mathcal{L}_e(f)$ (for details on $T\mathcal{R}_e(f), T\mathcal{L}_e(f)$, and $T\mathcal{A}_e(f)$, see [**8**]). The set of C^∞ or holomorphic function germs $(\mathbb{K}^n, S) \to \mathbb{K}$ (resp., $(\mathbb{K}^p, 0) \to \mathbb{K}$), such that the terms of their Taylor series up to $(i-1)$ are zero, is denoted by m_S^i (resp., m_0^i) for any non-negative integer i. Therefore, m_S^0 (resp., m_0^0) is exactly the same as

---

2010 *Mathematics Subject Classification.* Primary 57R45; Secondary 58K20, 58K45.

Key words and phrases. Multicusp, reduced Kodaira-Spencer-Mather map, higher version of the reduced Kodaira-Spencer-Mather map, liftable vector field.

C_S (resp., C_0). A C^∞ or holomorphic multigerm $f : (\mathbb{K}^n, S) \to (\mathbb{K}^p, 0)$ is said to be *finitely determined* if there exists a positive integer k such that the inclusion $m_S^k \theta_S(f) \subset T\mathcal{A}_e(f)$ holds.

For a given C^∞ or holomorphic multigerm $f : (\mathbb{K}^n, S) \to (\mathbb{K}^p, 0)$, a vector field $\xi \in \theta_0(p)$ is said to be *liftable over* f if $\xi \circ f$ belongs to $T\mathcal{L}_e(f) \cap T\mathcal{R}_e(f)$. The set of vector fields liftable over f naturally has a C_0-module structure. In [**6**], in order to express the minimal number of generators for the module of vector fields liftable over f, the second author introduced the following homomorphism ${}_i\overline{\omega}f$ for a given C^∞ or holomorphic multigerm f and a non-negative integer i.

$$\begin{array}{rcl} {}_i\overline{\omega}f : \dfrac{m_0^i \theta_0(p)}{m_0^{i+1}\theta_0(p)} & \to & \dfrac{f^* m_0^i \theta_S(f)}{T\mathcal{R}_e(f) \cap f^* m_0^i \theta_S(f) + f^* m_0^{i+1} \theta_S(f)}, \\ {}_i\overline{\omega}f([\xi]) & = & [\omega f(\xi)]. \end{array}$$

Note that ${}_0\overline{\omega}f$ is identical to the map $\overline{\omega}f$ defined in [**4**], i.e.,

$$\begin{array}{rcl} \overline{\omega}f : \dfrac{\theta_0(p)}{m_0 \theta_0(p)} & \to & \dfrac{\theta_S(f)}{T\mathcal{R}_e(f) + f^* m_0 \theta_S(f)}, \\ \overline{\omega}f([\xi]) & = & [\omega f(\xi)]. \end{array}$$

Suppose that $\mathbb{K} = \mathbb{C}, n \geq p$ and $S = \{\text{one point } x\}$. Then, the map $\overline{\omega}f$ is called the *reduced Kodaira-Spencer map*, and it is denoted by $\rho_f(x)$ in [**2**]. In this paper, we are mainly interested in the case $n < p$ and S is not a set of only one point; hence, we call $\overline{\omega}f$ (resp., ${}_i\overline{\omega}f$) the *reduced Kodaira-Spencer-Mather map* (resp., a *higher version of the reduced Kodaira-Spencer-Mather map*). When $n \leq p$, the module of vector fields liftable over f can be investigated by using higher versions of the reduced Kodaira-Spencer-Mather map as follows.

THEOREM 1 ([**6**]). *Let $f : (\mathbb{K}^n, S) \to (\mathbb{K}^p, 0)$ be a finitely determined multigerm of corank at most one. Suppose that there exists a non-negative integer i such that ${}_i\overline{\omega}f$ is bijective. Then, the minimal number of generators for the module of vector fields liftable over f is exactly* $\dim_{\mathbb{K}} \ker({}_{i+1}\overline{\omega}f)$.

Here, *corank at most one* for a multigerm $f : (\mathbb{K}^n, S) \to (\mathbb{K}^p, 0)$ implies that $n \leq p$ and $\max\{n - \text{rank} Jf(s_j) \mid 1 \leq j \leq |S|\} \leq 1$ holds, where $Jf(s_j)$ is the Jacobian matrix of f at $s_j \in S$.

In [**6**], we may find several examples satisfying the assumption of Theorem 1. Unfortunately, there are no examples satisfying the condition that ${}_i\overline{\omega}f$ is bijective for some i such that $2 \leq i$. In this paper, for any $i \in \mathbb{N}$, we give a concrete multigerm f such that ${}_i\overline{\omega}f$ is bijective. M. A. S. Ruas asked for such examples during the 11th International Workshop on Real and Complex Singularities at the University of São Paulo in São Carlos (2010). Thus, this paper answers her question affirmatively.

Let $c : \mathbb{K} \to \mathbb{K}^2$ be the map defined by $c(x) = (x^2, x^3)$, and for any real number θ, let $R_\theta : \mathbb{K}^2 \to \mathbb{K}^2$ be the linear map that gives the rotation of $\mathbb{K}^2$ about the origin with respect to the angle θ.

$$R_\theta \begin{pmatrix} X \\ Y \end{pmatrix} = \begin{pmatrix} \cos\theta & -\sin\theta \\ \sin\theta & \cos\theta \end{pmatrix} \begin{pmatrix} X \\ Y \end{pmatrix}.$$

Let $\theta_0, \ldots, \theta_i$ be real numbers satisfying $0 \leq \theta_j < 2\pi$ $(0 \leq j \leq i)$ and $0 \neq |\theta_j - \theta_k| \neq \pi$ $(j \neq k)$. Put $S = \{s_0, \ldots, s_i\}$ $(s_j \neq s_k$ if $j \neq k)$ and define $c_{\theta_j} : (\mathbb{K}, s_j) \to (\mathbb{K}^2, 0)$ as $c_{\theta_j}(x) = R_{\theta_j} \circ c(x_j)$, where $x_j = x - s_j$. A multigerm $\{c_{\theta_0}, \ldots, c_{\theta_i}\} : (\mathbb{K}, S) \to$

$(\mathbb{K}^2, 0)$ is called a *multicusp*, and it is denoted by $c_{(\theta_0,\dots,\theta_i)}$. In the case $i = 0$ (resp., $i = 1$), it is called a *cusp* (resp., *double cusp*). It is known that double cusps are open $\mathcal{A}$-simple multigerms $(\mathbb{K}, S) \to (\mathbb{K}^2, 0)$ (see [**1**]), and to the best of authors' knowledge, there is no literature on multicusps for the case $2 \leq i$.

For a given multicusp $c_{(\theta_0,\dots,\theta_i)}$ such that $i \in \mathbb{N}$, put

$$c_{\hat{\theta}_j} = \{c_{\theta_0}, \dots, \hat{c_{\theta_j}}, \dots, c_{\theta_i}\} : (\mathbb{K}, S - \{s_j\}) \to (\mathbb{K}^2, 0),$$

where $\hat{c_{\theta_j}}$ denotes the removal of the cusp c_{θ_j}. Note that in order to define $c_{\hat{\theta}_j}$, i must be positive. The main result of this paper can be stated in terms of ${}_i\overline{\omega}c_{\hat{\theta}_j}$ as follows.

THEOREM 2. *For any $i \in \mathbb{N}$ and any multicusp $c_{(\theta_0,\dots,\theta_i)}$, the following equality holds.*

$$\frac{m_0^i\theta_0(2)}{m_0^{i+1}\theta_0(2)} = \bigoplus_{j=0}^{i} \ker({}_i\overline{\omega}c_{\hat{\theta}_j}).$$

COROLLARY 1. *For any $i \in \mathbb{N}$ and any multicusp $c_{(\theta_0,\dots,\theta_i)}$, ${}_i\overline{\omega}c_{(\theta_0,\dots,\theta_i)}$ is bijective.*

Note that the reduced Kodaira-Spencer-Mather map $\overline{\omega}c_\theta = {}_0\overline{\omega}c_\theta$ is not surjective for any cusp c_θ [**4**] because $\theta_S(f) \neq T\mathcal{A}_e(f)$ for $f = c_\theta$ and $S = \{0\}$. Thus, Corollary 1 does not hold in the case $i = 0$. Since the image of $c_{(\theta_0,\dots,\theta_i)}$ is a plane curve, the minimal number of generators for the module of vector fields liftable over $c_{(\theta_0,\dots,\theta_i)}$ is always 2 in the complex case [**7**]. This implies that for any j $(0 \leq j \leq i)$,

$$\dim_{\mathbb{C}}\left(\ker\left({}_i\overline{\omega}c_{\hat{\theta}_j}\right)\right) = 2.$$

Nevertheless, since we want to study both real and complex cases in this paper, we directly show that for any j $(0 \leq j \leq i)$,

$$\dim_{\mathbb{K}}\left(\ker\left({}_i\overline{\omega}c_{\hat{\theta}_j}\right)\right) = 2$$

by using linear algebra (see Lemma 1 in §2).

For any $i \in \mathbb{N}$, let j be an integer such that $0 \leq j \leq i$. Then, since the kernel of ${}_i\overline{\omega}c_{\hat{\theta}_i}$ is a 2-dimensional vector space, there exists a non-zero vector field $\xi_j \in m_0^i\theta_0(2)$ such that $\omega c_{\hat{\theta}_j}(\xi_j) \in T\mathcal{R}_e(c_{\hat{\theta}_j}) + c_{\hat{\theta}_j}^* m_0^{i+1}\theta_S(c_{\hat{\theta}_j})$. The map ${}_{i+1}\overline{\omega}c_{\hat{\theta}_j}$ is surjective because the map ${}_i\overline{\omega}c_{\hat{\theta}_j}$ is surjective by Corollary 1. Thus, by the preparation theorem, there must exist a vector field $\widetilde{\xi_j} \in m_0^{i+1}\theta_0(2)$ such that $\omega c_{\hat{\theta}_j}(\xi_j + \widetilde{\xi_j}) \in T\mathcal{R}_e(c_{\hat{\theta}_j})$. On the other hand, $\omega c_{\theta_j}(\xi_j + \widetilde{\xi_j}) \notin T\mathcal{R}_e(c_{\theta_j})$ by Theorem 2. Therefore, by integrating $\xi_j + \widetilde{\xi_j} + \frac{\partial}{\partial t}$, we obtain a diffeomorphism germ $H_j : (\mathbb{K}^2, 0) \to (\mathbb{K}^2, 0)$ satisfying the following 4 conditions:

(1) The $(i-1)$-jet of H_j at the origin is equal to the $(i-1)$-jet of $id._{(\mathbb{K}^2,0)}$ at the origin.
(2) The i-jet of H_j at the origin is slightly different from the i-jet of $id._{(\mathbb{K}^2,0)}$ at the origin.
(3) The image of $H_j \circ c_{\hat{\theta}_j}$ is equal to the image of $c_{\hat{\theta}_j}$.
(4) The image of $H_j \circ c_{\theta_j}$ is not equal to the image of c_{θ_j}.

Similarly, we obtain a diffeomorphism germ $\widetilde{H} : (\mathbb{K}^2, 0) \to (\mathbb{K}^2, 0)$ satisfying the following 4 conditions:

(1) The i-jet of $\widetilde{H}$ at the origin is equal to the i-jet of $id._{(\mathbb{K}^2,0)}$ at the origin.
(2) The $(i+1)$-jet of $\widetilde{H}$ at the origin is slightly different from the $(i+1)$-jet of $id._{(\mathbb{K}^2,0)}$ at the origin.
(3) The image of $\widetilde{H} \circ c_{(\theta_0,\ldots,\theta_i)}$ is equal to the image of $c_{(\theta_0,\ldots,\theta_i)}$.
(4) There exists a sufficiently small $x \in \mathbb{K}$ such that $\widetilde{H} \circ c_{(\theta_0,\ldots,\theta_i)}(x) \neq c_{(\theta_0,\ldots,\theta_i)}(x)$.

PROOF OF COROLLARY 1. Since $1 \leq i$, by Theorem 2, we have

$$\ker({}_i\overline{\omega}c_{(\theta_0,\ldots,\theta_i)}) \subset \bigcap_{j=0}^{i} \ker({}_i\overline{\omega}c_{\hat{\theta}_j}) = \{0\}.$$

Hence, ${}_i\overline{\omega}c_{(\theta_0,\ldots,\theta_i)}$ is injective.

Since

$$\dim_{\mathbb{K}}\left(\frac{m_0^i\theta_0(2)}{m_0^{i+1}\theta_0(2)}\right) = 2(i+1) \qquad \text{and}$$

$$\dim_{\mathbb{K}}\left(\frac{f^*m_0^i\theta_S(f)}{T\mathcal{R}_e(f)\cap f^*m_0^i\theta_S(f) + f^*m_0^{i+1}\theta_S(f)}\right) = 2(i+1),$$

where we have put $f = c_{(\theta_0,\ldots,\theta_i)}$, ${}_i\overline{\omega}c_{(\theta_0,\ldots,\theta_i)}$ is an injective linear map between equidimensional vector spaces; this implies that it must be bijective. □

Theorem 2 is proved in Section 2.

2. Proof of Theorem 2

Theorem 2 is proved by induction with respect to i.

The case $i = 1$. Without loss of generality, we may assume that $\theta_0 < \theta_1$. Then, it is sufficient to prove Theorem 2 for $R_{-\theta_0} \circ c_{(\theta_0,\theta_1)}$. Thus, in the following, we assume that $0 = \theta_0 < \theta_1 < 2\pi$ and $\theta_1 \neq \pi$.

It is easily seen that

$$\left\{\left[\begin{pmatrix} 2X \\ 3Y \end{pmatrix}\right], \left[\begin{pmatrix} 2Y \\ 0 \end{pmatrix}\right]\right\}$$

is a basis of $\ker({}_1\overline{\omega}c_0)$. Therefore, two vectors

$$\left[\begin{pmatrix} (2+\sin^2\theta_1)X - (\cos\theta_1\sin\theta_1)Y \\ -(\cos\theta_1\sin\theta_1)X + (2+\cos^2\theta_1)Y \end{pmatrix}\right],$$
$$\left[\begin{pmatrix} -(2\cos\theta_1\sin\theta_1)X + (2\cos^2\theta_1)Y \\ -(2\sin^2\theta_1)X + (2\cos\theta_1\sin\theta_1)Y \end{pmatrix}\right]$$

constitute a basis of $\ker({}_1\overline{\omega}c_{\theta_1})$.

Since

$$\dim_{\mathbb{K}}\left(\frac{m_0\theta_0(2)}{m_0^2\theta_0(2)}\right) = 4,$$

it is sufficient to show that $\ker({}_1\overline{\omega}c_0) \cap \ker({}_1\overline{\omega}c_{\theta_1}) = \{0\}$. In order to show this equality, it is sufficient to show that the determinant of the following matrix is not zero.

$$M = \begin{pmatrix} 2 & 0 & 2+\sin^2\theta_1 & -2\cos\theta_1\sin\theta_1 \\ 0 & 2 & -\cos\theta_1\sin\theta_1 & 2\cos^2\theta_1 \\ 0 & 0 & -\cos\theta_1\sin\theta_1 & -2\sin^2\theta_1 \\ 3 & 0 & 2+\cos^2\theta_1 & 2\cos\theta_1\sin\theta_1 \end{pmatrix}.$$

Since $\det M = -20\sin^2\theta_1$, it is not zero by the assumption of θ_1.

The case $i = k+1$ under the assumption that Theorem 2 holds in the case $i = k$. Since it has been assumed that Theorem 2 holds in the case $i = k$, Corollary 1 holds in the case $i = k$. It follows that for any j such that $0 \le j \le k+1$, ${}_{k+1}\overline{\omega}c_{\hat{\theta_j}}$ is surjective.

LEMMA 1. *For any j such that $0 \le j \le k+1$, the following holds:*

$$\dim_{\mathbb{K}}\left(\ker\left({}_{k+1}\overline{\omega}c_{\hat{\theta_j}}\right)\right) = 2.$$

Although Lemma 1 is nothing but a special case of Proposition 4 in [**6**], we give the proof of Lemma 1 here for the sake of clarity for the readers.

PROOF OF LEMMA 1. Take any j $(0 \le j \le k+1)$ and fix it. Put $f = c_{\hat{\theta_j}}$. For f, we need several notions defined in [**5**]. For any non-negative integer ℓ, put ${}_\ell Q(f) = f^*m_0^\ell C_S / f^*m_0^{\ell+1}C_S$, ${}_\ell\delta(f) = \dim_{\mathbb{K}} {}_\ell Q(f)$, and ${}_\ell\gamma(f) = \dim_{\mathbb{K}}\ker({}_\ell\bar{t}f)$, where ${}_\ell\bar{t}f : {}_\ell Q(f) \to {}_\ell Q(f)^2$ is defined by ${}_\ell\bar{t}f([\eta]) = [tf(\eta)]$. Then, it is easily seen that ${}_\ell\delta(f) = 2(k+1)$ and ${}_\ell\gamma(f) = 1$.

For f, tf is injective and ${}_{k+1}\overline{\omega}f$ is surjective. Therefore,

$$\begin{aligned} & \dim_{\mathbb{K}}\ker({}_{k+1}\overline{\omega}f) \\ = \;& \dim_{\mathbb{K}}\frac{m_0^{k+1}\theta_0(2)}{m_0^{k+2}\theta_0(2)} - \dim_{\mathbb{K}}\frac{f^*m_0^{k+1}\theta_S(f)}{T\mathcal{R}_e(f)\cap f^*m_0^{k+1}\theta_S(f) + f^*m_0^{k+2}\theta_S(f)} \\ = \;& 2(k+2) - \dim_{\mathbb{K}}\frac{\frac{f^*m_0^{k+1}\theta_S(f)}{f^*m_0^{k+2}\theta_S(f)}}{\frac{T\mathcal{R}_e(f)\cap f^*m_0^{k+1}\theta_S(f)}{T\mathcal{R}_e(f)\cap f^*m_0^{k+2}\theta_S(f)}} \\ = \;& 2(k+2) - (2\cdot{}_{k+1}\delta(f) - ({}_{k+1}\delta(f) - {}_{k+1}\gamma(f) + {}_k\gamma(f))) \\ = \;& 2(k+2) - 2(k+1) = 2. \end{aligned}$$

□

By Lemma 1, it is sufficient to show that

$$\frac{m_0^{k+1}\theta_0(2)}{m_0^{k+2}\theta_0(2)} = \sum_{j=0}^{k+1}\ker\left({}_{k+1}\overline{\omega}c_{\hat{\theta_j}}\right).$$

Let $(U,(X,Y))$ be a local coordinate system of $\mathbb{K}^2$ at the origin. Since X^{k+1}, $X^kY, \ldots, Y^{k+1}$ are generators of m_0^{k+1}, for any $[\xi] \in \frac{m_0^{k+1}\theta_0(2)}{m_0^{k+2}\theta_0(2)}$, there must exist $\xi_1, \xi_2 \in m_0^k\theta_0(2)$ such that $\xi = X\xi_1 + Y\xi_2$. Then, by the assumption of induction we have the following:

$$[\xi_1], [\xi_2] \in \bigoplus_{j=0}^{k}\ker\left({}_k\overline{\omega}\left(c_{\hat{\theta_{k+1}}}\right)_{\hat{\theta_j}}\right).$$

Therefore, the following holds:

$$[\xi] \in \sum_{j=0}^{k} \ker\left({}_{k+1}\overline{\omega}\left(c_{\theta_{\hat{k+1}}}\right)_{\hat{\theta}_j}\right).$$

Hence, we may put

$$[\xi] = \sum_{j=0}^{k} [\widetilde{\xi}_j],$$

where $[\widetilde{\xi}_j] \in \ker\left({}_{k+1}\overline{\omega}\left(c_{\theta_{\hat{k+1}}}\right)_{\hat{\theta}_j}\right)$. Then, we have the following lemma.

LEMMA 2. *For any j such that $0 \leq j \leq k$, the following holds:*

$$[\widetilde{\xi}_j] \in \ker\left({}_{k+1}\overline{\omega}c_{\hat{\theta}_j}\right) + \ker\left({}_{k+1}\overline{\omega}c_{\theta_{\hat{k+1}}}\right).$$

PROOF OF LEMMA 2. Since $1 \leq k$, the set $\{0, 1, \ldots, k+1\} - \{j, k+1\}$ is not empty. Thus we may choose an element $\widetilde{\ell} \in \{0, 1, \ldots, k+1\} - \{j, k+1\}$. For $\widetilde{\ell}$, put $L = \{0, 1, \ldots, k+1\} - \{\widetilde{\ell}\}$. Then, from the assumption of induction, we have the following:

$$\frac{m_0^k \theta_0(2)}{m_0^{k+1}\theta_0(2)} = \bigoplus_{\ell \in L} \ker\left({}_k\overline{\omega}\left(c_{\hat{\theta}_{\widetilde{\ell}}}\right)_{\hat{\theta}_\ell}\right).$$

Hence, and since $[\widetilde{\xi}_j] \in \ker\left({}_{k+1}\overline{\omega}\left(c_{\theta_{\hat{k+1}}}\right)_{\hat{\theta}_j}\right)$, the following holds:

$$[\widetilde{\xi}_j] \in \left(\sum_{\ell \in L} \ker\left({}_{k+1}\overline{\omega}\left(c_{\hat{\theta}_{\widetilde{\ell}}}\right)_{\hat{\theta}_\ell}\right)\right) \bigcap \ker\left({}_{k+1}\overline{\omega}\left(c_{\theta_{\hat{k+1}}}\right)_{\hat{\theta}_j}\right).$$

Since $\widetilde{\ell} \neq j$ and $\widetilde{\ell} \neq k+1$, the intersection

$$\ker\left({}_{k+1}\overline{\omega}\left(c_{\hat{\theta}_{\widetilde{\ell}}}\right)_{\hat{\theta}_\ell}\right) \bigcap \ker\left({}_{k+1}\overline{\omega}\left(c_{\theta_{\hat{k+1}}}\right)_{\hat{\theta}_j}\right)$$

is contained in

$$\ker\left({}_{k+1}\overline{\omega}c_{\hat{\theta}_j}\right) + \ker\left({}_{k+1}\overline{\omega}c_{\theta_{\hat{k+1}}}\right)$$

for any $\ell \in L$. Thus, $[\widetilde{\xi}_j]$ must belong to $\ker\left({}_{k+1}\overline{\omega}c_{\hat{\theta}_j}\right) + \ker\left({}_{k+1}\overline{\omega}c_{\theta_{\hat{k+1}}}\right)$. □

By Lemma 2, we have the following:

$$[\xi] = \sum_{j=0}^{k} [\widetilde{\xi}_j] \in \sum_{m=0}^{k+1} \ker\left({}_{k+1}\overline{\omega}c_{\theta_{\hat{m}}}\right).$$

Therefore, we have

$$\frac{m_0^{k+1}\theta_0(2)}{m_0^{k+2}\theta_0(2)} = \sum_{m=0}^{k+1} \ker\left({}_{k+1}\overline{\omega}c_{\theta_{\hat{m}}}\right).$$ □

References

1. P. A. Kolgushkin and R. R. Sadykov, *Simple singularities of multigerms of curves*, Rev. Mat. Complut. **14** (2001), 311–344. MR1871300 (2002h:14002)
2. E. J. N. Looijenga, *Isolated Singular Points on Complete Intersections*, London Mathematical Society Lecture Note Series, **77**, Cambridge University Press, Cambridge, 1984. MR747303 (86a:32021)
3. J. Mather, *Stability of C^∞ mappings,* III. *Finitely determined map-germs*, Publ. Math. Inst. Hautes Études Sci. **35** (1969), 127–156. MR0275459 (43:1215a)
4. J. Mather, *Stability of C^∞ mappings,* IV, *Classification of stable map-germs by $\mathbb{R}$-algebras*, Publ. Math. Inst. Hautes Études Sci. **37** (1970), 223–248. MR0275460 (43:1215b)
5. T. Nishimura, *$\mathcal{A}$-simple multigerms and $\mathcal{L}$-simple multigerms*, Yokohama Math. J. **55** (2010), 93–104. MR2666140 (2011i:58066)
6. T. Nishimura, *Vector fields liftable over finitely determined multigerms of corank at most one*, arXiv:1112.1214.
7. K. Saito, *Theory of logarithmic differential forms and logarithmic vector fields*, J. Fac. Sci. Univ. Tokyo Sect. IA Math. **27** (1980), 265–291. MR586450 (83h:32023)
8. C. T. C. Wall, *Finite determinacy of smooth map-germs*, Bull. London Math. Soc. **13** (1981), 481–539. MR634595 (83i:58020)

Graduate School of Mathematical Sciences, Kyushu University, 744, Motooka, Nishi-ku, Fukuoka 819-0395, Japan
E-mail address: ma210046@math.kyushu-u.ac.jp

Research Group of Mathematical Sciences, Research Institute of Environment and Information Sciences, Yokohama National University, Yokohama 240-8501, Japan
E-mail address: nishimura-takashi-yx@ynu.jp

Contemporary Mathematics
Volume **569**, 2012
http://dx.doi.org/10.1090/conm/569/11247

Small growth vectors of the compactifications of the contact systems on $J^r(1,1)$

Piotr Mormul

ABSTRACT. It is well known that the compactifications of the canonical contact systems living on real jet spaces $J^r(1,1)$, $r \geq 2$, are locally universal Goursat distributions, Δ^r, living on compact manifolds (called Goursat monsters) having open dense jet-like ($J^r(1,1)$-like) parts.
By virtue of the results of Jean (1996), one was able, for each $r \geq 2$, to recursively compute the *small growth vector* of Δ^r at any point of the r-th monster. The result was got by performing series of r operations taken, in function of the local geometry of Δ^r in question, from the set of fixed recursive rules $\{1, 2, 3\}$ in Jean's notation (called in the present text S, T, G, respectively). By the local universality of Δ^r one was thus able to compute *all* small growth vectors of all existing Goursat distributions.
In the work of Mormul (2004) proposed were explicit solutions to the series of Jean's recurrences. Yet, those formulas appeared as if from thin air, and proofs were postponed to another publication.
In the present contribution we submit proofs of the formulas from the year 2004. The paper is not a plain check that our candidates satisfy Jean's recurrences. We effectively retrieve the very formulas – they gradually (and naturally) emerge in the steps of an inductive proof.

1. Goursat distributions and their small growth vectors

Goursat flags are certain special nested sequences, in general of variable length r $(2 < r < n-2)$ of subbundles in the tangent bundle TM to a smooth or analytic n-dimensional manifold M: $D^r \subset D^{r-1} \subset \cdots \subset D^1 \subset D^0 = TM$. Namely, one demands, for $l = r, r-1, \ldots, 1$, that (a) the Lie square of D^l be D^{l-1} and (b) cork $D^l = l$. In other words, that the *big growth vector* of the distribution D^r be $[n-r,\, n-r+1,\, n-r+2, \ldots,\, n-1,\, n]$ at each point of M.

This very restricted class of objects was being investigated (intermittently) over the last 120 years, with contributions, among others, by E. von Weber [**8**] and E. Cartan [**1**]. They independently proved that every corank-r Goursat distribution D^r on M around a generic point locally behaves in a unique way visualised by the *chained model* – the germ at $0 \in \mathbb{R}^n(x^1, \ldots, x^{r+2}; x^{r+3}, \ldots, x^n)$ of

$$\big(\partial_n, \ldots, \partial_{r+3};\ \partial_{r+2},\ \partial_1 + x^3\partial_2 + x^4\partial_3 + \cdots + x^{r+2}\partial_{r+1}\big)$$

2010 *Mathematics Subject Classification.* Primary 58A17, 30.
Key words and phrases. Goursat flag, small growth vector, solution of Jean's recurrences, encoding of letter words by sequences of integers.
Supported by Polish Government MNiSzW Grant N N201 397937.

(these are vector fields generators; effectively used are only the first $r+2$ coordinates). Chained models can be viewed as the simplest instance of a family of local writings (preliminary normal forms) of Goursat distributions, obtained much later by Kumpera and Ruiz. As a matter of fact, Kumpera and Ruiz discovered, to the big surprise of the mathematical community, *singularities* hidden in flags, and their pseudo-normal forms were merely a by-product – visualisations of those singularities.

When $n = r+2$, a Goursat distribution D^r on M has around generic points a unique local description as the germ at $0 \in \mathbb{R}^{r+2}(t,\, x,\, x_1, \dots,\, x_r)$ of

$$\text{(A)} \qquad \big(\partial/\partial x_r,\ \partial/\partial t + x_1\partial/\partial x + x_2\partial/\partial x_1 + \cdots + x_r\partial/\partial x_{r-1}\big)\,.$$

That is, of the canonical contact system on $J^r(1,1)$. (The independent variable is t, the dependent one is x, and $x_1,\, x_2, \dots,\, x_r$ can be viewed as the consecutive derivatives of x with respect to t.) This is precisely the result obtained independently by von Weber and Cartan.

For a theory of Goursat flags it has been essential to have a flexible length r that could be increased without changing a manifold, allowing for flags to be prolonged in length on one and the same manifold. In what follows, however, we deal only with the most classical situation $n = r+2$, when the initial Goursat distribution D^r is of rank 2 and its flag is already of maximal length $r = n-2$. For, as long as Goursat distributions are considered only *locally*, this causes no loss of generality. Compare, for instance, Corollary 1.3 in [**6**].

By definition, the rank-2 and corank-r Goursat distributions have the big growth vector $[2,\, 3,\, 4, \dots,\, r+1,\, r+2]$ at each point.

DEFINITION 1. Small growth vector (sgrv in all what follows) $\{n_j(p)\}$ of a distribution D at a point $p \in M$ is the sequence of dimensions at p of the (local) modules of vector fields V_j, $V_1 = D$, $V_{j+1} = V_j + [D,\, V_j]$,

$$n_j(p) = \dim V_j(p)\,, \qquad j = 1,\, 2,\, 3,\, \dots$$

We are interested only in distributions for which this sequence, for every $p \in M$, attains (sooner or later) the value $\dim M$, with 'sooner or later' underscored. Such distributions are called *completely nonholonomic.* When, for such a distribution D, there happens $n_{l-1}(p) < n_l(p) = \dim M$, then l is called the *nonholonomy degree* of D at p.

It is a short Lie algebra exercise to show that the Goursat distributions are completely nonholonomic. Around jet-like points p where the visualisation (A) is effective, there holds $n_j(p) = 1+j$ for $j = 1, 2, \dots, n-1$, and then the nonholonomy degree, minimal possible for Goursat in dimension n, is $n-1 = r+1$, the number r being the order of jets in the visualisation. For non-jet-like points, *much* slower sgrv's (and hence bigger nonholonomy degrees) are possible, as is announced in [**6**] and will be eventually clear from the present work. In any fixed dimension $n \geq 4$, the biggest nonholonomy degree will turn out to be F_n, the n-th Fibonacci number.

A given Goursat distribution D^r may not feature all possible singularities in corank r, nor all possible sgrv's of Goursat distributions of that corank. But in each corank r there do exist manifolds with Goursat structures of corank r on them that are locally universal – feature all possible singularities, hence also all possible sgrv's of Goursat distributions of corank r. Such are, for instance, the kinematical models extensively used and analysed in [**3**], and especially *monster*

Goursat manifolds $\mathbb{P}^r(\mathbb{R}^2)$, with locally universal Goursat structures Δ^r living on them, constructed in [**4**].[1]

A natural question is what are all occurring sgrv's of Goursat distributions of a fixed corank $r \geq 2$. Or, which is the same for that fixed r, what are all sgrv's of the distribution Δ^r living on $\mathbb{P}^r(\mathbb{R}^2)$.

We are going to recapitulate the information necessary for the present contribution. First of all, the germs of Goursat flags of length r are partitioned into F_{2r-3} invariant *geometric classes* labelled (or: encoded) by words of length r over the alphabet $\{\mathrm{G}, \mathrm{S}, \mathrm{T}\}$ starting with GG and such that GT is not allowed.[2] We call such words *admissible.* The roots of this concept can be traced back to the paper [**3**], where the prototypes of geometric classes were called *regions.* By virtue of Jean's results [**3**], the sgrv of a Goursat germ depends solely on its geometric class. Hence, in length r there are not more than F_{2r-3} different sgrv's of Goursat germs.

Remark 1. *It was not explained in* [**3**] *whether the small growth vectors attached to the geometric classes, in any fixed length, were all different. It had remained a fine point for quite a time, having become completely clear only on the solutions' side of Jean's recurrences. See in this respect in* [**6**]*: the discussion in Section 2.1 and Theorem 3.5.*

Those who prefer the universal objects may think of the geometric classes in length r as just the strata of a [very regular, besides] stratification of the r-th monster $\mathbb{P}^r(\mathbb{R}^2)$. Those strata are embedded submanifolds of $\mathbb{P}^r(\mathbb{R}^2)$ of codimensions equal to the numbers of non-G letters in their codes. (As a matter of fact, Jean's stratification of $\mathbb{P}^r(\mathbb{R})^2$ is labelled in [**4**] by r-letter words over a newer alphabet $\{\mathrm{R}, \mathrm{V}, \mathrm{T}\}$, with R replacing the previous G, V replacing S, and T replacing T.) Repeating, then, within each stratum, or geometric class, the sgrv of Δ^r is constant, not depending on points.

In fact, it is a nondecreasing sequence of integers that starts with 2 and takes on, for the Goursat distributions, all intermediate integer values up to $r+2$ inclusively. The key issue is to ascertain with what multiplicities do these integers appear. We recall that the sequence of multiplicities of integers in the sgrv of a given geometric class has been called in [**6**] the **derived vector** of that class. We stick to this terminology in the present paper. Sometimes, for brevity's sake, we will speak about the derived vector of an admissible word (= the code of a geometric class).

The recipes to compute the derived vectors out of admissible (G,S,T) words which encode the geometric classes, were given in [**6**]: firstly on a recursive basis rephrasing the results of [**3**], then in closed form formulas, which explicitly solved the recurrences for derived vectors. The first way did not need a proof, because Jean's recurrences, for the functions *beta* he used, automatically implied the recurrences for derived vectors being nothing but the sequences of *increments* of the beta functions.

The second way necessitated a proof, or proofs of three separate statements in [**6**]: Proposition 3.2, Theorem 3.3, and Theorem 3.4. The purpose of the present

[1]Goursat monsters could be alternatively built over, say, S^2 or T^2 or yet another compact 2-dimensional manifold, instead of $\mathbb{R}^2$. But they would serve just the same objective of local universality, while it is much easier to work with $\mathbb{P}^r(\mathbb{R}^2)$, notwithstanding $\mathbb{R}^2$ is not compact.

[2]As concerns the count of such words, for the number F_{2r-3} see Observation 1.10 in [**6**].

text is to furnish such proofs, after a slight reformatting of the statements under consideration: a part of old Proposition 3.2 and Theorem 3.3 are now merged into Theorem 2, while the rest of Proposition 3.2 and Theorem 3.4 are merged into Theorem 3.

In [**6**], for each geometric class $\mathcal{C}$ of length r, the derived vector of $\mathcal{C}$, denoted by der($\mathcal{C}$), is a function $\{2, 3, \ldots, r+1\} \to \mathbb{N}$ in which der($\mathcal{C}$)(j) is the multiplicity of j in the sgrv of $\mathcal{C}$, for every integer $2 \le j \le r+1$. Eventually the sgrv ends with the ceiling value $r+2$.

So how does Jean recursively arrive (modulo the mentioned passing to the increments of his original functions beta) at the functions der($\mathcal{C}$)?

THEOREM 1 ([**3**]). *For each geometric class* $\mathcal{C}$ *in length* r, *the function* der($\mathcal{C}$) *equals the last term* d^r *in the sequence of functions* $d^1, d^2, \ldots, d^r$ *that are constructed, by means of operations* G, S, T *defined below, as follows.*

One defines, recursively for $j = 1, 2, \ldots, r$, functions $d^j\colon \{2, \ldots, j+1\} \to \mathbb{N}$. One starts by declaring $d^1 = (1)$ and $d^2 = (1, 1)$, and then continues

$$d^{j+2} = \begin{cases} \mathrm{G}\left(d^{j+1}\right), & \text{when the } (j+2)\text{-nd letter in the code of } \mathcal{C} \text{ is G,} \\ \mathrm{S}\left(d^j,\, d^{j+1}\right), & \text{when the } (j+2)\text{-nd letter in the code of } \mathcal{C} \text{ is S,} \\ \mathrm{T}\left(d^j,\, d^{j+1}\right), & \text{when the } (j+2)\text{-nd letter in the code of } \mathcal{C} \text{ is T.} \end{cases}$$

Concerning the operations in use, the simplest among them is the operation G,

$$\mathrm{G}(\alpha) = \begin{pmatrix} 2 & 3 & 4 & 5 & \ldots \\ \downarrow & \downarrow & \downarrow & \downarrow & \ldots \\ 1 & \alpha(2) & \alpha(3) & \alpha(4) & \ldots \end{pmatrix}.$$

So it is an "insert a 1 on the left and shift the argument vector α by one slot to the right" operation. Clearly, G(α) is a sequence (function) by one entry longer than α (with the domain by one element bigger than the domain of α).
The remaining operations S and T are more involved. They are two-argument, not just one-argument like G. Their first argument, say α, is a sequence by one entry shorter than the second argument, say β, and their outputs are yet by one entry longer than β,

$$\mathrm{S}(\alpha,\, \beta) = \begin{pmatrix} 2 & 3 & 4 & 5 & 6 & \ldots \\ \downarrow & \downarrow & \downarrow & \downarrow & \downarrow & \ldots \\ 1 & 1 & \alpha(2)+\beta(3) & \alpha(3)+\beta(4) & \alpha(4)+\beta(5) & \ldots \end{pmatrix},$$

$$\mathrm{T}(\alpha,\, \beta) = \begin{pmatrix} 2 & 3 & 4 & 5 & 6 & \ldots \\ \downarrow & \downarrow & \downarrow & \downarrow & \downarrow & \ldots \\ 1 & 1 & 2\beta(3)-\alpha(2) & 2\beta(4)-\alpha(3) & 2\beta(5)-\alpha(4) & \ldots \end{pmatrix}.$$

So, in perhaps more catching terms, S is a *Fibonacci-like rule*, with appropriate shifts of the input vectors α and β, while T is an *arithmetic progression rule*, with likewise shifts of the inputs. In fact, when the two inputs α and β are identically indented on the left, then their S (T, resp.) output starts, one row below, by inserting two 1's on the left and then performing the Fibonacci rule (arithmetic progression rule, resp.) on α and β skew-wise from the NW to the SE. To give an instance of each of these two-argument operations in action, and skipping brackets for bigger transparence,

$$\begin{array}{rcllllll} \alpha & = & 1 & 1 & 2 & 3 & & \\ \beta & = & 1 & 1 & 1 & 2 & 3 & \\ \mathrm{S}(\alpha,\beta) & = & \mathbf{1} & \mathbf{1} & 2 & 2 & 4 & 6 \end{array}$$

and

$$\begin{array}{rcllllll} \alpha & = & 1 & 1 & 2 & & \\ \beta & = & 1 & 1 & 2 & 3 & \\ \mathrm{T}(\alpha,\beta) & = & \mathbf{1} & \mathbf{1} & 1 & 3 & 4 \end{array}$$

(the two initial 1's inserted in the outputs on the left are written in bold).

REMARK 2. *When, for any fixed length* r, *does the sum of values of* $\mathrm{der}(\mathcal{C})$ *attain the maximal value, or: when the nonholonomy degree* $1+\sum_{i=2}^{r+1}\mathrm{der}(\mathcal{C})(i) = \max$? *It is an exercise in the* G, S, *and* T *operations to show that, for each* r, *the maximal nonholonomy degree is realized only by the [most singular] class* $\mathcal{C} = \mathrm{G\,G}\underbrace{\mathrm{S\,S}\ldots\mathrm{S}}_{r-2}$. *Then* $\mathrm{der}(\mathcal{C})(i) = F_{i-1}$ *for* $i = 2, 3, \ldots, r+1$, *and*

$$1+\sum_{i=2}^{r+1}\mathrm{der}(\mathcal{C})(i) = 1+F_1+F_2+\cdots+F_r = F_{r+2}\,,$$

as invoiced earlier in this section. (Thus the nonholonomy degree of the distribution Δ^r *always varies between* $r+1$ *and* F_{r+2} *and these bounds are sharp for each* $r \geq 2$. *For instance, for* Δ^3, *it varies between* 4 *and* 5. *In general, however, not all intermediate integer values are realized by the nonholonomy degrees of* Δ^r. *The smallest length with this happening is* 6: *the value* 20 *located between* 7 *and* $21 = F_{6+2}$ *is not realized, cf.* Theorem 23 in [**2**].)

1.1. Useful aggregates of geometric classes. In what follows, the length (r) of a geometric class will not be essential. Much more important will be the number, say $s+1 \geq 1$, of letters S in class' code that can otherwise be arbitrarily long.[3] To word our results, we arbitrarily **fix this discrete parameter** $s \geq 0$.

We have to introduce some more, still necessary, parameters – the numbers of letters T and G in classes' words going past the letters S, those letters S being run *backwards* from code's right end to left.

Namely, agree that the last S in a code is followed by $k_0 \geq 0$ letters T, and then by $l_0 \geq 0$ letters G. Agree also that the one before last S is followed by $k_1 \geq 0$ letters T and then by $l_1 \geq 0$ letters G, and so on backwards until the first appearing letter S being followed by $k_s \geq 0$ letters T, then by $l_s \geq 0$ letters G.

On top of that, let a code start with $l_{s+1} \geq 2$ letters G. (Except for this last quantity, all introduced integer parameters may even vanish, as it happens in the 'Fibonacci' classes GGS_{s+1} discussed in Remark 2 above.)

REMARK 3. *The indexation of integer parameters* k_j *and* l_j *proposed here slightly differs, by a backward shift of indices by one, from the indexation in* [**6**]. *The same applies to the, central for the contribution, family (depending on the*

[3]By Theorem 1, in any fixed length $r \geq 2$, the generic jet-like geometric class with no letters S, written shortly G_r, and only this class, has the simplest derived vector $(1, 1, \ldots, 1)$ (r 1's). Within G_r, and only within it (in that length), the small growth vector of a Goursat distribution is the same as its big growth vector. In the paper we only consider the *remaining* geometric classes featuring certain letters S in their codes. That is, we consider the singular geometric classes.

parameters k_j) of sequences (A_0, A_1, A_2, ...) introduced below before Theorem 2. The formulas, especially in Theorem 3 below, assume thus a more compact view than in [**6**].

In terms of these parameters, what can be said about the derived vectors emerging from Theorem 1?

We will first answer the question for the geometric classes having $l_1 = l_2 = \cdots = l_s = 0$. That is, for classes in whose codes the letters G occur only in the beginning (in number $l_{s+1} \geq 2$) and possibly in the end (in number $l_0 \geq 0$). The union of such geometric classes has been called in [**6**] the hyperclass 0_s. In the present text we use the concept of a hyperclass only implicitly, not explicitly.

To word the relevant theorem, let us introduce a *family* of integer sequences

- $A_0 = 1, \quad A_1 = 2+k_0, \quad A_j = A_{j-2}+(1+k_{j-1})A_{j-1} \quad \text{for} \quad 2 \leq j \leq s+1\,.$

(This family is parametrized by the non-negative integers k_0, $k_1, \ldots, k_s$.)

2. Main theorems

THEOREM 2 ([**6**]). *In the derived vectors of the geometric classes with $s+1$ letters* S *and such that $l_1 = l_2 = \cdots = l_s = 0$, there appear the following $s+2$ different values, in the growing order: A_0, $A_1, \ldots, A_{s+1}$. All these derived vectors are non-decreasing and, concerning the multiplicities of the listed values:*

- A_0 appears $2 + k_0 + l_0$ times in row;
- for $1 \leq j \leq s$, the number A_j appears $1 + k_j$ times in row;
- A_{s+1} appears $l_{s+1} - 1$ times in row.

Proof of this theorem is given in Section 3.

Before describing the sgrv's of the remaining geometric classes, we need an infinite series of sequences of the type A. This series will be parametrized by yet another natural number N. (N is subject to some restrictions in function of s, but s is not a priori bounded from above.)

- $A_0^{+N} = 1, \quad A_1^{+N} = 2 + k_N \;(1 \leq N \leq s)\,,$
- $A_j^{+N} = A_{j-2}^{+N} + (1 + k_{j-1+N})A_{j-1}^{+N} \quad \text{for} \quad 2 \leq j,\; N+j \leq s+1\,.$

We also have to specifically put in relief those letters S in a class' code that are followed by **non-zero numbers** of letters G (going past some letters T, if any). They will play a central role in the characterization of the derived vector of such a class.

Let, for a given geometric class not served by Theorem 2 (hence, in particular, having $s+1 \geq 2$ letters S in its code),

$$1 \leq n_1 < n_2 < \cdots < n_q \leq s$$

be *all* indices i between 1 and s such that $l_i > 0$. (The index n_1 exists for each such class. The index n_2 – already not for each such class. For its existence, the parameter s should be at least 2 and, among the discrete parameters l_1, $l_2, \ldots, l_s$ at least two should be non-zero, etc.) With these notations, we are now ready to complete the information given in Theorem 2 by the following

THEOREM 3 ([**6**]). *In the derived vectors of the geometric classes with $s+1$ letters* S *and such that, among the parameters $\{l_1, l_2, \ldots, l_s\}$, exactly $l_{n_1}, l_{n_2}, \ldots, l_{n_q}$ are non-zero (positive), $1 \leq n_1 < n_2 < \cdots < n_q \leq s$, $1 \leq q \leq s$, there appear $s+2$*

different values, listed below in the growing order, in the following $q+1$ separate rows:

$$A_0,\ A_1,\ \dots,\ A_{n_1-1}\,;$$
$$A_{n_1}\big(A_0^{+n_1},\ A_1^{+n_1},\ \dots,\ A_{n_2-n_1-1}^{+n_1}\big)\,;$$
$$A_{n_1}A_{n_2-n_1}^{+n_1}\big(A_0^{+n_2},\ A_1^{+n_2},\ \dots,\ A_{n_3-n_2-1}^{+n_2}\big)\,;$$
$$\dots\quad\dots\quad\dots\quad\dots\quad\dots$$
$$A_{n_1}\prod_{j=1}^{q-2}A_{n_{j+1}-n_j}^{+n_j}\Big(A_0^{+n_{q-1}},\ A_1^{+n_{q-1}},\ \dots,\ A_{n_q-n_{q-1}-1}^{+n_{q-1}}\Big)\,;$$
$$A_{n_1}\prod_{j=1}^{q-1}A_{n_{j+1}-n_j}^{+n_j}\Big(A_0^{+n_q},\ A_1^{+n_q},\ \dots,\ A_{s-n_q+1}^{+n_q}\Big)\,.$$

These derived vectors are always non-decreasing and, among the above-listed values:

- The 0-th of them (A_0) appears $2+k_0+l_0$ times in row;
- for $1\le j\le s$, the j-th value appears $1+k_j+l_j$ times in row (that is, $1+k_j$ times, when $j\notin\{n_1,\,n_2,\dots,\,n_q\}$);
- the last and biggest $(s+1)$-st value appears $l_{s+1}-1$ times in row.

Observations. (i) The table of $s+2$ different values in $\operatorname{der}(\mathcal{C})$ in Theorem 3 is now formatted slightly differently than in Theorem 3.4 in [**6**]. The last entries in the rows (except the last row) *there* are *now* shifted to the front of the following rows. This is done by purely technical reasons of making the proof (in Section 4) more transparent.

(ii) Also, to simplify the presentation in Theorem 3, the numbers standing *before* the brackets in the all but first rows in the above list do multiply all number entries standing *inside* the relevant brackets.

(iii) When $n_1=1$, or when $n_{i+1}-n_i=1$, then there is only one entry in the relevant row in the list. When $n_q=s$, there are only two entries in the last row.

Proof of Theorem 3 is given in Section 4.

EXAMPLE 1. We want to illustrate Theorem 3 on one geometric class present on the list of 93 *orbits* of the local classification of Δ^7, made in [**5**]. (The language of [**5**] was different, but the very objective of that paper was to locally classify Δ^7.) Namely, for $\mathcal{C}=$ GGSTSGS (called 3.1.3.2.3 in [**5**]; it *is* a single orbit in $\mathbb{P}^7\mathbb{R}^2$) we have $s=2$, $q=1$, $n_1=1$ and $k_0=l_0=k_1=0$, $l_1=k_2=1$, $l_2=0$, $l_3=2$. By Theorem 3, the table of different values appearing in $\operatorname{der}(\mathcal{C})$ (consisting of $q+1=2$ rows) is

$$A_0\,;$$
$$A_1\Big(A_0^{+1},\ A_1^{+1},\ A_2^{+1}\Big)\,,$$

or else, after the due substitutions,

$$1\,;$$
$$2(1,\,2,\,5)\,.$$

By the same Theorem 3, the multiplicities of these values are $2+k_0+l_0=2$, $1+k_1+l_1=2$, $1+k_2+l_2=2$, $l_3-1=1$, respectively. Hence $\operatorname{der}(\mathcal{C})=$

(1, 1, 2, 2, 4, 4, 10). Now let us look how Jean's recurrences lead to the same result:

$$\begin{array}{llllllll} \mathrm{G} & 1 \\ \mathrm{G} & 1 & 1 \\ \mathrm{S} & 1 & 1 & 2 \\ \mathrm{T} & 1 & 1 & 1 & 3 \\ \mathrm{S} & 1 & 1 & 2 & 2 & 5 \\ \mathrm{G} & 1 & 1 & 1 & 2 & 2 & 5 \\ \mathrm{S} & 1 & 1 & 2 & 2 & 4 & 4 & 10\,. \end{array}$$

EXAMPLE 2. The next illustration also comes from the list of 93 local models for Δ^7 in dimension 9 mentioned in Example 1. The geometric class $\mathcal{C} = $ GGSGSSG (called 3.2.3.3.2± in [**5**]) is one of the four exceptional classes in that length – it is not an orbit, but the union of *two* orbits in $\mathbb{P}^7\mathbb{R}^2$. For $\mathcal{C}$, we have $s = 2, q = 1, n_1 = 2$, and $k_0 = 0$, $l_0 = 1$, $k_1 = l_1 = k_2 = 0$, $l_2 = 1$, $l_3 = 2$. By Theorem 3 again, the table of different values appearing in $\mathrm{der}(\mathcal{C})$ is

$$A_0,\ A_1\,;$$
$$A_2\left(A_0^{+2},\ A_1^{+2}\right),$$

or, after the substitutions,

$$1,\ 2\,;$$
$$3(1,\ 2)\,.$$

By Theorem 3, the multiplicities of these values are $2+k_0+l_0 = 3$, $1+k_1+l_1 = 1$, $1+k_2+l_2 = 2$, $l_3 - 1 = 1$, respectively. Hence $\mathrm{der}(\mathcal{C}) = (1,\ 1,\ 1,\ 2,\ 3,\ 3,\ 6)$. And, as a matter of record, Jean's recurrences lead to the same result:

$$\begin{array}{llllllll} \mathrm{G} & 1 \\ \mathrm{G} & 1 & 1 \\ \mathrm{S} & 1 & 1 & 2 \\ \mathrm{G} & 1 & 1 & 1 & 2 \\ \mathrm{S} & 1 & 1 & 2 & 2 & 4 \\ \mathrm{S} & 1 & 1 & 2 & 3 & 3 & 6 \\ \mathrm{G} & 1 & 1 & 1 & 2 & 3 & 3 & 6\,. \end{array}$$

EXAMPLE 3. Our last example deals with a [unimodal] class in length 8 (i. e., in dimension 10), discussed in [**5**] as well (Remark 3, pp. 110-111 there). We take $\mathcal{C} = $ GGSGSGSG (called 3.2.3.2.3.2 in [**5**]) which is a 1-parameter union of orbits in $\mathbb{P}^8\mathbb{R}^2$. In this situation $s = 2, q = 2, n_1 = 1, n_2 = 2$, and $k_0 = 0$, $l_0 = 1$, $k_1 = 0$, $l_1 = 1$, $k_2 = 0$, $l_2 = 1$, $l_3 = 2$. The table of different values appearing in $\mathrm{der}(\mathcal{C})$ (consisting this time of $q+1 = 3$ rows) is, always by Theorem 3,

$$A_0\,;$$
$$A_1\left(A_0^{+1}\right);$$
$$A_1A_1^{+1}\left(A_0^{+2},\ A_1^{+2}\right),$$

or else,

$$1\,;$$
$$2(1)\,;$$
$$2\cdot 2(1,\,2)\,.$$

Theorem 3 says that the multiplicities of these values are $2+k_0+l_0=3$, $1+k_1+l_1=2$, $1+k_2+l_2=2$, $l_3-1=1$, respectively. Hence $\mathrm{der}(\mathcal{C})=(1,\,1,\,1,\,2,\,2,\,4,\,4,\,8)$. And, needless to say, Jean's recurrences produce the same vector:

G	1							
G	1	1						
S	1	1	2					
G	1	1	1	2				
S	1	1	2	2	4			
G	1	1	1	2	2	4		
S	1	1	2	2	4	4	8	
G	1	1	1	2	2	4	4	8.

REMARK 4. *In* [**6**] *there is Remark 4.1 which says that, in the end of the day, the two-step recurrences from* [**3**] *recalled in Theorem* 1, *can be replaced by [much less transparent, however] one-step recurrences. We want to underline that it is but an immediate consequence of Theorems* 2 *and* 3, *for the derived vectors emerging from these theorems are just functions of the codes of geometric classes under consideration. Functions of the extensible set of arguments* $s,\,k_0,\,l_0,\,k_1,\,l_1,\ldots,\,k_s,\,l_s,\,l_{s+1}$. *In fact,*

- *prolonging a geometric class by* G *(increasing* l_0 *by* 1*) results in an extremely simple operation on the derived vector,*
- *prolonging it by* T *(meaning that* $l_0=0$*) is just increasing* k_0 *by* 1, *with the resulting changes in the derived vector,*
- *while prolonging it by* S *means: - increasing the parameter* s *by* 1, *- the related shift of indices in the parameters* k *and* l, *- the consequent changes in the derived vector, all of them governed by Theorems* 2 *and/or* 3.

3. Proof of Theorem 2

The proof is by induction on $s\ge 0$. For $s=0$, i. e., for the geometric classes with just one letter S in their codes ($s+1=1$), the derived vectors can be computed directly from Theorem 1. Indeed, for the classes $\mathrm{G}_{l_1}\mathrm{S}\,\mathrm{T}_{k_0}\mathrm{G}_{l_0}$, where $l_1\ge 2$, we are to perform $l_1-2\ge 0$ operations G starting from the function d^2, then one operation S, then $k_0\ge 0$ operations T, and eventually $l_0\ge 0$ operations G.

We know from the beginning that $d^1=(1)$, $d^2=(1,\,1)$. So, after l_1-2 operations G,

$$d^{l_1}=\big(\underbrace{1,\,1,\ldots,\,1}_{l_1}\big)=\big(1_{l_1}\big)\,.$$

Now, by the Fibonacci-like rule governing the operation S,

$$d^{l_1+1}=\mathrm{S}(d^{l_1-1},\,d^{l_1})=\big(1,\,1,\,2_{l_1-1}\big)\,. \tag{3.1}$$

When $k_0 > 0$, we next perform k_0 operations T, that is, compute k_0 subsequent terms in arithmetic progressions with the two initial terms either $(1, 1)$ or $(1, 2)$:

$$d^{l_1+2} = \mathrm{T}(d^{l_1}, d^{l_1+1}) = \big(1,\, 1,\, 1,\, 3_{l_1-1}\big)$$

$$\dots \quad \dots \quad \dots \quad \dots$$

$$d^{l_1+1+k_0} = \big(1_{2+k_0},\, (2+k_0)_{l_1-1}\big)\,. \tag{3.2}$$

Observe that, when $k_0 = 0$, the expression on the right in (3.1) coincides with the expression on the right in (3.2). So (3.2) is also valid for $k_0 = 0$. Now, applying the operation G l_0 times to the sequence on the RHS of (3.2), we get

$$d^{l_1+1+k_0+l_0} = \big(1_{2+k_0+l_0},\, (2+k_0)_{l_1-1}\big) = \Big(\underbrace{A_0, \dots, A_0}_{2+k_0+l_0},\, \underbrace{A_1, \dots, A_1}_{l_1-1}\Big),$$

as stated in theorem's statement, irrespectively of whether $l_0 > 0$ or $l_0 = 0$. The beginning of the induction is done.

We are now to justify, for any fixed integer value $s \geq 0$, the induction step '$s \Rightarrow s+1$'. That is, we assume theorem's statement for classes with **not more than** $s+1$ letters S in their codes and work towards the analogous statement for an arbitrary class $\mathcal{C}$ having $s+2$ letters S and with such discrete parameters that $l_1 = l_2 = \cdots = l_{s+1} = 0$.

It is natural to truncate the code of $\mathcal{C}$ just before the last from the left (the first from the right!) letter S in it. The inductive hypothesis will be applied to this truncated word, with the up-shift of indices in theorem's parameters by 1 and – attention – with the last parameter l_0 in the very wording of theorem (its role played now by l_1) vanishing. In parallel, the hypothesis will also be applied to this truncated word with *its* last (or: first from the right) letter *deleted.* We are going to treat their derived vectors as known and explicitly described by the theorem under proof.

Observe that the latter (doubly truncated) word, may have either $s+1$ or s letters S. The first case occurs, clearly, when there is a gap in the code of $\mathcal{C}$ between its $(s+1)$-st and $(s+2)$-nd letter S counted from the left (between the 1st and 2nd letter S counted from the right). That is, when $k_1 \geq 1$. The second case – when there is no mentioned gap, or else when $k_1 = 0$. We will argue completely separately in each case, for the information concerning the derived vector of the doubly truncated word depends critically on the case occurring.

3.1. Case $k_1 \geq 1$. In this case, the above-mentioned truncated words are of the form $\mathbf{W}\mathrm{ST}_{k_1}$ and $\mathbf{W}\mathrm{ST}_{k_1-1}$, where $\mathbf{W}$ is an admissible word with s letters S. So, by the inductive assumption, the vector $\mathrm{der}(\mathbf{W}\mathrm{ST}_{k_1})$ is non-decreasing and features (with due multiplicities precised in the inductive assumption) the numbers

$$A_0^{+1},\, A_1^{+1},\, \dots,\, A_s^{+1},\, A_{s+1}^{+1},$$

while the vector $\mathrm{der}(\mathbf{W}\mathrm{ST}_{k_1-1})$ is also non-decreasing and features formally 'similar' numbers (with their proper multiplicities also precised in the inductive assumption), yet built on the basis of the initial data $k_1 - 1, k_2, \dots, k_{s+1}$ instead of $k_1, k_2, \dots, k_{s+1}$ in the previous case. Let us write those numbers down as

$$\bar{A}_0^{+1},\, \bar{A}_1^{+1},\, \dots,\, \bar{A}_s^{+1},\, \bar{A}_{s+1}^{+1}.$$

These are our basic construction bricks. With their help we will firstly express the vector

$$\mathrm{S}\Big(\mathrm{der}(\mathbf{W}\mathrm{ST}_{k_1-1}), \mathrm{der}\big(\mathbf{W}\mathrm{ST}_{k_1}\big)\Big)\,. \tag{3.3}$$

In order to perform this operation S, one writes its vector arguments one above the other, identically indented on the left.
So, in one row, called $(*)$, from the left to right:

- $2 + k_1 - 1 = 1 + k_1$ times $\bar{A}_0^{+1}$, then
- $1 + k_2$ times $\bar{A}_1^{+1}$, then
- $\ldots\ \ \ldots\ \ \ldots\ \ \ldots\ \ \ldots$
- $1 + k_{s+1}$ times $\bar{A}_s^{+1}$, and then
- $l_{s+2} - 1$ times $\bar{A}_{s+1}^{+1}$.

And then in the following row, called $(**)$, also from the left to right:

- $2 + k_1$ times A_0^{+1}, then
- $1 + k_2$ times A_1^{+1}, then
- $\ldots\ \ \ldots\ \ \ldots\ \ \ldots$
- $1 + k_{s+1}$ times A_s^{+1}, and, eventually,
- $l_{s+2} - 1$ times A_{s+1}^{+1}.

In the subsequent row, indented on the left exactly as $(*)$ and $(**)$ are, we ought, by the definition of S, to start with two 1's and then perform the Fibonacci rule skew-wards from NW to SE. Let us present this procedure in detail in the initial parts of the row $(*)$ (with its $1 + k_1$ initial entries) and $(**)$ (with its $2 + k_1$ initial entries):

$$\begin{array}{cccccc}
\bar{A}_0^{+1} & \ldots & \bar{A}_0^{+1} & \bar{A}_0^{+1} & & \\
A_0^{+1} & A_0^{+1} & \ldots & A_0^{+1} & A_0^{+1} & \\
\mathbf{1} & \mathbf{1} & A_0^{+1} + \bar{A}_0^{+1} & \ldots & A_0^{+1} + \bar{A}_0^{+1} & A_0^{+1} + \bar{A}_0^{+1}
\end{array}$$

Given that we know the entire rows $(*)$ and $(**)$, it is now visible that the overall result of the operation S is a row, called $(***)$, by one entry longer than $(**)$, in which there go, from the left to right,

- 2 times 1, then
- $1 + k_1$ times $A_0^{+1} + \bar{A}_0^{+1}$, then
- $1 + k_2$ times $A_1^{+1} + \bar{A}_1^{+1}$, then
- $\ldots\ \ \ldots\ \ \ldots\ \ \ldots\ \ \ldots$
- $1 + k_{s+1}$ times $A_s^{+1} + \bar{A}_s^{+1}$, and then
- $l_{s+2} - 1$ times $A_{s+1}^{+1} + \bar{A}_{s+1}^{+1}$.

Now we are to perform the operation T certain ($k_0 \geq 0$) number of times. When $k_0 > 0$, we apply it for the first time to the rows $(**)$ and $(***)$ (these two are its vector arguments). Let us present this procedure in detail, like with the preceding operation, in the initial parts of these rows, focusing on the $2 + k_1$ initial entries in the row $(**)$ and on the $3 + k_1$ initial entries in the row $(***)$:

$$\begin{array}{ccccccc}
1 & A_0^{+1} & A_0^{+1} & \ldots & A_0^{+1} & & \\
1 & 1 & A_0^{+1} + \bar{A}_0^{+1} & A_0^{+1} + \bar{A}_0^{+1} & \ldots & A_0^{+1} + \bar{A}_0^{+1} & \\
\mathbf{1} & \mathbf{1} & 1 & A_0^{+1} + 2\bar{A}_0^{+1} & A_0^{+1} + 2\bar{A}_0^{+1} & \ldots & A_0^{+1} + 2\bar{A}_0^{+1}
\end{array}$$

Given that we know the entire rows $(**)$ and $(***)$, it now becomes clear (a) how the rest of the output of this first operation T looks like, and (b) how the entire continuation with the consecutive $k_0 - 1$ operations T looks like. Namely, the resulting eventual derived vector consists of the entries

- $2 + k_0$ times $1(= A_0)$, then
- $1 + k_1$ times $A_0^{+1} + (1 + k_0)\bar{A}_0^{+1}$, then
- $1 + k_2$ times $A_1^{+1} + (1 + k_0)\bar{A}_1^{+1}$, then
- $\dots \quad \dots \quad \dots \quad \dots \quad \dots$
- $1 + k_{s+1}$ times $A_s^{+1} + (1 + k_0)\bar{A}_s^{+1}$, and then
- $l_{s+2} - 1$ times $A_{s+1}^{+1} + (1 + k_0)\bar{A}_{s+1}^{+1}$.

Observe that, when $k_0 = 0$ (i.e., when there is no operation T after the S) this result also takes effect, because it coincides with the row $(***)$.

What we still need in this case $k_1 \geq 1$ is

LEMMA 1. $A_j^{+1} + (1 + k_0)\bar{A}_j^{+1} = A_{j+1}$ *for* $j = 0, 1, \dots, s, s+1$.

Proof by induction on j. The beginning of induction, for $j = 0$ and $j = 1$, is immediate. Indeed, $A_0^{+1} + (1 + k_0)\bar{A}_0^{+1} = 1 + (1 + k_0) \cdot 1 = A_1$ and

$$A_1^{+1}+(1+k_0)\bar{A}_1^{+1} = 2+k_1+(1+k_0)(2+k_1-1) = 1+(2+k_0)(1+k_1) = A_0+(1+k_1)A_1 = A_2\,.$$

The induction step, in the form $(j - 1, j) \Rightarrow j + 1$ for any fixed j, $1 \leq j \leq s$, is only a bit longer. From the definition of the A sequences

$$A_{j+1}^{+1} = A_{j-1}^{+1} + (1 + k_{j+1})A_j^{+1}, \tag{3.4}$$

$$\bar{A}_{j+1}^{+1} = \bar{A}_{j-1}^{+1} + (1 + k_{j+1})\bar{A}_j^{+1}. \tag{3.5}$$

Now we check if the lemma's statement holds for $j + 1$:

$$\begin{aligned}
A_{j+1}^{+1}+(1 + k_0)\bar{A}_{j+1}^{+1} &\overset{\text{by (3.4) and (3.5)}}{=} A_{j-1}^{+1}+(1 + k_{j+1})A_j^{+1}+(1 + k_0)\big(\bar{A}_{j-1}^{+1}+(1 + k_{j+1})\bar{A}_j^{+1}\big)\\
&= A_{j-1}^{+1} + (1 + k_0)\bar{A}_{j-1}^{+1} + (1 + k_{j+1})\big(A_j^{+1} + (1 + k_0)\bar{A}_j^{+1}\big)\\
&\overset{\text{by ind. ass.}}{=} A_j + (1 + k_{j+1})A_{j+1} = A_{j+2}\,.
\end{aligned}$$

Taking into account this lemma, after the operations S and $k_0 \geq 0$ times T we have the result row in a clearer form

- $2 + k_0$ times A_0, then
- $1 + k_1$ times A_1, then
- $1 + k_2$ times A_2, then
- $\dots \quad \dots \quad \dots \quad \dots$
- $1 + k_{s+1}$ times A_{s+1}, and then
- $l_{s+2} - 1$ times A_{s+2}.

This is the vector $\mathrm{der}\big(\mathbf{W}\mathrm{ST}_{k_1}\mathrm{ST}_{k_0}\big)$. To finish the case $k_1 \geq 1$, we remember that there still are $l_0 \geq 0$ letters G in the end of the code of $\mathcal{C}$. But the operation G is extremely simple: one just adds, from the left, one entry 1. Hence the eventual vector $\mathrm{der}\big(\mathbf{W}\mathrm{ST}_{k_1}\mathrm{ST}_{k_0}\mathrm{G}_{l_0}\big)$ has just l_0 entries 1 $(= A_0)$ more in the beginning. That is, it consists of

- $2 + k_0 + l_0$ times A_0, then
- $1 + k_1$ times A_1, then
- $1 + k_2$ times A_2, then
- $\dots \quad \dots \quad \dots \quad \dots$
- $1 + k_{s+1}$ times A_{s+1}, and then
- $l_{s+2} - 1$ times A_{s+2}.

Case $k_1 \geq 1$ [in the inductive proof of Theorem 2] is now proved.

3.2. Case $k_1 = 0$. Now the two truncated words are just of the form **WS** and **W**, where **W** is an admissible word with s letters S.
The word **WS** is admissible, too, and has $s+1$ letters S. So, by the inductive assumption, the vector der(**WS**) is non-decreasing and features, with due multiplicities precised in the inductive assumption, the numbers, written in the growing order:

$$A_0^{+1},\ A_1^{+1},\ \ldots,\ A_s^{+1},\ A_{s+1}^{+1},$$

now with an important restriction $k_1 = 0$, hence $A_1^{+1} = 2 + k_1 = 2$ etc. The word **W** having s letters S, by the same assumption, the vector der(**W**) is non-decreasing and features the numbers, in the growing order (with their proper multiplicities also precised in the inductive assumption),

$$A_0^{+2},\ A_1^{+2},\ \ldots,\ A_s^{+2}.$$

These are our construction bricks this time. We first have to compute the derived vector $\mathrm{S}\big(\mathrm{der}(\mathbf{W}), \mathrm{der}(\mathbf{WS})\big)$. To this end we expand der(**W**) in a row, called $(*')$, consisting of

- $2 + k_2$ times A_0^{+2}, then
- $1 + k_3$ times A_1^{+2}, then
- $\ldots\quad\ldots\quad\ldots\quad\ldots$
- $1 + k_{s+1}$ times A_{s-1}^{+2}, and, eventually,
- $l_{s+2} - 1$ times A_s^{+2}.

(Clearly, when $s = 0$, then only the last group is present here and der(**W**) consists of $l_2 - 1$ values $A_0^{+2} = 1$.)
And we also expand der(**WS**), in a row called $(**')$, only slightly differing from the row $(**)$ in the previous case. The difference resides in the current specification $k_1 = 0$ (which generates, naturally, a different sequence of values A than when $k_1 > 0$):

- 2 times A_0^{+1}, then
- $1 + k_2$ times A_1^{+1}, then
- $\ldots\quad\ldots\quad\ldots\quad\ldots$
- $1 + k_{s+1}$ times A_s^{+1}, and, eventually,
- $l_{s+2} - 1$ times A_{s+1}^{+1}.

The continuation is, basically, already known. The row $(*')$ is written above row $(**')$, both identically indented on the left, and in the following row below them, called $(***')$ and identically indented on the left, we start with two 1's written in bold to put them in relief, then apply the Fibonacci-like rule from the NW to SE. Let us trace down this procedure in the initial parts of the rows: first $2 + k_2$ terms in $(*')$ and first $3 + k_2$ terms in $(**')$:

$$\begin{array}{rcccccccc}
(*') & 1 & A_0^{+2} & \ldots & & A_0^{+2} & & & \\
(**') & 1 & 1 & A_1^{+1} & & \ldots & & A_1^{+1} & \\
(***') & \mathbf{1} & \mathbf{1} & 2 & A_1^{+1} + A_0^{+2} & & \ldots & & A_1^{+1} + A_0^{+2}
\end{array}$$

(the first entry in $(*')$ is written as 1, and not A_0^{+2}). Given the above-listed multiplicities of different values in the rows $(*')$ and $(**')$, the pattern observed on the initial terms of both the arguments and result smoothly extends to the entire rows. The outcome row $(***') = \mathrm{S}\big((*'),\, (**')\big)$ consists of entries, in the order of their appearing:

- 2 times 1, then
- 1 time 2, then
- $1 + k_2$ times $A_1^{+1} + A_0^{+2}$, then
- $1 + k_3$ times $A_2^{+1} + A_1^{+2}$, then
- ... ...
- $1 + k_{s+1}$ times $A_s^{+1} + A_{s-1}^{+2}$, and then
- $l_{s+2} - 1$ times $A_{s+1}^{+1} + A_s^{+2}$.

Now we are to continue with the operation T performed k_0 times. Since T is a (skew-wards NW to SE) arithmetic-progression-like operation which is initially being applied after the operation S acting on the rows: $(*')$ as the first argument in S and $(**')$ as second argument, the arithmetic progressions obtained in the outcome of k_0 operations T are either constant and equal to 1, or have their *differences* in the row $(*')$ – just entries in that row. In fact, these are the sequences

- $1, 1, 1, \ldots, 1$ (altogether k_0 sequences of lengths $3, 4, \ldots, 2 + k_0$),
- $1, 2, 3, \ldots, 2 + k_0$ (1 time),
- $A_1^{+1}, A_1^{+1} + A_0^{+2}, A_1^{+1} + 2A_0^{+2}, \ldots, A_1^{+1} + (1 + k_0)A_0^{+2}$ ($1 + k_2$ times),
- $A_2^{+1}, A_2^{+1} + A_1^{+2}, A_2^{+1} + 2A_1^{+2}, \ldots, A_2^{+1} + (1 + k_0)A_1^{+2}$ ($1 + k_3$ times),
- ... ...
- $A_s^{+1}, A_s^{+1} + A_{s-1}^{+2}, A_1^{+1} + 2A_{s-1}^{+2}, \ldots, A_1^{+1} + (1 + k_0)A_{s-1}^{+2}$ ($1 + k_{s+1}$ times),
- $A_{s+1}^{+1}, A_{s+1}^{+1} + A_s^{+2}, A_{s+1}^{+1} + 2A_s^{+2}, \ldots, A_{s+1}^{+1} + (1 + k_0)A_s^{+2}$ ($l_{s+2} - 1$ times)

(except for the irregular sequences in the beginning, the arithmetic sequences here have length $2 + k_0$, and their first (resp., second) terms are always in the row $(**')$ (resp., $(***')$).

Summing up, the vector $\operatorname{der}\big(\mathbf{W}\mathrm{SST}_{k_0}\big)$ is non-decreasing and consists of

- $2 + k_0$ times $1 = A_0$, then
- 1 time $2 + k_0 = A_1$, then
- $1 + k_2$ times $A_1^{+1} + (1 + k_0)A_0^{+2}$, then
- $1 + k_3$ times $A_2^{+1} + (1 + k_0)A_1^{+2}$, then
- ... ...
- $1 + k_{s+1}$ times $A_s^{+1} + (1 + k_0)A_{s-1}^{+2}$, and eventually
- $l_{s+2} - 1$ times $A_{s+1}^{+1} + (1 + k_0)A_s^{+2}$.

To proceed, analogously as in the previous case $k_1 > 0$, we need

LEMMA 2. *When* $k_1 = 0$, *then* $A_j^{+1} + (1 + k_0)A_{j-1}^{+2} = A_{j+1}$ *for* $j = 1, 2, \ldots, s, s + 1$.

Attention. There is no use to directly compare this lemma with Lemma 1, because now we are under the restriction $k_1 = 0$. The two lemmas deal with *disjoint* families of integer sequences A.

Proof of lemma is inductive on j. For $j = 1$ it is quick, $A_1^{+1} + (1 + k_0)A_0^{+2} = 2 + k_1 + 1 + k_0 = 1 + 2 + k_0 = A_0 + A_1 = A_0 + (1 + k_1)A_1 = A_2$.
For $j = 2$ it is a bit longer,

$$\begin{aligned} A_2^{+1} + (1 + k_0)A_1^{+2} &= 1 + (1 + k_2)A_1^{+1} + (1 + k_0)(2 + k_2) \\ &= 1 + 2(1 + k_2) + (1 + k_0)(1 + 1 + k_2) = 2 + k_0 + (1 + k_2)(1 + 2 + k_0) \\ &= A_1 + (1 + k_2)(A_0 + A_1) = A_1 + (1 + k_2)(A_0 + (1 + k_1)A_1) \\ &= A_1 + (1 + k_2)A_2 = A_3\,. \end{aligned}$$

As for the inductive step, again in the form $(j-1,\, j) \Rightarrow j+1$, with, this time, $j \in \{2, 3, \ldots, s\}$, we will nearly mimick the proof of Lemma 1. We will again make use of identity (3.4) and of a similar identity

$$A_j^{+2} = A_{j-2}^{+2} + (1 + k_{j+1})A_{j-1}^{+2} \tag{3.6}$$

which follows from the definition of the A sequences as well.

Let us check now if the lemma's statement holds for $j+1$:

$$\begin{aligned} A_{j+1}^{+1} + (1+k_0)A_j^{+2} &\overset{\text{by (3.4) and (3.6)}}{=} A_{j-1}^{+1} + (1+k_{j+1})A_j^{+1} + (1+k_0)\big(A_{j-2}^{+2} + (1+k_{j+1})A_{j-1}^{+2}\big) \\ &= A_{j-1}^{+1} + (1+k_0)A_{j-2}^{+2} + (1+k_{j+1})\big(A_j^{+1} + (1+k_0)A_{j-1}^{+2}\big) \\ &\overset{\text{by ind. ass.}}{=} A_j + (1+k_{j+1})A_{j+1} = A_{j+2}\,. \end{aligned}$$

Lemma 2 is now proved by induction.

We are about to finish the proof of the case $k_1 = 0$ in the theorem. Indeed, Lemma 2 simplifies the expressions for the entries in the vector $\operatorname{der}\big(\mathbf{W}\mathrm{SST}_{k_0}\big)$, while the l_0 operations G that are still to be performed add l_0 entries 1 on the left. Hence the vector $\operatorname{der}\big(\mathbf{W}\mathrm{SST}_{k_0}\mathrm{G}_{l_0}\big)$, is non-decreasing and consists of

- $2 + k_0 + l_0$ times A_0, then
- $1 + k_1 = 1$ time A_1, then
- $1 + k_2$ times A_2, then
- $1 + k_3$ times A_3, then
- $\ldots \quad \ldots \quad \ldots$
- $1 + k_{s+1}$ times A_{s+1}, and eventually
- $l_{s+2} - 1$ times A_{s+2}.

Case $k_1 = 0$ [in the inductive proof of Theorem 2] is finished.

Theorem 2 is proved.

3.3. Reformulations of Theorem 2. In order to simplify our to-be-done work in proving Theorem 3 in Section 4, let us slightly reformulate Theorem 2, then draw a corollary from that reformulation, and then yet another, more far-reaching, corollary.

OBSERVATION 1. *In the notation from* Theorem 2, *for an admissible word with* $s+1$ *letters* S *and such that* $l_1 = l_2 = \cdots = l_s = 0$, *the sequence of operations*

$$\mathrm{S},\ \mathrm{T}^{k_s},\ \mathrm{S},\ \mathrm{T}^{k_{s-1}},\ \ldots,\ \mathrm{S},\ \mathrm{T}^{k_1},\ \mathrm{S},\ \mathrm{T}^{k_0},\ \mathrm{G}^{l_0}$$

started on the pair of arguments $\big(\underbrace{1,\, 1, \ldots,\, 1}_{l_{s+1}-1}\big)$ *and* $\big(\underbrace{1,\, 1, \ldots,\, 1}_{l_{s+1}}\big)$ *produces as the eventual result the vector described in the statement of* Theorem 2.

This is just a reformulation of Theorem 2, compare the beginning of its proof.

COROLLARY 1. *The sequence of operations*

$$\mathrm{S},\ \mathrm{T}^{k_{n-1}},\ \mathrm{S},\ \mathrm{T}^{k_{n-2}},\ \ldots,\ \mathrm{S},\ \mathrm{T}^{k_1},\ \mathrm{S},\ \mathrm{T}^{k_0},\ \mathrm{G}^{l_0}$$

(n is a natural integer not exceeding s) started on the pair of vector arguments $\big(\underbrace{w_0,\, w_0, \ldots,\, w_0}_{m_0}\big)$ *and* $\big(1,\, \underbrace{w_0,\, w_0, \ldots,\, w_0}_{m_0}\big)$, *where* w_0 *is a positive integer, produces in the result the vector with the entries, consecutively:*

- $2 + k_0 + l_0$ times A_0, then
- $1 + k_1$ times A_1, then
- ... ...
- $1 + k_{n-1}$ times A_{n-1}, and eventually
- m_0 times $A_n w_0$.

This corollary is only slightly more general than Observation 1 (instead of s there is $n-1$ now, m_0 instead of $l_{s+1} - 1$, and w_0 instead of 1) and its method of proof is the same as that in Theorem 2.

COROLLARY 2. *The same sequence of operations as in* Corollary 1, *when started on the pair of longer vector arguments*

$$\big(\underbrace{w_0,\, w_0, \dots,\, w_0}_{m_0},\, \underbrace{w_1,\, w_1, \dots,\, w_1}_{m_1},\, \dots,\, \underbrace{w_N,\, w_N, \dots,\, w_N}_{m_N}\big)$$

and

$$\big(1,\, \underbrace{w_0,\, w_0, \dots,\, w_0}_{m_0},\, \underbrace{w_1,\, w_1, \dots,\, w_1}_{m_1},\, \dots,\, \underbrace{w_N,\, w_N, \dots,\, w_N}_{m_N}\big)\,,$$

where $w_0,\, w_1, \dots,\, w_N$ *are positive integers, produces in the result the vector with the entries, consecutively:*

- $2 + k_0 + l_0$ times A_0, then
- $1 + k_1$ times A_1, then
- ... ...
- $1 + k_{n-1}$ times A_{n-1}, then
- m_0 times $A_n w_0$, then
- m_1 times $A_n w_1$, then
- ... ...
- m_N times $A_n w_N$.

This is a direct consequence of Corollary 1. Indeed, replacing in that corollary w_0 and m_0 by w_i and m_i, respectively, $i \in \{1,\, 2, \dots,\, N\}$, and then *concatenating* the vector results for $i = 1,\, 2, \dots,\, N$ with the initial vector result in Corollary 1, we get the statement in Corollary 2.

4. Proof of Theorem 3

The proof of Theorem 3 will be by induction on $q \in \{0,\, 1, \dots,\, s\}$ in all the pairs of integers $(q,\, s)$, $0 \le q \le s$. The size of $s \ge q$ will not play any role in the arguments.

Before doing this we want to formulate and prove a statement that would greatly simplify the inductive step in that announced induction. Note that the notation used in Proposition 1 below is fully compatible with that of Corollary 2. It is done purposefully – in order for the reader to see that it is basically Corollary 2 again, only given in other terms.

PROPOSITION 1. *Let the code of a geometric class* $\mathcal{C}$ *have the form*

$$\mathbf{W}\mathrm{GST}_{k_{n-1}}\mathrm{ST}_{k_{n-2}} \dots \mathrm{ST}_{k_1}\mathrm{ST}_{k_0}\mathrm{G}_{l_0}\,,$$

$\mathbf{W}$ – *the code of an ancestor of* $\mathcal{C}$, *ending on* S *or* T, *whose derived vector* $\mathrm{der}(\mathbf{W})$ *is non-decreasing and has the different values* $1 = w_0 < w_1 < \cdots < w_N$ *appearing in it with multiplicities* $m_0,\, m_1, \dots,\, m_N$, *respectively. Then* $\mathrm{der}(\mathcal{C})$ *is non-decreasing as well and consists of*

- $2 + k_0 + l_0$ times A_0, then
- $1 + k_1$ times A_1, then
- $\dots \quad \dots \quad \dots \quad \dots$
- $1 + k_{n-1}$ times A_{n-1}, then
- m_0 times A_n $(= A_n w_0)$, then
- m_1 times $A_n w_1$, then
- $\dots \quad \dots \quad \dots$
- m_N times $A_n w_N$.

Proof of Proposition 1. It is a direct consequence of Corollary 2, because the derived vector of $\mathbf{W}$ can be taken as the first vector argument in that corollary, upon which the derived vector of $\mathbf{W}$G becomes the second vector argument in that corollary. (The operation G adds a single 1 on the left, while the particular value $w_0 = 1$ in the proposition does not interfere with this action.)

The beginning of induction for $q = 0$ **in Theorem 3** is already done – it is Theorem 2, proved already for all $s \geq 0$.

The inductive step '$q - 1 \Rightarrow q$' **in Theorem 3** for any fixed $q \geq 1$ and arbitrary $s \geq q$, assuming the due information for $q - 1$ and all $s' \geq q - 1$.

Let us focus our attention on the one before last from the left (or last, if $l_0 = 0$) segment of G's in the word of $\mathcal{C}$ and truncate this word, calling $\mathbf{W}$ the result of truncation, *before* the last letter G in that segment. That is, take

$$\mathbf{W} = \mathrm{G}_{l_{s+1}} \mathrm{S}\,\mathrm{T}_{k_s} \dots \mathrm{S}\,\mathrm{T}_{k_{n_1}} \mathrm{G}_{l_{n_1}-1}$$

where $l_{n_1} - 1 \geq 0$ (the value 0 is not excluded). Then the vector der($\mathbf{W}$) is known from the inductive assumption, for there are only $q - 1$ positive values among the parameters $l_{n_1+1}, l_{n_1+2}, \dots, l_s$ related to the word $\mathbf{W}$, and $s' = s - n_1 \geq q - 1$. (Indeed, from our construction, $s - n_1 \geq n_q - n_1 > n_{q-1} - n_1 > \cdots > n_2 - n_1 \geq 1$.) So what is precisely der($\mathbf{W}$)?

The inductive assumption *formally* would have yielded a description in terms of the set of letters (as if a formal alphabet)

$$\{n_1, n_2, \dots, n_{q-1}, s - n_1, k_0, l_0, \dots, k_{s-n_1}, l_{s-n_1}, l_{s-n_1+1}\},$$

because of the relevant bare cardinalities related to the word $\mathbf{W}$. Yet, in the concrete case of $\mathbf{W}$, all *indices* in the parameters k and l are naturally raised by $+n_1$, ranging from n_1 to $s + 1$ (and not from 0 to $s - n_1 + 1$). And the *distances* in indices are $n_2 - n_1, n_3 - n_1, \dots, n_q - n_1$ instead of $n_1 - 0, n_2 - 0, \dots, n_{q-1} - 0$, respectively (the distances in indices in $\mathbf{W}$ are now counted from n_1, not from 0).

Taking into account all these specific features, the data about $\mathbf{W}$ which follow from the inductive assumption, and are about to be plugged into Proposition 1, read:

In the vector der($\mathbf{W}$) there appear $s - n_1 + 2$ different values $1 = w_0 < w_1 < w_2 < \cdots < w_{s-n_1+1}$, listed below in the growing order, in the following $(q-1)+1 = q$

separate rows:

$$
\begin{gathered}
A_0^{+n_1},\ A_1^{+n_1},\ \dots,\ A_{n_2-n_1-1}^{+n_1}\,;\\
A_{n_2-n_1}^{+n_1}\big(A_0^{+n_2},\ A_1^{+n_2},\ \dots,\ A_{n_3-n_2-1}^{+n_2}\big)\,;\\
\dots\quad \dots\quad \dots\quad \dots\quad \dots\\
\prod_{j=1}^{q-2} A_{n_{j+1}-n_j}^{+n_j}\left(A_0^{+n_{q-1}},\ A_1^{+n_{q-1}},\ \dots,\ A_{n_q-n_{q-1}-1}^{+n_{q-1}}\right);\\
\prod_{j=1}^{q-1} A_{n_{j+1}-n_j}^{+n_j}\left(A_0^{+n_q},\ A_1^{+n_q},\ \dots,\ A_{s-n_q+1}^{+n_q}\right).
\end{gathered}
$$

Moreover, der(**W**) is non-decreasing and the multiplicities of its above-listed different values are as follows.

- $w_0 = A_0^{+n_1}$ appears $m_0 = 2 + k_{n_1} + (l_{n_1} - 1) = 1 + k_{n_1} + l_{n_1}$ times in row (remember that the last group of G's in **W** consists, by construction, of $l_{n_1} - 1$ letters);
- for $1 \le j \le s - n_1$, w_j appears $m_j = 1 + k_{n_1+j} + l_{n_1+j}$ times in row (that is, $1 + k_{n_1+j}$ times, when $j \notin \{n_2 - n_1,\ n_3 - n_1,\ \dots,\ n_q - n_1\}$);
- the last and biggest w_{s-n_1+1} appears $m_{s-n_1+1} = l_{s+1} - 1$ times in row.

We are about to apply Proposition 1 to the code of the class $\mathcal{C}$ under the induction step procedure and to its truncation **W**, for $n = n_1$, $N = s - n_1 + 1$, the different values w_i in der(**W**) and their multiplicities m_i ($i = 0,\ 1, \dots,\ s-n_1+1$), as specified above.

So Proposition 1 says that the table of different values in der($\mathcal{C}$) is being obtained by just multiplying by A_{n_1} all the entries in the table of q rows for der(**W**) above *and* by inserting on top of them the single row $A_0,\ A_1, \dots,\ A_{n_1-1}$. That is, the table of different values showing up in the vector der($\mathcal{C}$) is of the form

$$
\begin{gathered}
A_0,\ A_1,\ \dots,\ A_{n_1-1}\,;\\
A_{n_1}\big(A_0^{+n_1},\ A_1^{+n_1},\ \dots,\ A_{n_2-n_1-1}^{+n_1}\big)\,;\\
A_{n_1}A_{n_2-n_1}^{+n_1}\big(A_0^{+n_2},\ A_1^{+n_2},\ \dots,\ A_{n_3-n_2-1}^{+n_2}\big)\,;\\
\dots\quad \dots\quad \dots\quad \dots\quad \dots\\
A_{n_1}\prod_{j=1}^{q-2} A_{n_{j+1}-n_j}^{+n_j}\left(A_0^{+n_{q-1}},\ A_1^{+n_{q-1}},\ \dots,\ A_{n_q-n_{q-1}-1}^{+n_{q-1}}\right);\\
A_{n_1}\prod_{j=1}^{q-1} A_{n_{j+1}-n_j}^{+n_j}\left(A_0^{+n_q},\ A_1^{+n_q},\ \dots,\ A_{s-n_q+1}^{+n_q}\right),
\end{gathered}
$$

precisely as expected. Moreover, also by Propositiopn 1, the vector der($\mathcal{C}$) is non-decreasing and the multiplicities of its listed different values which emerge from Proposition 1 are identical (recall that $n = n_1$) with those in the statement of Theorem 3. Theorem's statements for q positive parameters among $\{l_1,\ l_2, \dots,\ l_s\}$ are now justified and the step of induction is completed.

Theorem 3 is now proved.

Afterword. In the contribution [**4**], among many an issue raised, there is a far-reaching Question 9.19, consisting of parts 1, 2, and 3.

Part 1, especially when understood *sensu largo* [is it possible to produce Jean's recurrences in pure Goursat monster terms?], is a true challenge.

Part 2 is elementary and covered by the present work (which, as a matter of record, completes [**6**]).

Part 3 is an invitation to research, for the mapping Pc (*Puiseux characteristic* being associated to a geometric class = a word over $\{R, V, T\}$) is explicit in [**4**], while the mapping denoted by GW in [**4**] ([small] growth vector being associated to a geometric class = a word over $\{G, S, T\}$) is recursive in [**3**], and explicit in [**6**] and in the present contribution.

Added in proof. A solution to Part 3 has been recently given by Shanbrom in [**7**]. Using the formulae from [**6**], he neatly describes the mapping $\mathrm{Pc} \circ \mathrm{GW}^{-1}$ attaching the Puiseux characteristic to any given small growth vector in the Goursat monster.

References

[1] E. Cartan, *Sur l'équivalence absolue de certains systèmes d'équations différentielles et sur certaines familles de courbes.* Bull. Soc. Math. France **XLII** (1914), 12–48. MR1504722

[2] M. Cheaito and P. Mormul, *Rank–2 distributions satisfying the Goursat condition: all their local models in dimension* 7 *and* 8. ESAIM: Control, Optimisation and Calculus of Variations **4** (1999), 137–158 (electronic). MR1816509 (2001j:58003)

[3] F. Jean, *The car with N trailers: characterisation of the singular configurations.* ESAIM: Control, Optimisation and Calculus of Variations **1** (1996), 241–266 (electronic). MR1411581 (97j:70008)

[4] R. Montgomery and M. Zhitomirskii, *Points and Curves in the Monster Tower.* Memoirs of the AMS **956** (2010). MR2599043 (2011d:58009)

[5] P. Mormul, *Local classification of rank–2 distributions satisfying the Goursat condition in dimension* 9. In: P. Orro and F. Pelletier (Eds), Singularités et géométrie sous-riemannienne, Chambéry 1997; Travaux en cours vol. 62, Hermann, Paris 2000, pp. 89–119.

[6] P. Mormul, *Geometric classes of Goursat flags and the arithmetics of their encoding by small growth vectors.* Central European J. Math. **2** (2004), 859–883. MR2172057 (2006j:58004)

[7] C. Shanbrom, *The Puiseux characteristic of a small growth vector,* arXiv:1103.2999v2 [math.DG].

[8] E. von Weber, *Zur Invariantentheorie der Systeme Pfaff'scher Gleichungen.* Berichte Ges. Leipzig, Math–Phys. Classe **L** (1898), 207–229.

Institute of Mathematics, University of Warsaw, Banach str. 2, 02-097 Warsaw, Poland.

Current address: Mathematical Institute, Polish Academy of Sciences, Śniadeckich str. 8, 00-956 Warsaw, Poland.

E-mail address: mormul@mimuw.edu.pl

Contemporary Mathematics
Volume **569**, 2012
http://dx.doi.org/10.1090/conm/569/11246

Vassiliev type invariants for generic mappings, revisited

Toru Ohmoto

ABSTRACT. We discuss Vassiliev-type invariants for isotopy classes of generic maps $M^m \to \mathbb{R}^n$ $(m, n \geq 2)$ using a kind of infinite dimensional Alexander duality. First we describe a general framework based on Mather's stratification of the space $C^\infty(M, \mathbb{R}^n)$ of smooth mappings, and then deal with *finite type invariants* for generic maps in an axiomatic way. Also we mention about a relation with characteristic classes of manifold-bundles with fiber M as a sort of degeneracy loci problem.

Introduction

In this short note, we revisit an 'old and new' theme on *the topology of the discriminant hypersurface* Γ *in the infinite dimensional space* $\mathcal{M}$ *of smooth mappings from an* m*-dimensional manifold* M *to Euclidean space*, initiated by R. Thom [**27, 28**],

$$\mathcal{M} := C^\infty(M, \mathbb{R}^n) \supset \Gamma := \{C^\infty \text{ unstable maps}\}.$$

In particular we are interested in

(A) $H^0(\mathcal{M} - \Gamma)$ as the space of all isotopy invariants of generic maps;
(B) the space $\mathcal{M}$ as a linear representation of the group $\mathrm{Diff}\, M$;
(C) the space of all invariants of smooth oriented closed d-manifolds belonging to a fixed cobordism class.

The topic (A) leads us to the theory of **Vassiliev-type invariants for generic maps** [**30, 3, 6, 7, 20, 31, 22, 5, 9, 13, 14**]. We focus on an elementary common feature of the theory in general dimensions m, n of the source and target manifolds greater than or equal to 2. That would be much different from the theory of knot invariants. For instance, according to the $\mathcal{A}$-classification theory of map-germs, *first order invariants* of generic maps $M^m \to \mathbb{R}^n$ $(m, n \geq 2)$ can be quite rich, while the first order invariant for knots $S^1 \to \mathbb{R}^3$ is trivial.

First we overview a general framework based on an invariant stratification of the mapping space $\mathcal{M}$ due to J. Mather [**18**] and a characteristic spectral sequence due to V. A. Vassiliev [**29**]. Second, in case of $m, n \geq 2$, *naïve finite type invariants for generic maps* are introduced in an axiomatic way similar to the case of knots (Definition 3.2). We then show that such naïve finite type invariants are basically reduced to polynomials of first order invariants (Theorem 3.7). In some particular

2010 *Mathematics Subject Classification.* Primary 57R45, 55R40; Secondary 55R80.
Key words and phrases. Singularity Theory, Vassiliev invariants, characteristic classes.
This was supported in part by JSPS grant No.21540057.

dimensions (e.g. in case of generic immersions of a surface into $\mathbb{R}^4$), similar results have been reported in [**14, 13, 9**], but the common essential reason has not been quite clear so far. Our purpose is to figure it out in most general form.

The main point is as follows: (i) The *transverse k-th self-intersection locus* of Γ consists of irreducible components $\Gamma(\boldsymbol{a})$ labeled by k-tuples $\boldsymbol{a} = (\alpha_j)$ of local singularity types of generic 1-parameter families of maps, and each $\Gamma(\boldsymbol{a})$ turns to be *connected but not simply-connected* in general; (ii) the value of the jump $\nabla^k v$ of any invariant v of order k vanishes on *coherence relations for local invariants*, i.e., the relations determined by local data of adjacencies around singularity types of codimension 2. Contrarily, in case of knots [**30**], (i) the k-th self-intersection locus consists of many contractible components labeled by different 'k-chord diagrams', and (ii) the value of the jump $\nabla^k v$ vanishes on the so-called 1- *and* 4-*term relations*, which are in fact the coherence relations associated to adjacencies to cusps and triple points, respectively. That is to say, in our case, the complexity of combinatorics among local singularity types much increases, while the complexity of chord diagrams (=configurations of unstable singular points) is much simplified, because such a configuration space on M is connected. The lost information must be hidden beyond the first cohomology groups $H^1(\Gamma(\boldsymbol{a}))$, non-transverse self-intersection loci, some combinatorial data on the global configuration of critical point sets, ... etc, which may suggest further studies on true finite type invariants for generic maps.

As for (B) we describe an equivariant version of some tools used in (A) in order to obtain a geometric presentation of **characteristic classes of manifold-bundles** with fiber M and group Diff(M), based on Kazarian [**15, 16, 17**]. We will briefly discuss a few examples and proposals.

Finally, we remark about a counterpart under **global contact equivalence** (C) which has already been suggested in Thom [**27**]. For instance, consider $\mathcal{M}_0 = C^\infty(S^{n+d}, S^n)_{base}$ $(n \gg 0)$ and the discriminant $\mathcal{D}$ consisting of maps having critical value at a fixed point 0. Then $H^0(\mathcal{M}_0 - \mathcal{D})$ may be regarded as the space of invariants of null-cobordant smooth closed manifolds of dimension d. In case of $d = 1$ Saeki [**23**] has worked out some details on the classification of singular fibers. In case of $d = 3$, Otsuki's theory of finite type invariants [**21**] for $\mathbb{Z}$-homology 3-spheres is the case when restricted to an open set of $\mathcal{M}_0 - \mathcal{D}$ so that $f^{-1}(0)$ have the fixed Betti numbers. There still remain further researches in this area.

All spaces and mappings are of class C^∞ throughout this paper.

The author would like to thank the organizers of the 11th International Workshop on Real and Complex Singularities at ICMC-USP, Sao Carlos in 2010. In fact this work was prepared for the conference.

1. Mapping space and Discriminant

1.1. $\mathcal{A}$-equivalence and invariant stratification. Let M be a compact m-dimensional manifold without boundary, and N an n-dimensional manifold without boundary. Denote the space of smooth maps, equipped with C^∞ topology, by $\mathcal{M} := C^\infty(M, N)$.

We say a map $\varphi : U \to \mathcal{M}$ from a finite dimensional manifold to $\mathcal{M}$ is *smooth* if the evaluation map $M \times U \to N$ is a C^∞ map. This gives a Frechet manifold structure on $\mathcal{M}$.

The $\mathcal{A}$-*equivalence group* or the *right-left group* is the direct product of diffeomorphism groups, $\mathcal{A}_{M,N} := \mathrm{Diff}(M) \times \mathrm{Diff}(N)$, which acts on $\mathcal{M}$ by $(\varphi, \tau).f := \tau \circ f \circ \varphi^{-1}$. Put $\mathcal{A}^0_{M,N}$ to be the connected component containing (id_M, id_N).

DEFINITION 1.1. C^∞ maps $f, g : M \to N$ are $\mathcal{A}$-*equivalent* if $\mathcal{A}_{M,N}.f = \mathcal{A}_{M,N}.g$, i.e., there exists $(\varphi, \tau) \in \mathcal{A}_{M,N}$ so that the following diagram commutes:

$$\begin{array}{ccc} M & \xrightarrow{\ f\ } & N \\ {\scriptstyle \varphi}\downarrow \simeq & & \simeq \downarrow{\scriptstyle \tau} \\ M & \xrightarrow{\ g\ } & N \end{array}$$

Further, f, g are C^∞-*isotopic* if $\mathcal{A}^0_{M,N}.f = \mathcal{A}^0_{M,N}.g$.

DEFINITION 1.2. We say f is a C^∞-*structurally stable map* if the orbit $\mathcal{A}_{M,N}.f$ is an open set in $\mathcal{M}$. If the pair of dimension (m, n) is in so-called *Mather's nice range*, C^∞-structurally stable maps form a residual open subset in $\mathcal{M}$, that is known as the C^∞-*Structural Stability Theorem* proved by J. Mather. In this paper we often call such f by a *generic map* for short.

We are interested in $\mathcal{A}_{M,N}$-orbits (or families of orbits) of finite codimension in $\mathcal{M}$, which are invariant Frechet submanifold of $\mathcal{M}$. By definition, f is C^∞-stable if and only if codim $\mathcal{A}_{M,N}.f = 0$. Put

$$\Gamma_\infty := \{\, f \in \mathcal{M} \mid \mathrm{codim}\, \mathcal{A}_{M,N}.f = \infty \,\}, \quad \mathcal{U}_0 := \mathcal{M} - \Gamma_\infty.$$

An $\mathcal{A}_{M,N}$-invariant subset K of $\mathcal{M}$ is called to be *pseudo-algebraic* if for any $f \in K \cap \mathcal{U}_0$ and for any smooth map $\varphi : U \to \mathcal{M}$ with $\varphi(u_0) = f$ which is transverse to the $\mathcal{A}_{M,N}$-orbit of f at u_0, the preimage $\varphi^{-1}(K) \subset U$ is a semi-algebraic subset in some local coordinate centered at u_0. A pseudo-algebraic set K is said to be of codimension s if the semi-algebraic subset $\varphi^{-1}(K)$ is of codimension s around u_0.

THEOREM 1.3 (Mather [**18**]). *Assume that (m, n) belongs to the nice range. Then Γ_∞ has infinite codimension, and there exists a filtration*

$$\mathcal{M} \supset \Gamma := \Gamma_1 \supset \Gamma_2 \supset \cdots \supset \Gamma_s \supset \cdots \supset \Gamma_\infty,$$

by $\mathcal{A}_{M,N}$-invariant closed pseudo-algebraic subsets Γ_s of codimension s such that it admits a topologically locally trivial fibration $\pi_s : \Gamma_s - \Gamma_{s+1} \to Y_s$ so that each fiber is an $\mathcal{A}_{M,N}$-orbit and Y_s is a finite dimensional manifold.

REMARK 1.4. 1) In Theorem 1.3, $\Gamma_s - \Gamma_{s+1}$ is a disjoint union of single $\mathcal{A}_{M,N}$-orbits of codimension s or moduli strata of $\mathcal{A}_{M,N}$-orbits whose base (parameter) spaces are components of Y_s, see §1.2 below.
2) By the definition, the complement $\mathcal{M} - \Gamma$ consists of all stable maps. An isotopy invariant of stable maps is a locally constant function over $\mathcal{M} - \Gamma$, thus we regard the 0-th singular cohomology group $H^0(\mathcal{M} - \Gamma)$ as the space of all isotopy invariants of stable maps.
3) The ranks of $H^0(\mathcal{M} - \Gamma)$ and of $H^0(\Gamma_s - \Gamma_{s+1})$ are at most countable.

1.2. Multi-singularity types. We outline the construction of the $\mathcal{A}_{M,N}$-invariant filtration in Theorem 1.3.

A *multi-germ* is a map-germ $\varphi : M, S \to N, p$ at finite points S mapped to a single point $p = \varphi(S)$ (including a *mono*-germ as the case when S consists of a single point). We denote by $\alpha, \beta, \cdots$, $\mathcal{A}$-equivalent classes (or $\mathcal{A}$-moduli families) of multi-germs, and also denote by $\boldsymbol{a} = (\alpha_1, \cdots, \alpha_l)$ an l-tuple of those $\mathcal{A}$-classes of multi-germs. The *codimension* $|\boldsymbol{a}|$ is given by the sum of $\mathcal{A}_e$-codimension of α_j. Note that α is a stable (multi-)singularity if and only if its $\mathcal{A}_e$-codimension $= 0$.

We say $f : M \to N$ *has multi-singularities of type* $\boldsymbol{a}$ (at S) if there are mutually disjoint finite subsets $S_1, \cdots, S_l$ in M ($S = \coprod S_j$) and l distinct points $p_1, \cdots, p_l \in N$ so that $f(S_j) = p_j$ and the germ $f : M, S_j \to N, p_j$ is of type α_j for each j.

For a multi-singularity type α of codimension ≥ 1, we put

$$\Gamma(\alpha) := \mathrm{Cl}\{\, f \in \Gamma \mid f \text{ has a multi-singularity of type } \alpha \,\}$$

(Cl stands for the closure), and for a l-tuple $\boldsymbol{a} = (\alpha_1, \cdots, \alpha_l)$ with $|\alpha_j| \geq 1$

$$\Gamma(\boldsymbol{a}) := \mathrm{Cl}\{\, f \in \Gamma \mid f \text{ has multi-singularities of type } \boldsymbol{a} \,\} = \bigcap_{j=1}^{l} \Gamma(\alpha_j).$$

The locus Γ_s in Theorem 1.3 is given by

$$\Gamma_s := \bigcup \Gamma(\boldsymbol{a})$$

taken over all multi-singularity types $\boldsymbol{a}$ of codimension $\geq s$. In particular we obtain a stratification

$$\Gamma_s - \Gamma_{s+1} = \bigsqcup_{|\boldsymbol{a}|=s} (\Gamma(\boldsymbol{a}) - \Gamma_{s+1}).$$

Note that

- $\Gamma(\boldsymbol{a})$ is a closed pseudo-algebraic subset of codimension s;
- $f \in \Gamma(\boldsymbol{a}) - \Gamma_{s+1}$ if and only if f has multi-singularities of type $\boldsymbol{a}$ at some S, any point in $f^{-1}(f(S)) - S$ is not critical, and f is infinitesimally C^∞-stable at any finite points in $M - S$;
- $\Gamma(\boldsymbol{a}) - \Gamma_{s+1}$ is a Frechet submanifold of codimension s in $\mathcal{M}$ having possibly countably many connected components: Each component is a single $\mathcal{A}^0_{M,N}$-orbit (if $\boldsymbol{a}$ consists of simple singularity types), or a moduli family of orbits (if $\boldsymbol{a}$ contains a moduli singularity type);
- $\Gamma(\boldsymbol{a}) - \Gamma_{s+1}$ is the transverse intersection of submanifolds $\Gamma(\alpha_j) - \Gamma_{|\alpha_j|+1}$ in $\mathcal{M}$.

The filtration in Theorem 1.3 is obtained by the above definitions and properties. A technical detail in the proof belongs to the geometry of real algebraic group action [**18**].

DEFINITION 1.5. We say a multi-singularity type α is *coorientable* if the normal bundle of $\Gamma(\alpha) - \Gamma_{|\alpha|+1}$ is orientable. Also $\boldsymbol{a} = (\alpha_1, \cdots, \alpha_l)$ is coorientable if each α_j is so.

DEFINITION 1.6. We define the *s-th transverse self-intersection locus of the discriminant hypersurface* Γ to be a union of $\Gamma(\boldsymbol{a})$ of all s-tuples $\boldsymbol{a} = (\alpha_1, \cdots, \alpha_s)$ with $|\alpha_j| = 1$ ($\forall j$); The rest is the *non-transverse* self-intersection locus of Γ, that is a union of $\Gamma(\boldsymbol{a})$ for $\boldsymbol{a}$ where some α_j has codimension greater than one.

EXAMPLE 1.7. Any $f \in \Gamma_2 - \Gamma_3$ has either
- only a couple of multi-singularities of codim $= 1$ with distinct two critical values (i.e., f belongs to the transverse double point locus of Γ); or
- only one multi-singularity of codim $= 2$ with a single critical value (i.e., f belongs to the non-transverse intersection locus of Γ).

For instance, if $m = 1$ and $n = 3$, a map in the former case is an immersion with two double points (two crossing changes), and a map in the latter case is an embedding except for one ordinary cusp point.

2. Vassiliev complex

2.1. Characteristic spectral sequence. Let us think of the situation as in Theorem 1.3. Put

$$\mathcal{U}_s := \mathcal{M} - \Gamma_{s+1},$$

then we have invariant open subsets of $\mathcal{M}$:

$$\mathcal{M} - \Gamma = \mathcal{U}_0 \subset \mathcal{U}_1 \subset \mathcal{U}_2 \subset \cdots \subset \mathcal{U}_s \subset \cdots \subset \mathcal{U}_\infty \subset \mathcal{M}.$$

The cohomology spectral sequence associated to this filtration is called the *Vassiliev-Kazarian spectral sequence for* $\mathcal{M}$ [**15, 29, 30**]. The E_1-term is

$$E_1^{s,t} := H^{s+t}(\mathcal{U}_s, \mathcal{U}_{s-1}; R) \xleftarrow{\simeq} H^t(\Gamma_s - \Gamma_{s+1}; R)$$

with a coefficient ring R. The arrow indicates the *Alexander duality for functional spaces* in the sense of Eells [**8**], that is, the Thom isomorphism for *coorientable* components of the s-codimensional Frechet manifold $\Gamma_s - \Gamma_{s+1}$ (for a non-coorientable connected component the Thom class within integer coefficients vanishes, but it works within $\mathbb{Z}_2$-coefficients). Thus $E_1^{s,0}$ is the R-module generated by coorientable connected components in $\Gamma_s - \Gamma_{s+1}$, especially,

$$E_1^{0,0} = H^0(\mathcal{U}_0) = H^0(\mathcal{M} - \Gamma).$$

The first cochain complex is

$$0 \to E_1^{0,t} \to E_1^{1,t} \to E_1^{2,t} \to \cdots$$

where the operator $d_1 : E_1^{s,t} \to E_1^{s+1,t}$ is the connection homomorphism ∂ of cohomology exact sequence for the triple $(\mathcal{U}_{s+1}, \mathcal{U}_s, \mathcal{U}_{s-1})$. As usual, we put for $r \geq 1$,

$$E_{r+1}^{s,t} := \frac{\ker[d_r : E_r^{s,t} \to E_r^{s+r,t-r+1}]}{\operatorname{Im}[d_r : E_r^{s-r,t+r-1} \to E_r^{s,t}]}$$

with d_{r+1}, and we have a natural homomorphism $E_\infty^{s,t} \to H^*(\mathcal{M})$.

Instead, we may take

$$0 \to E_1^{1,t} \to E_1^{2,t} \to E_1^{3,t} \to \cdots$$

and $E_r^{s,t}$ $(s \geq 1)$, that approximates $H^*(\mathcal{U}, \mathcal{U}_0) = H^*(\mathcal{M}, \mathcal{M} - \Gamma)$, the cohomology with support on Γ.

REMARK 2.1. The main technical invention in Vassiliev's original approach [**30**] is a *simplicial resolution* $\Gamma' \to \Gamma$ of the discriminant hypersurface, constructed in $\mathcal{M} \times \mathbb{R}^\infty$ in a purely combinatorial way. A natural filtration of Γ' produces a spectral sequence converging to a subspace in $H^0(\mathcal{M} - \Gamma)$. That spectral sequence is an entirely different one from the characteristic spectral sequence described above, so the readers should not confuse them. In this paper we wouldn't develop such a simplicial resolution technique, although it might be possible theoretically. Instead we use a more easier cochain subcomplex of our $E_1^{s,t}$ given by data of local singularity types and some additional semi-global data of configurations of singular points/values, [**29, 19, 23**]. A substitute to the 'E_1-term' A_* for the simplical resolution Γ' will be given in §3.3 by (the dual to) some cochain complex of singularities, cf. [**30, 2**].

2.2. Local Vassiliev complex of multi-singularities. Let $C^0(\mathcal{A}) = 0$ and for $s \geq 1$ let

$$C^s(\mathcal{A}) := \bigoplus R \cdot \boldsymbol{a}$$

the R-module generated by *coorientable* $\mathcal{A}$-classes $\boldsymbol{a}$ of multi-singularities of codimension s (for a technical detail, see [**29, 19**]). This is regarded as a submodule

$$C^s(\mathcal{A}) \subset E_1^{s,0} = H^0(\Gamma_s - \Gamma_{s+1}; R)$$

by identifying $\boldsymbol{a}$ with the constant function on $\Gamma(\boldsymbol{a}) - \Gamma_{s+1}$ (taking 1, otherwise 0). The coboundary operator $\partial : C^s(\mathcal{A}) \to C^{s+1}(\mathcal{A})$ is induced from d_1.

DEFINITION 2.2. The cochain complex $(C^*(\mathcal{A}), \partial)$ is called the *local Vassiliev complex for $\mathcal{A}$-classification of multi-singularities.*

The operator $\partial : C^s(\mathcal{A}) \to C^{s+1}(\mathcal{A})$ can explicitly be written down as follows. Let $\boldsymbol{a} \in C^s(\mathcal{A})$ and $\boldsymbol{b} \in C^{s+1}(\mathcal{A})$ (we fix coorientations of $\boldsymbol{a}$, $\boldsymbol{b}$ in advance). Take a versal deformation of $\boldsymbol{b}$. On the parameter space, the bifurcation diagram $\Psi(\boldsymbol{a})$ of type $\boldsymbol{a}$ is defined: It is either empty or 1-dimensional semi-algebraic curves approaching the origin (That corresponds to that the stratum $\Gamma(\boldsymbol{a})$ is pseudo-algebraic and of codimension s). Count the *incidence coefficient* $[\boldsymbol{a}; \boldsymbol{b}]$, defined by the algebraic intersection number of $\Psi(\boldsymbol{a})$ with a small oriented sphere centered at the origin. Then $\partial \boldsymbol{a} = \sum [\boldsymbol{a}; \boldsymbol{b}]\, \boldsymbol{b}$, the sum taken over all generators $\boldsymbol{b}$.

We use the $\mathcal{A}^0$-classification when the source and target manifolds M, N are oriented. Its Vassiliev complex is denoted by $C(\mathcal{A}^0)$.

REMARK 2.3. Notice that the local Vassiliev complex is determined only by the local classification of singularities. In fact, although there are possibly many connected components in each $\Gamma(\boldsymbol{a}) - \Gamma_{s+1}$, they are regarded as just 'one stratum' in the complex $C^*(\mathcal{A})$. Instead, a more finer subcomplex can be considered by adding some 'non-local' data to $C^*(\mathcal{A})$, which we provisionally call an *enriched Vassiliev complex*,

$$C^s_{en}(\mathcal{A}) \subset E_1^{s,0}.$$

The additional data in consideration would be, for instance, types of configurations of subsets $S_1, \cdots, S_k$ on M at each of which a local bifurcation occurs (in case of knots in 3-space, these data are called *chord diagrams* or *weight systems*), the placement of the singular set in M and of the singular value sets in N, the topological types of singular fibers (if $m \geq n$), and so on. $E_1^{s,0}$ itself can be regarded as the finest refinement among all the enriched complexes.

2.3. Local invariants for generic maps. There is a natural homomorphism

$$H^s(C(\mathcal{A})) \to E_2^{s,0} \to E_\infty^{s,0} \to H^s(\mathcal{M}, \mathcal{M} - \Gamma) \to H^s(\mathcal{M}).$$

In particular, if $H^{k+1}(\mathcal{M}) = 0$, we have

$$\rho : H^{k+1}(C(\mathcal{A})) \to H^k(\mathcal{M} - \Gamma) \quad \text{modulo } H^k(\mathcal{M}).$$

This map is explained as follows. For simplicity, assume that $\mathcal{M}$ is contractible (e.g., M is connected and $N = \mathbb{R}^n$). Let $c = \sum \lambda_{\boldsymbol{a}} \cdot \boldsymbol{a}$ be a cocycle of $C^{k+1}(\mathcal{A})$, and D^{k+1} and $S^k = \partial D^{k+1}$ denote the standard oriented disk and the sphere. Given any smooth map $f : S^k \to \mathcal{M} - \Gamma$, we can take a smooth map $\gamma : D^{k+1} \to \mathcal{M}$ which is an extension of f, i.e., $\gamma|_{\partial D^{k+1}} = f$, and is transverse to Γ_{k+1}. Then the value of $\rho([c])$ on the cycle $[f]$ is given by

$$\rho([c])([f]) = \sum \lambda_{\boldsymbol{a}} \cdot \sharp\,[\Gamma(\boldsymbol{a}) \cap \gamma].$$

Here $\sharp\,[\Gamma(\boldsymbol{a}) \cap \gamma]$ stands for the intersection number, that is, the number of points $b \in D^{k+1}$ taking into account the sign ± 1 so that the C^∞-map $\gamma_b : M \to N$ admits multi-singularities of type $\boldsymbol{a}$ at some finite points in M. The sign means the coincidence between a given orientation of D^{k+1} and the coorientation of $\boldsymbol{a}$.

DEFINITION 2.4. An element $\rho(c) \in H^k(\mathcal{M} - \Gamma)$ (determined up to $H^k(\mathcal{M})$) is called a *local invariant* for k-cycles in $\mathcal{M} - \Gamma$. In particular, in case of $k = 0$, a *local invariant for generic maps* is a locally constant function $v : \mathcal{M} - \Gamma \to R$ determined by some element of $H^1(C(\mathcal{A})) = \ker \partial_1$ up to constants (Goryunov [**6**]). If we take some enriched Vassiliev complex instead, then we say $v = \rho(c)$ is *semi-local* or *enriched-local.*

EXAMPLE 2.5. For generic plane curves, there are Arnold's basic invariants: $J^\pm$ are local, while St is *enriched-local* in our sense, because its definition involves the cyclic order of the preimage of a triple point. There are several higher dimensional analogies to J^+, J^-, St: local invariants for $M^2 \to \mathbb{R}^3$ are related to inverse self-tangencies and quadric points [**6**]; for generic immersions $M^3 \to \mathbb{R}^5$ the Smale invariant is related [**9**], and so on. A list of some works is noted below (this list is not complete!).

(m,n)	object/singularities
$(1,1)$	functions on circle bundles/global max.-min. [**15**]
$(1,2)$	generic immersed curves, wavefronts [**2, 1**]
$(1,3)$	knots and links [**30**]
$(2,1)$	Morse function/singular fibers [**23**]
$(2,2)$	generic maps [**20, 12**]
$(2,3)$	generic maps [**6**]
$(2,4)$	generic immersions [**14, 13**], ribbon knots [**11**]
$(3,2)$	generic maps [**31**]
$(3,3)$	geneirc maps [**22, 7**]
$(3,4)$	generic immersions [**5**]
$(3,5)$	generic immersions [**9**]
$(4,3)$	generic maps/singular fibers [**23, 32**]

EXAMPLE 2.6. For instance, look at the case of $m = n = 2$, studied in [**20**]. In that case, stable singularity types (codimension 0) are of fold, cusps and double folds, and there are 10 multi-singularities of codimension one and 20 multi-singularities of codimension two (Here couples of codimension one singularities are omitted, since they do not effect coherence relations for local invariants):

$$0 \longrightarrow C^1(\mathcal{A}_{2,2}) \simeq \mathbb{Z}^{10} \xrightarrow{\ \partial\ } C^2(\mathcal{A}_{2,2}) \simeq \mathbb{Z}^{20}$$

It turns out that $H^1(C(\mathcal{A}); \mathbb{Z}) = \ker \partial$ has rank 3 (the generators are denoted by ΔI_i $(i = 1, 2, 3)$ in [**20**]) and local invariants (over $R \ni \frac{1}{4}$) are generated modulo constants by

$$\begin{array}{ll} \Delta c = 2\Delta I_1 : & \sharp \text{ of cusps;} \\ \Delta d = \Delta I_2 + \Delta I_3 : & \sharp \text{ of double folds;} \\ \Delta \nu = \Delta I_1 - \Delta I_2 + \Delta I_3 : & \text{projective Bennequin invariant.} \end{array}$$

The last one is an interesting invariant for apparent contours obtained as a variant of the Bennequin number for Legendre curves (e.g., [**4**] for a computer program to compute $\Delta\nu$).

When we take the $\mathcal{A}^0$-equivalence instead of $\mathcal{A}$, Δc breaks into two independent invariants $\Delta c_\pm$ for the numbers of positive and negative cusps.

Further, note that type B (beak-to-beak) of codimension one can be separated into two types according to how components of contour curves are mutually connected. This yields an *enriched-local* invariant [**12**]:

$$\Delta I_4 = \Delta l - \Delta b_1 + \Delta b_2 : \ \sharp \text{ of components of critical set } C(f).$$

3. Finite type invariants for generic maps

3.1. A naïve approach. A path transverse to the discriminant Γ is a smooth homotopy which causes a generic local bifurcation, say of type α. We denote such a path by

$$\Xi^{(\alpha)} : [-1,1] \longrightarrow \mathcal{M}, \ \epsilon_1 \mapsto \Xi^{(\alpha)}_{\epsilon_1},$$

so that it is transverse to $\Gamma(\alpha) - \Gamma_2$ at $\epsilon = 0$. For *cooriented* α, we always assume that $\Xi^{(\alpha)}$ is *compatible with the coorientation*, i.e., the path is directed to the given normal orientation.

Let $\boldsymbol{a} = (\alpha_1, \cdots, \alpha_s)$, all α_j being of codimension one. A normal slice to $\Gamma(\boldsymbol{a}) - \Gamma_{s+1}$ is a smooth family of maps which causes local bifurcations of type α_j $(j = 1, \cdots, s)$ around s distinct points in N independently. We denote it by

$$\Xi^{\boldsymbol{a}} : [-1,1]^s \longrightarrow \mathcal{M}, \ (\epsilon_1, \cdots, \epsilon_s) \mapsto \Xi^{\boldsymbol{a}}_{\epsilon_1\cdots\epsilon_s}$$

so that $\Xi^{\boldsymbol{a}}_{\epsilon_1\cdots\epsilon_s} \in \Gamma(\alpha_i)$ if and only if $\epsilon_i = 0$. If $\boldsymbol{a}$ is cooriented, $\Xi^{\boldsymbol{a}}$ is assumed to be compatible with the coorientation for each α_j.

EXAMPLE 3.1. In case of maps $S^1 \to \mathbb{R}^3$, there is only one singularity type of codimension one, $\delta :=$ *crossing change.* Then Ξ^{δ^s} for the s-tuple of δ means a family which causes crossing changes at s distinct points in $\mathbb{R}^3$ independently; In particular $\Xi^{\delta^s}_0 : S^1 \to \mathbb{R}^3$ is an immersion with s double points. In case of maps $M^2 \to \mathbb{R}^2$, there arise 10 different types of singularities of codimension one (Example 2.5).

To define 'finite type invariants' in general case, we take in mind an analogy to *partial derivatives* of functions in several variables.

DEFINITION 3.2. For a locally constant function $v : \mathcal{M} - \Gamma \to R$, we define the *$s$-th partial derivative* $\nabla^{\boldsymbol{a}} v := \nabla^{\alpha_1} \cdots \nabla^{\alpha_s} v$ *with respect to* $\boldsymbol{a} = (\alpha_1, \cdots, \alpha_s)$, $|\alpha_j| = 1$, to be a locally constant function over $\Gamma(\boldsymbol{a}) - \Gamma_{s+1}$ given by

$$\nabla^{\boldsymbol{a}} v\,(\Xi^{\boldsymbol{a}}_{0\cdots 0}) := \sum \epsilon_1 \cdots \epsilon_s \cdot v(\Xi^{\boldsymbol{a}}_{\epsilon_1\cdots\epsilon_s}),$$

the sum taken over 2^s combinations of $\epsilon_i = \pm 1$. We say that v is an invariant *of order at most* r if the following condition (T) is satisfied:

$$(T) \qquad \nabla^{\boldsymbol{a}} v = 0 \qquad (\forall \boldsymbol{a} \text{ of codimension } s \geq r+1).$$

In particular, in case that $m \geq 2$, such an invariant is called a *naïve finite type invariant.*

REMARK 3.3. The partial derivative

$$\nabla^\alpha v(\Xi_0) = v(\Xi_+) - v(\Xi_-)$$

defines an operator $\nabla^\alpha : H^0(\Gamma(\boldsymbol{a}) - \Gamma_{s+1}) \to H^0(\Gamma(\boldsymbol{a} \cup \alpha) - \Gamma_{s+2})$.

Denote by V_r the R-module generated by invariants of order $\leq r$:

$$V_0 \subset V_1 \subset \cdots \subset V_\infty := \bigcup_{r=0}^{\infty} V_r \subset H^0(\mathcal{M} - \Gamma; R).$$

Obviously, an invariant of order zero is a function which is constant over each connected component of $\mathcal{M}$, therefore $V_0 = H^0(\mathcal{M}; R)$.

LEMMA 3.4. *V_∞ is a graded R-algebra: If $v_1 \in V_r$ and $v_2 \in V_\ell$, then $v_1 v_2 \in V_{r+\ell}$, where $v_1 v_2(f) := v_1(f) v_2(f)$ for $f \in \mathcal{M} - \Gamma$.*

PROOF. Similarly as in the case of knots, a Leibniz type formula holds:

$$\begin{aligned} \nabla^\alpha(v_1 v_2)(\Xi_0) &= v_1(\Xi_+)v_2(\Xi_+) - v_1(\Xi_-)v_2(\Xi_-) \\ &= v_1(\Xi_+)(v_2(\Xi_+) - v_2(\Xi_-)) + (v_1(\Xi_+) - v_1(\Xi_-))v_2(\Xi_-) \\ &= v_1(\Xi_+) \cdot \nabla^\alpha v_2(\Xi_0) + \nabla^\alpha v_1(\Xi_0) \cdot v_2(\Xi_-). \end{aligned}$$

By the induction, we observe that any value of $\nabla^{\boldsymbol{a}} v_1 v_2$ is written by a linear combination $\nabla^{\boldsymbol{a}'} v_1 \nabla^{\boldsymbol{a}''} v_2$ with $\boldsymbol{a} = \boldsymbol{a}' \cup \boldsymbol{a}''$. □

3.2. (Non-)transverse self-intersection. Let $\boldsymbol{a} = (\alpha_j)$, $|\alpha_j| = 1$. In general the manifold $\Gamma(\boldsymbol{a}) - \Gamma_{s+1}$ has many connected components separated by the discriminant hypersurface $\Gamma(\boldsymbol{a}) \cap \Gamma_{s+1}$. Any transverse self-intersection locus $\Gamma(\boldsymbol{a} \cup \alpha)$, where $|\alpha| = 1$, is a top stratum of $\Gamma(\boldsymbol{a}) \cap \Gamma_{s+1}$, but it can happen that some non-transverse self-intersection locus also becomes a top stratum. Provisionally we make an additional strong condition:

DEFINITION 3.5. We say that an invariant $v : \mathcal{M} - \Gamma \to R$ satisfies the *non-transverse loci condition* (NT) if for any $\boldsymbol{a} = (\alpha_1, \cdots, \alpha_s)$ with $|\alpha_j| = 1$ the value of $\nabla^{\boldsymbol{a}} v$ does not change when passing through any *non-transverse* self-intersection loci of codimension one in $\Gamma(\boldsymbol{a})$. We denote by V'_r $(r \geq 1)$ the subspace of V_r consisting of invariants satisfying (NT) (for $r = 0$, $V'_0 = V_0$).

LEMMA 3.6. *Any polynomial of local invariants is a naïve finite type invariant with (NT).*

PROOF. By Lemma 3.4, it is enough to see the case $r = 1$. If $v_c : \mathcal{M} - \Gamma \to R$ is a local invariant corresponding to a Vassiliev 1-cocycle $c = \sum_\alpha c_\alpha \cdot \alpha$, then $\nabla^\alpha v_c$ is a *constant function* with value c_α over $\Gamma(\alpha) - \Gamma_2$. Obviously v_c is of first order and satisfies (NT), since any jump of $\nabla^\alpha v_c$ on Γ_2 is zero. □

The converse is also true: We will show that *our operators ∇^α are linearly dependent* in the relation corresponding to the coherence condition $\partial = 0$ of local invariants: Let $R = \mathbb{Q}$ for simplicity.

THEOREM 3.7. *Let $m, n \geq 2$ and the pair (m, n) be in Mather's nice range. Let M be connected, oriented and of dimension m, and $N = \mathbb{R}^n$, oriented. Then,*

- *If $n > m+1$, naïve finite type invariants are polynomials of local invariants modulo constants:*

$$V_r/V_{r-1} = \mathrm{Sym}^r(V_1/V_0), \qquad V_1/V_0 = H^1(C(\mathcal{A}^0)).$$

- *If $n \le m+1$, naïve finite type invariants with non-transverse loci condition are polynomials of local invariants modulo constants.*

REMARK 3.8. A similar theorem has already been known by different arguments for generic immersions in cases of $n = 2m(\ge 4)$ in [**14, 13**], and for $n = 2m-1(\ge 5)$ in [**9**]: For instance, naïve finite type invariants of generic immersions $M^2 \to \mathbb{R}^4$ are generated by the number of double points and the normal Euler number [**14**]. Notice that when taking n much less than $2m-1$, the number of unavoidable singularities rapidly increases and the situation would be much more complicated.

3.3. Irreducibility. An elementary fact is that *the locus $\Gamma(\boldsymbol{a})$ is irreducible* in the sense that any $f_0, f_1 \in \Gamma(\boldsymbol{a}) - \Gamma_{s+1}$ are joined by generic paths within the locus $\Gamma(\boldsymbol{a})$:

LEMMA 3.9. *Under the same assumption as in Theorem 3.7, the following properties hold:*

(1) *Let $v \in V_r'$, then $\nabla^{\boldsymbol{a}} v$ is a constant for each r-tuple $\boldsymbol{a}$.*
(2) *If $n > m+1$, $V_r' = V_r$.*
(3) $V_1'/V_0' = H^1(C(\mathcal{A}^0))$.

PROOF. (1) Let $v \in V_r'$. By definition the r-th partial derivative $\nabla^{\boldsymbol{a}} v$ is a locally constant function $\Gamma(\boldsymbol{a}) - \Gamma_{r+1} \to \mathbb{Q}$. We show that $\nabla^{\boldsymbol{a}} v$ is constant.

Let $f_0 \in \Gamma(\boldsymbol{a}) - \Gamma_{r+1}$, which has multi-singularities of type $\boldsymbol{a}$ at $S \subset M$ and $T := f_0(S) \subset \mathbb{R}^n$ and no more complicated singularities. Put

$$\Gamma(\boldsymbol{a}; f_0) := \{\, g \in \mathcal{M} \mid g = f_0 \text{ on some neighborhood of } S \,\},$$

which is an affine subspace of $\mathcal{M}$ contained in $\Gamma(\boldsymbol{a})$. For any $g \in \Gamma(\boldsymbol{a}; f_0) - \Gamma_{r+1}$, we can take a generic path γ_t from f_0 to g in $\Gamma(\boldsymbol{a}; f_0)$. Since v satisfies conditions (NT) and (T), any jump of $\nabla^{\boldsymbol{a}} v$ along γ_t is zero, hence $\nabla^{\boldsymbol{a}} v(f_0) = \nabla^{\boldsymbol{a}} v(g)$. Now take another $f_1 \in \Gamma(\boldsymbol{a}) - \Gamma_{r+1}$ with multi-singularities of type $\boldsymbol{a}$ at some S_1 and $T_1 = f_1(S_1)$. Since M^m and $\mathbb{R}^n$ are connected and $m, n \ge 2$, there exists an 1-parameter family $(\sigma_t, \tau_t) \in \mathcal{A}^0_{M,N}$ so that

- $\sigma_0 = id_M$ and $\tau_0 = id_{\mathbb{R}^n}$ (trivial),
- $\sigma_1(S_1) = S$, $\tau_1(T_1) = T$,
- $g_1 := \tau_1 \circ f_1 \circ \sigma_1^{-1}$ coincides with f on a neighborhood of S.

Since f_1 and g_1 are isotopic and $g_1 \in \Gamma(\boldsymbol{a}; f_0) - \Gamma_{r+1}$, we have

$$\nabla^{\boldsymbol{a}} v(f_1) = \nabla^{\boldsymbol{a}} v(g_1) = \nabla^{\boldsymbol{a}} v(f_0).$$

Thus $\nabla^{\boldsymbol{a}} v$ is constant on $\Gamma(\boldsymbol{a}) - \Gamma_{r+1}$.
(2) Let $n > m+1$. Then, in the proof of (1) above, we can choose the path γ_t so that the image $\gamma_t(M - S)$ does not pass through any point of $T = f_0(S)$, i.e., γ_t does not create any new multi-singularity mapped to a point of T. Namely, the path γ_t meets only transverse self-intersection locus of Γ, thus $V_r = V_r'$.
(3) Let $v \in V_1'$. It follows from (1) that $\nabla^\alpha v$ is constant for each α of codimension one. Put

$$c_v := \sum_\alpha (\nabla^\alpha v) \cdot \alpha \in C^1(\mathcal{A}^0).$$

Since v is a locally constant function $\mathcal{M} - \Gamma \to \mathbb{Q}$, the coherence relation around each $\Gamma(\beta)$ of codimension 2 is automatically satisfied:

$$(*) \qquad \sum_{\alpha} [\alpha;\beta] \cdot \nabla^{\alpha} v = 0,$$

i.e., $\partial c_v = 0$, that means that v is a local invariant modulo constant. Lemma 3.6 shows local implies order one, thus the claim follows. □

PROOF. (Theorem 3.7) By Lemma 3.9 (1), one can define a functional $\nabla^{\boldsymbol{a}} : V'_r \to \mathbb{Q}$ for each r-tuple $\boldsymbol{a}$. Note that $v \in V'_{r-1}$ if and only if $\nabla^{\boldsymbol{a}} v = 0$ for all r-tuples $\boldsymbol{a}$. Given a $(r-1)$-tuple $\boldsymbol{a}'$ and a singularity type β of codimension 2, the relation $(*)$ in the proof of (3) generates a linear relation

$$(*_r) \qquad \sum_{\alpha} [\alpha;\beta] \cdot \nabla^{\boldsymbol{a}' \cup \alpha} = 0$$

among functionals $\nabla^{\boldsymbol{a}} = \nabla^{\boldsymbol{a}' \cup \alpha} = \nabla^{\alpha} \cdot \nabla^{\boldsymbol{a}'}$ (That means a coherence relation for locally constant functions $\nabla^{\boldsymbol{a}'} v$ on $\Gamma(\boldsymbol{a}') - \Gamma_r$ around $\Gamma(\boldsymbol{a}' \cup \beta)$, where $v \in V'_r$).

Consider the vector space spanned by the functionals indexed by $\boldsymbol{a}$ ($|\boldsymbol{a}| = r$), and let A_r denote the quotient modulo the relations $(*_r)$ for all $\boldsymbol{a}'$ and β:

$$A_r := \frac{\bigoplus_{|\boldsymbol{a}|=r} \mathbb{Q} \cdot \nabla^{\boldsymbol{a}}}{\text{relations } (*)_r}.$$

Then

- A natural pairing $A_r \times V'_r \to \mathbb{Q}$ induces an injective linear map
$$V'_r / V'_{r-1} \to \mathrm{Hom}(A_r, \mathbb{Q})$$
- A_r is naturally isomorphic to the $\mathbb{Q}$-vector space of homogeneous differential operators of order r on the space V'_1/V'_0.

So we have an injective linear map $V'_r/V'_{r-1} \to \mathrm{Sym}^r(V'_1/V'_0)$. It follows from Lemma 3.6 that this map is surjective, thus it is an isomorphism. This completes the proof. □

REMARK 3.10. Theorem 3.7 says that the naïve definition of finite type invariants for generic maps is somehow irrelevant. For improving the definition, there come up several difficulties which depend on the pair of dimensions m, n. In case of $n \le m+1$ $(m, n \ge 2)$, for instance, $(m,n) = (2,2)$ or $(2,3)$, the next task would be to define 'second order' invariants for stable maps from M^2 to $\mathbb{R}^2$ (or $\mathbb{R}^3$) involving non-transverse self-intersection loci. To define 'non-local' invariants, we must add some data on the global configuration of critical curves or double point curves of maps: Those data are used to identify connected components of $\Gamma(\boldsymbol{a}) - \Gamma_{s+1}$. In case of $n > m+1$ $(m \ge 2)$, for instance, $(m,n) = (2,4)$, the difficulty is different from the above, and we need some completely new idea for 'finite type'.

REMARK 3.11. (1) In case of knots in 3-space ($n = 1$, $m = 3$), we should replace r-tuples $\boldsymbol{a}$ in the above argument by (equivalent classes of) r-chord diagrams on S^1, and let A_r denote the space generated by r-chord diagrams modulo all 1- and 4-term relations, which are a sort of coherence relations like $(*_r)$ (see (2) below). As well-known, it holds that

$$V_r/V_{r-1} = \mathrm{Hom}(A_r, \mathbb{Q}),$$

proved by M. Kontsevich. In Vassiliev's paper [**30**] a purely combinatorial construction of knot invariants is given by using *Actuality Table*, that might be applicable to our setting.
(2) We have seen that if $n > m+1$ and $m \geq 2$, (NT) is automatically satisfied. For maps $S^1 \to \mathbb{R}^3$, it fails. Indeed, the *triple point stratum* $\Gamma(\tau)$, consisting of immersions with a triple point τ, belongs to the *non-transverse* self-intersection locus of Γ in our sense (Definition 1.6). Any path on the strata $\Gamma(\delta^2)$ of maps with two double points may come across $\Gamma(\tau)$ in general, so we can not ignore this triple point stratum. In fact, the value of the jumps of invariants $\nabla^{\delta^2} v$ at $\Gamma(\tau)$ yields a coherence relation, i.e., the 4-term relation, see [**2, 30**].

4. Characteristic classes for fiber bundles

4.1. Classifying space. Let M be a compact, connected oriented manifold. We regard the affine space $\mathcal{M} = C^\infty(M, \mathbb{R}^n)$ *as a representation of the diffeomorphism group* $G = \mathrm{Diff}\, M$.

First, we shall recall a well-known construction (e.g., [**8**]) of the classifying space of the topological group $G = \mathrm{Diff}\, M$ of orientation preserving diffeomorphisms. If n is quite high, $C^\infty(M, \mathbb{R}^n) - \Gamma = \mathrm{Emb}(M, \mathbb{R}^n)$, the space of all embeddings of M in $\mathbb{R}^n$. Sending $n \to \infty$, we may identify the classifying space of G with the topological quotient

$$BG = B\,\mathrm{Diff}\, M = \mathrm{Emb}(M, \mathbb{R}^\infty)/\mathrm{Diff}\, M.$$

Put $EG = \mathrm{Emb}(M, \mathbb{R}^\infty)$, then EG is highly connected, thus the canonical map $EG \to BG$ gives the universal principal bundle for the group G. Let $BM := (EG \times M)/G$, the associated manifold-bundle with fiber M, then any manifold-bundle $E \to B$ (B paracompact), with fiber M and structure group G, can be obtained up to isomorphisms from the universal bundle $BM \to BG$ via a classifying map $\rho : B \to BG$. Any element of $H^*(BG)$ is called a universal G-characteristic class: G-characteristic classes of $E \to B$ are defined by their ρ^*-image in $H^*(B)$.

4.2. Degeneracy loci and universal polynomials. We are concerned with the "*geometric realization problem*" of a given characteristic class for a manifold-bundle $E \to B$.

A Singularity Theory approach to this problem is as follows: Let us think of the composition of an embedding of M and a projection onto $\mathbb{R}^n$ for some n small enough,

$$\begin{array}{ccc} M & \xrightarrow{\ incl\ } & \mathbb{R}^\infty \\ & \searrow^{f} & \downarrow{\scriptstyle proj} \\ & & \mathbb{R}^n \end{array}$$

Put $\mathcal{M} = C^\infty(M, \mathbb{R}^n)$ and $B\mathcal{M} \to BG$ to be the associated bundle with fiber $\mathcal{M}$ and group G. Now $\mathcal{M}$ is a contractible space, hence the Borel cohomology $H^*_G(\mathcal{M}) := H^*(B\mathcal{M})$ is isomorphic to $H^*_G(pt) = H^*(BG)$. The Vassiliev complex has much meanings in this equivariant setting: There is a natural homomorphism

$$Tp : H^s(C(\mathcal{A})) \to E^{s,0}_\infty \to H^s_G(\mathcal{M}) \simeq H^*(BG).$$

We denote by $Tp_c \in H^*(BG)$ the G-characteristic class associated to a Vassiliev cocycle $c = \sum \lambda_i \boldsymbol{a}_i \in C^s(\mathcal{A})$, $\partial c = 0$. This means the following: Suppose that we are given a bundle $\pi : E \to B$ with fiber M over a manifold B and a smooth map

$f : E \to \mathbb{R}^n$ over the total space of the bundle. Denote by $\rho = \rho_\pi : B \to BG$ the classifying map associated to the bundle.

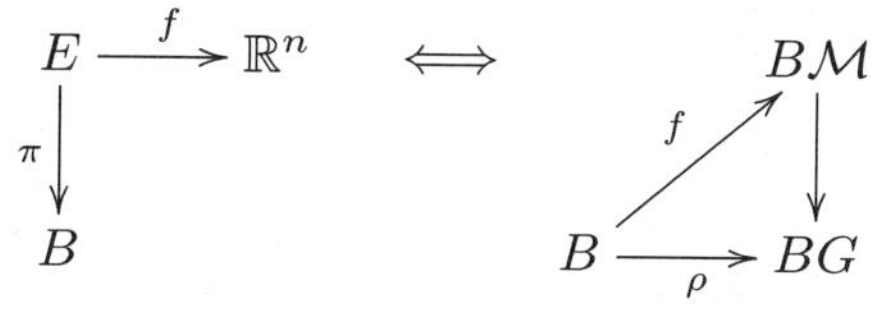

To any multi-singularity type $\boldsymbol{a}$ we associate the *bifurcation locus* $B_{\boldsymbol{a}}(f) \subset B$, which is a locally closed submanifold consisting of points $b \in B$ so that the map $f_b : E_b \simeq M \to \mathbb{R}^n$ obtained by the restriction to a fiber $E_b = \pi^{-1}(b)$ admits the multi-singularity of type $\boldsymbol{a}$ at some finite points in E_b. Given a Vassiliev cocycle $c := \sum \lambda_i \boldsymbol{a}_i$ and appropriately generic $f : E \to \mathbb{R}^n$, we can define the *bifurcation cycle* $B_c(f)$ to be the geometric cycle $\sum \lambda_i B_{\boldsymbol{a}_i}(f)$ in B: It is a geometric presentation of the G-characteristic class so that

$$\mathrm{Dual}\,[\overline{B_c(f)}] = \rho^* Tp_c.$$

The Thom polynomial theory for $\mathcal{A}$-equivalence of map-germs (e.g. [**19, 16, 17, 29**]) says that for a Vassiliev cocycle of mono-singularities $c = \sum \lambda_\alpha \alpha$, Tp_c is written by a universal polynomial in the relative Novikov-Landweber classes

$$\pi_* cl^I(T_\pi) = \pi_*(cl_1^{i_1}(T_\pi) \cdots cl_k^{i_k}(T_\pi))$$

where $T_\pi (= \ker d\pi)$ is the relative tangent bundle of $\pi : E \to B$ and cl means a certain characteristic class of vector bundles, e.g., Pontrjagin class, Euler class (with $\mathbb{Z}$-coefficients), Stiefel-Whitney class (with $\mathbb{Z}_2$-coefficients). For cocycles of $\mathcal{A}$-multi-singularities, the same conjecturally holds, cf. Kazarian [**16**].

Here n should be reasonably small: For if we take n to ∞, any cocycles of BG live in $\mathcal{M} - \Gamma$, thus there is no chance to obtain nontrivial geometric presentations by bifurcation loci.

We may think of the same story for some enriched Vassiliev complex instead of the local complex of singularities. Some natural questions are, e.g., to find

- the precise forms of Tp_c for given classes $[c] \in H^*(C(\mathcal{A}))$,
- nontrivial relations among those Tp_c's,
- elements in $H^*(C_{en}(\mathcal{A}))$ representing torsion elements of $H^*(BG;\mathbb{Z})$.

EXAMPLE 4.1. For example, in case that M is oriented circle S^1,

$$H^*(BS^1) = H^*(BU(1)) = \mathbb{Z}[c_1]$$

where c_1 is the first Chern class of complex line bundles. In [**15**] it is shown that the class c_1 can be realized by some bifurcation locus of functions $E \to \mathbb{R}$ or maps $E \to \mathbb{R}^2$ over total space E of a S^1-bundle, and also universal polynomials Tp_Σ in c_1 for several types Σ are computed. When taking the target space $\mathbb{R}^n$ $(n \geq 3)$, it seems that c_1 can not be realized by any bifurcation points, i.e., c_1 lives in the space of embeddings.

EXAMPLE 4.2. Recall that for an oriented C^∞-surface bundle $\pi : E \to B$ with fiber a closed oriented surface M, the r-th *Morita-Miller-Munford class* $e_r(E) \in H^{2r}(B;\mathbb{Z})$ is defined to be the pushforward $\pi_* e(T_\pi)^{r+1}$ where T_π is the relative tangent bundle over the total space E and $e(T_\pi) \in H^2(E;\mathbb{Z})$ is the Euler class. It is easy to see that for $r \geq 1$

$$Tp_{\Sigma^2}[f : E \to \mathbb{R}^{r+1}] = e_r(E),$$

that is, the class $e_r(E)$ is realized by the Σ^2-bifurcation locus of generic maps $E \to \mathbb{R}^{r+1}$, where Σ^2 means corank 2 singularities of germs $\mathbb{R}^2, 0 \to \mathbb{R}^{r+1}, 0$ (the differential has kernel dimension ≥ 2):

$$[\overline{B_{\Sigma^2}(f)}] = \pi_*[\overline{\Sigma^2(f)}] = \pi_* e(T^*_\pi \otimes f^*\epsilon^{r+1}) = \pi_* e(T^*_\pi)^{r+1} = e_r(E).$$

For instance, the first Morita-Miller-Mumford class $e_1(E)$ is geometrically observed by using generic maps $f : E \to \mathbb{R}^2$. If B is a closed surface, $B_{\Sigma^2}(f)$ consists of finite points in B, over which $f : E \to \mathbb{R}^2$ has the singularity of type

$$I_{22} + II_{22} : (x, y; a, b) \mapsto (x^2 \pm y^2 + x^3 + ay, xy + bx)$$

where x, y are local coordinates of fiber M and a, b are some local coordinates of the base space B. As another example, there is a work by Saeki-Yamamoto [**24**] which shows that $e_1(E)$ is realized by the codimension 2 bifurcation locus corresponding to a special topological type of singular fiber of generic functions $f : E \to \mathbb{R}$: The singular fiber contains three circle components each two of which meet at a nodal point.

5. Contact equivalence for mappings

5.1. Global contact equivalence. Let M be a compact manifold of dimension $m = n+d$ and N of dimension $n+k$. Let B be a k-dimensional compact closed (regular) submanifold of N and $p : M \times N \to M$ the projection to the first factor. If $f : M \to N$ is transverse to B, then $f^{-1}(B)$ is a d-dimensional submanifold of M; If f is tangent to B at some points, then the preimage $f^{-1}(B)$ has singularities of contact ($\mathcal{K}$-)type at those points. Let $p \in f^{-1}(B)$ be a singular point, then the contact singularity type at p is determined by the contact equivalence class of the germ

$$\pi \circ f : \mathbb{R}^{n+d}, 0 \to \mathbb{R}^n, 0$$

where we take local coordinates of M and N centered at p and $f(p)$ respectively, and $\pi : \mathbb{R}^{n+k}, 0 \to \mathbb{R}^n, 0$ denotes a submersion-germ defining $B = \pi^{-1}(0)$ locally.

We introduce an equivalence relation on $\mathcal{M} = C^\infty(M, N)$ to measure how a map $f : M \to N$ is tangent to B: Define $\mathcal{K}_{M,N,B}$ (or $\mathcal{K}_B$ for short) to be the subgroup of Diff$(M \times N)$ consisting of diffeomorphisms H which preserves both the submanifold $M \times B$ and the fiber structure of $p : M \times N \to M$. C^∞-maps $f, g : M \to N$ are said to be $\mathcal{K}_B$-*equivalent* if there exists $H \in \mathcal{K}_B$ such that H sends the graph of f to the graph of g. Then, in particular, $f^{-1}(B)$ and $g^{-1}(B)$ are isomorphic via some diffeomorphism of the ambient manifold M.

Let $\mathcal{D}_\infty := \{f \in \mathcal{M}, \text{codim } \mathcal{K}_B.f = \infty\}$ and $\mathcal{U}_0 := \mathcal{M} - \mathcal{D}_\infty$. $f \in \mathcal{U}_0$ if and only if $f^{-1}(B)$ has finitely many singular points, each of which is of type $\mathcal{K}$-finite. In entirely the same way as in the case of Theorem 1.3, we have

THEOREM 5.1. *$\mathcal{D}_\infty$ has infinite codimension in $\mathcal{M}$, and there exists a filtration*

$$\mathcal{M} \supset \mathcal{D} := \mathcal{D}_1 \supset \mathcal{D}_2 \supset \cdots \supset \mathcal{D}_s \supset \cdots \supset \mathcal{D}_\infty,$$

by $\mathcal{K}_B$-invariant closed pseudo-algebraic subsets $\mathcal{D}_s$ of codimension s such that it admits a topologically locally trival fibration $\pi_s : \mathcal{D}_s - \mathcal{D}_{s+1} \to Y'_s$ so that each fiber is an $\mathcal{K}_B$-orbit and Y'_s is a finite dimensional manifold.

EXAMPLE 5.2. 0) $f \in \mathcal{M} - \mathcal{D}$ if and only if f is transverse to B.
1) $\mathcal{D}_1 - \mathcal{D}_2$ consists of maps f having one Morse singularity A_1 on $f^{-1}(B)$, whose contact type is $\mathcal{K}$-equivalent to

$$A_{1,k} : (x_1, \cdots, x_{d+1}, z) \mapsto (-x_1^2 - \cdots - x_k^2 + x_{k+1}^2 + \cdots + x_{d+1}^2, z),$$

for some $0 \leq k \leq [\frac{d+1}{2}]$. $A_{1,k}$ is *coorientable*, except for d odd, $k = [\frac{d+1}{2}]$. This is the surgery to attach a k-handle to a smooth fiber.
2) Also $\mathcal{D}_2 - \mathcal{D}_3$ consists of maps f having either of
- two Morse singularities, or
- one A_2-singularity (=cancelation of handle surgeries).

$$A_{2,k} : (x_1, \cdots, x_{d+1}, y, z) \mapsto (x_1^3 + yx_1 \pm x_2^2 \pm \cdots \pm x_{d+1}^2, y, z).$$

5.2. Thom-Pontrjagin construction. The classical construction is as follows: Now let M be a compact oriented d-manifold, and embed it in Euclidean space $\mathbb{R}^{d+n}$ ($n \gg 0$). Let $B := BSO(n)$, the Grassmannian of oriented n-planes in $\mathbb{R}^{d+n}$, and let $\rho : M \to B$ the map sending $p \in M$ to the orthogonal plane to TM_p, i.e., the classifying map of the normal bundle of rank n. ρ is extended to a C^∞-map f from $S^{d+n} = \mathbb{R}^{d+n} \cup \infty$ to the Thom space $N_n := MSO(n)$ (a neighborhood of ∞ mapped to the base point of N_n), then $M = f^{-1}(B)$. Let us consider the space of such maps $S^{d+n} \to N_n$ (stationary around ∞), and denote it by

$$\mathcal{M} := C^\infty(S^{d+n}, N_n)_{base}, \quad B := BSO(n) \subset N_n \qquad (n \gg 0).$$

It follows from Theorem 5.1 that there is a $\mathcal{K}_B$-invariant stratification

$$\mathcal{M} \supset \mathcal{D} := \mathcal{D}_1 \supset \mathcal{D}_2 \supset \cdots.$$

If two points $f, g \in \mathcal{M} - \mathcal{D}$ are joined by a generic path γ in $\mathcal{M}$, then d-manifolds $f^{-1}(B)$ and $g^{-1}(B)$ are cobordant, that is, each connected component of $\mathcal{M}$ corresponds to a cobordism class: $\Omega^{ori}(d) = \pi_0(\mathcal{M})$ ($n \to \infty$). That was a basic theorem of R. Thom.

More generally, a sort of global version of *Martinet's versality theorem* holds, see Kazarian [**16, 17**] (also [**24**]): Let $e : S^{d+n} \times \mathcal{M} \to N_n$ be the evaluation map $e(p, f) := f(p)$, and $\mathcal{H} := e^{-1}(B)$ the preimage of B. Denote by $\pi_B : \mathcal{H} \to \mathcal{M}$ the restriction of the second factor projection. Then, we may regard π_B as the "universal C^∞ stable map", and $\mathcal{D}$ as the "discriminant set of π_B":

$$\begin{array}{ccccccc} Q^{d+p} & \longrightarrow & \mathcal{H} & \xrightarrow{incl} & S^{d+n} \times \mathcal{M} & \xrightarrow{e} & N_n \\ {\scriptstyle g}\downarrow & & \downarrow{\scriptstyle \pi_B} & & & & \\ P^p & \longrightarrow & \mathcal{M} & & & & \end{array}$$

That is, a suitably generic map $g : Q \to P$ corresponds to a smooth map $P \to \mathcal{M}$ which is transverse to each $\mathcal{D}_s$.

In case of $d = 1$, Saeki [**23**] has studied a cochain complex for topological types of singular 1-dimensional fiber, which is an enriched Vassiliev complex for $\mathcal{K}_B$-invariant filtration of $\mathcal{D}$.

5.3. Naïve invariants. Let us take a connected component $\mathcal{M}_1$ of $\mathcal{M}$, which corresponds to a cobordism class of oriented closed d-manifolds. Then $H^0(\mathcal{M}_1 - \mathcal{D})$ is regarded as the space of invariants of all such d-manifolds belonging to the fixed cobordism class.

EXAMPLE 5.3. The 'null-cobordant' component in $\mathcal{M}$ can be replaced by

$$\mathcal{M}_0 = C^\infty(S^{d+n}, S^n)_{base}, \qquad B = \{0\} \subset S^n, \quad n \gg 0.$$

Note $\Omega^{ori}(1) = \Omega^{ori}(2) = \Omega^{ori}(3) = 0$. So, in these cases, $\mathcal{M} = \mathcal{M}_0$.

Naïve finite type invariants can be defined in the same way as in §3.1. However, by entirely the same reason as seen in Theorem 3.7, the naïve finite type invariants are reduced to polynomials of local invariants, such as the Euler characteristics $\chi : \mathcal{M} - \mathcal{D} \to \mathbb{Z}$, $f \mapsto \chi(f^{-1}(B))$, cf. [**25**].

Instead, in order to keep the information of glueing maps of handle surgeries, we need more restrictions, i.e., not to be allowed to make other surgeries freely. A choice is to restrict us to a smaller mapping space: For instance, take an open subset $\mathcal{M}_{betti}$ of $\mathcal{M}_0$ so that $f^{-1}(B)$ has *fixed Betti numbers*. In case of $d = 3$, it is nothing but the theory of finite type invariants for homology 3-spheres (Ohtsuki [**21**]) and its generalization. We don't know any result at all in this direction for other dimensional cases.

References

[1] Aicardi, F., *Dicsriminants and Local Invariants of Planar Fronts,* The Arnold-Gelfand math. seminars : geometry and singularity theory, edited by V.I.Arnold etc, Birkhäuser, 1997, 1–76. MR1429884 (99k:58021)

[2] Arnol'd, V. I. *Topological invariants of plane curves and caustics,* Univ. Lect. Series, vol. 5, AMS, (1994). MR1286249 (95h:57003)

[3] Arnol'd, V. I. *The Vassiliev Theory of Discriminants and Knots,* First European Congress of Math., Paris, July 1992, Vol. 1, Birkhäuser, (1994), 3–29. MR1341819 (96m:57010)

[4] Bellettini, G., Beorchia, V. and Paollini, M., *An explicit formula for a Bennequin-type invariant of apparent contours* Topology and its Applications vol.156, 4, (2009), 747–760. MR2492959 (2010e:57013)

[5] Casonatto, C., Romero Fuster, M. C. and Wik Atique, R., *Topological invariants of stable immersions of oriented 3-manifolds in $\mathbb{R}^4$,* to appear in Proceedings of the Workshop on Singularities in Generic Geometry and Applications, Topology and its Applications (2011).

[6] Goryunov, V. V., *Local invariants of mappings of surfaces into three-space,* The Arnold-Gelfand math. seminors: geometry and singularity theory, edited by V.I.Arnold etc, Birkhäuser (1997), 223–255. MR1429894 (97m:57040)

[7] Goryunov, V. V., *Local invariants of maps between 3-manifolds,* Talk in Workshop on Singularities in Geometry and Application, Banach center, (Bedlewo, Poland) May 16-20, 2011.

[8] Eells, J., *A setting for global analysis,* Bull. AMS, 1966, 751–807. MR0203742 (34:3590)

[9] Ekholm, T., *Immersions and their self intersections,* Dissertation, Dept. Math., Uppsala University 1998. MR2715243

[10] Ekholm, T., *Regular homotopy and Vassiliev invariants of generic immersions $S^k \to \mathbb{R}^{2k-1}, k \geq 4$,* Jour. Knot Theory and its Ramifications, 1998. MR1671496 (2000d:57047)

[11] Habiro, K., Kanenobu, T. and Shima, A., *Finite type invariants of ribbon 2-knots,* In Low-dimensional topology (Funchal, 1998), Contemp. Math. 233, AMS (1999), 187–196. MR1701683 (2000e:57038)

[12] Hacon, D., Mendes de Jesus, C., Romero Fuster, M.C., *Global topological invariants of stable maps from a surface to the plane,* Proceedings of the 6th Workshop On Real and Complex Singularities, 2001. Lect. Notes in Pure Appl. Math., vol. 232, Marcel Dekker, (2003).

[13] Januszkiewicz, T, Swiatkowski, J., *Finite type invariants of generic immersions of M^n into R^{2n} are trivial.* Differential and symplectic topology of knots and curves, 61–76, Amer. Math. Soc. Transl. Ser. 2, 190, Amer. Math. Soc., Providence, RI, 1999. MR1738390 (2001b:57051)

[14] Kamada, S., *Vanishing of a certain kind of Vassiliev invariants of 2-knots,* Proc. AMS, 127, 11, (1999), 3421–3426. MR1610788 (2000b:57038)

[15] Kazarian, M. É., *Classifying spaces of singularities and Thom polynomials,* New developments in Singularity Theory (Cambridge 2000), NATO Sci.Ser. II, 21, Kluwer Acad. Publ.,Dordrecht, 2001, 117-134. MR1849306 (2002f:58072)
[16] Kazarian, M. É., *Multisingularities, cobordisms and enumerative geometry,* Russian Math. Survey 58:4 (2003), 665–724 (Uspekhi Mat. nauk 58, 29–88). MR2042262 (2005d:58070)
[17] Kazarian, M. É., *Thom polynomials,* Proc. sympo. “Singularity Theory and its application” (Sapporo, 2003), Adv. Stud. Pure Math. vol. 43 (2006), 85–136. MR2313410 (2008h:58074)
[18] Mather, J. N., *Infinite dimensional group actions,* Cartan Fest. Analyse et topologie, Astérisque 32, 33, Paris, Soc. Math. de France, 1976, 165–172. MR0426026 (54:13975)
[19] Ohmoto, T., *Vassiliev complex for contact classes of real smooth map-germs,* Rep. Fac. Sci. Kagoshima Univ., (1994), pp.1–12. MR1341346 (96m:57046)
[20] Ohmoto, T. and Aicardi, F., *First order local invariants of apparent contours,* Topology, vol. 45 (2006) pp.27-45. MR2170493 (2006f:57015)
[21] Ohtsuki, T., *Finite type invariants of integral homology 3-spheres,* Jour. Knot Theory and its Ramification, vol. 5, Issue 1, 1996, pp.101–115. MR1373813 (97i:57019)
[22] Oset-Sinha, R., *Topological invariants of stable maps from 3-manifolds to three-space,* Dissertation, Universitat de València, 2009.
[23] Saeki, O., *Topology of Singular Fibers of Differentiable Maps,* Lect. Notes Math. 1854, Springer, 2004. MR2106689 (2005m:58085)
[24] Saeki, O and Yamamoto, T., *Singular fibers and characteristic classes,* Topology and its Applications, vol. 155, 2 (2007), 112–120. MR2368204 (2008j:57049)
[25] Sirokova, N., *The space of 3-manifolds,* C. R. de l'Acad. Sci. 331, 2, 15 (2000), 131–136. MR1781529 (2001g:57026)
[26] Saeki, O, Szücs, A and Takase, M., *Regular homotopy classes of immersions of 3-manifolds into 5-space,* Manuscripta Math., 108 (2002), 1432–1785. MR1912945 (2004b:57030)
[27] Thom, R., *The bifurcation subset of a space of maps,* Manifolds - Amsterdam, Lect. Notes Math. 197 Springer, (1971), 202–208. MR0281222 (43:6941)
[28] Thom, R., *Structural Stability and Morphogenesis,* W. A. Benjam, (1972). MR0488156 (58:7722b)
[29] Vassiliev, V. A., *Lagrange and Legendre characteristic classes,* Gordon and Breach, 1988. MR1065996 (91k:57034)
[30] Vassiliev, V. A., *Cohomology of knot spaces,* Adv. Soviet. Math., vol. 21, AMS, Providence, RI. (1990), pp. 23–69. MR1089670 (92a:57016)
[31] Yamamoto, M., *First Order Semi-Local Invariants of Stable Maps of 3-Manifolds into the Plane,* Proc. London Math. Soc. 92 (2) (2006), 471–504.. MR2205725 (2006k:57079)
[32] Yamamoto, T., *Singular fibers of two-colored maps and cobordism invariants,* Pacific J. Math. 234 (2) (2008), 379–398. MR2373454 (2009f:57046)

Department of Mathematics, Faculty of Science, Hokkaido University, Sapporo 060-0810, Japan
E-mail address: ohmoto@math.sci.hokudai.ac.jp

Contemporary Mathematics
Volume **569**, 2012
http://dx.doi.org/10.1090/conm/569/11240

Sections of Analytic Variety

B. Oréfice and J. N. Tomazella

ABSTRACT. We present a study of sections of a germ of analytic variety V through relations of equivalences defined by geometric subgroups, introduced by J. Damon, that in some way preserve V and we show that under certain conditions these equivalences relate. We also obtain a classification under $\mathcal{R}_V$-equivalence of sections of S_k-singularities.

1. Introduction

Let $(V, 0)$ be the germ of an analytic subvariety of K^n ($K = \mathbb{R}$ or $\mathbb{C}$). The study of equivalence of the germ $f : (K^n, 0) \to (K^p, 0)$ by requiring that the germs of diffeomorphisms on K^n preserve V has been done through groups $\mathcal{R}_V$, $\mathcal{K}_{RV}$, $\mathcal{K}_V$ and $\mathcal{A}_V$ which are geometric subgroups of the classical groups of singularity theory $\mathcal{R}$, $\mathcal{K}$, $\mathcal{A}$ (see [**29**]). This study began in [**1**] with V the discriminant of a versal unfolding and has been followed by [**27**], [**31**], [**4**], [**5**], [**6**], [**11**], etc. Results for unfolding, versality, determinacy theorems and topological triviality for these subgroups were presented in [**10**], [**11**] and [**12**]. A specific study for $\mathcal{K}_V$ was given in [**8**] and [**9**], for $\mathcal{R}_V$ in [**5**]. Results about $\mathcal{R}_V$-topological triviality of families of function germs were given in [**28**] and [**29**].

In this paper we present the equivalence relations defined by these subgroups and show that under certain conditions these equivalences are related. For groups $\mathcal{R}_V$ and $\mathcal{K}_{RV}$ we give algebraic and geometric conditions necessary and sufficient for finite determinacy and we obtain a classification of sections of S_k-singularities (defined in [**22**]) given by submersions under $\mathcal{R}_V$-equivalence.

2. Equivalence of Sections

Let $\mathcal{O}_n$ be the ring of analytic function germs $f : (K^n, 0) \to K$, with $K = \mathbb{R}$ or $\mathbb{C}$ and let m_n be its maximal ideal.

A germ of a subset $(V, 0) \subset (K^n, 0)$ is the germ of analytic variety if there exist analytic function germs $f_1, ..., f_r$ such that $V = \{x \in K^n : f_1(x) = ... = f_r(x) = 0\}$. We denote by $I(V)$ the ideal in $\mathcal{O}_n$ consisting of germs vanishing on V, that is, $I(V) = \{\phi \in \mathcal{O}_n : \phi(x) = 0 \, \forall \, x \in V\}$.

DEFINITION 2.1. A section of V is a set C obtained by the intersection of V with the germ of a variety $(X, 0) \subset (K^n, 0)$.

2010 *Mathematics Subject Classification.* Primary 32S05, Secondary 58K40.

The first author has been partially supported by FAPESP Grant 2008/53944-8. and by CAPES Grant BEX 2820/10-2.

All germs which we consider will be analytic. Classically the three principal notions of equivalence for map germs are $\mathcal{R}$, $\mathcal{A}$ and $\mathcal{K}$-equivalence. We consider the space of analytic germs $f : (K^n, 0) \to (K^p, 0)$, which we denote by $\mathcal{O}_{n,p}$. Let $\mathcal{D}_s$ denote the group of germs of diffeomorphisms $\varphi : (K^s, 0) \to (K^s, 0)$. The group $\mathcal{R} = \mathcal{D}_n$ acts on $\mathcal{O}_{n,p}$ by $\varphi.f = f \circ \varphi^{-1}$. The group $\mathcal{A} = \mathcal{D}_n \times \mathcal{D}_p$ acts on $\mathcal{O}_{n,p}$ by $(\varphi, \phi).f = \phi \circ f \circ \varphi^{-1}$. The group $\mathcal{K}$, contact equivalence, consists of $H \in \mathcal{D}_{n+p}$ such that there is a $\varphi \in \mathcal{D}_n$ so that $H \circ i = i \circ \varphi$ and $\pi \circ H = \varphi \circ \pi$ where $i(x) = (x, 0)$ is the inclusion $i : K^n \to K^{n+p}$ and $\pi(x, y) = x$ is the projection $\pi : K^{n+p} \to K^p$, therefore $H(x, y) = (\varphi(x), \theta(x, y))$ with $\theta(x, 0) = 0$. Then $\mathcal{K}$ acts on $\mathcal{O}_{n,p}$ by $(\varphi(x), H.f(x)) = H(x, f(x))$ that is $\text{graph}(H.f) = H(\text{graph}(f))$. Germs are $\mathcal{R}$, $\mathcal{A}$ or $\mathcal{K}$-equivalent if they lie in common orbits of the group actions.

We can study sections of V in two ways: X is given by a parametrization $g : (K^m, 0) \to (K^n, 0)$ or as an inverse image, $X = l^{-1}(0)$, $l : (K^n, 0) \to (K^p, 0)$. We describe below these different approaches.

In the first case, i.e., taking the section of V as an image of a map germ $g : (K^m, 0) \to (K^n, 0)$, $m < n$, the important equivalence relations are:

DEFINITION 2.2. Two germs g and $g' : (K^m, 0) \to (K^n, 0)$ are:

(i) $\mathcal{A}_V$-equivalent if there exist germs $\varphi \in \mathcal{D}_m$ and $\phi \in \mathcal{D}_n$ with $g \circ \varphi = \phi \circ g'$ and $\phi(V) = V$.

(ii) $\mathcal{K}_V$-equivalent if there exists $H = (\varphi, \theta) \in \mathcal{K}$ such that $H(x, g(x)) = (\varphi(x), g'(\varphi(x)))$ and $H(K^m \times V) \subseteq K^m \times V$. That is,

$$\mathcal{K}_V = \{H = (\varphi, \theta) \in \mathcal{K} \; : \; \theta(K^m \times V) \subseteq V\}.$$

This subgroup of $\mathcal{K}$ can be seen as the semi-direct product of $\mathcal{R}$ and $\mathcal{C}_V$ where

$$\mathcal{C}_V = \{H = (id, \theta) \in \mathcal{K} : \theta(K^n \times V) \subseteq V\}$$

and id is the germ at 0 of the identity mapping on K^m.

The $\mathcal{K}$-equivalence is the one which shows the equivalence of the germs of varieties $g^{-1}(0)$, then the $\mathcal{K}_V$-equivalence shows the equivalence of the germs of varieties $g^{-1}(V)$. The group $\mathcal{K}_V$ is a geometric subgroup of $\mathcal{K}$ which satisfies the basic theorems of theory of singularities, see [**8**]. Relations between $\mathcal{A}$ and $\mathcal{K}_V$ are given in [**9**].

When $X = l^{-1}(0)$ for a germ $l : (K^n, 0) \to (K^p, 0)$, we consider the following relations of equivalence:

DEFINITION 2.3. Two germs l, $l' : (K^n, 0) \to (K^p, 0)$ are:

(i) $\mathcal{R}_V$-equivalent if there exists $\phi \in \mathcal{D}_n$ with $\phi(V) = V$ and $l \circ \phi = l'$. That is,

$$\mathcal{R}_V = \{\phi \in \mathcal{R} : \phi(V) = V\}$$

(ii) $\mathcal{K}_{RV}$-equivalent if there exists $H = (\varphi, \theta) \in \mathcal{K}$ with $\varphi(V) = V$ and $H(x, l(x)) = (\phi(x), l'(\phi(x)))$. That is,

$$\mathcal{K}_{RV} = \{H = (\varphi, \theta) \in \mathcal{K} \; : \; \varphi(V) = V\}$$

When $p = 1$, the group $\mathcal{R}_V$ has been studied by some authors, such as [**5**], [**13**] and [**26**].

The group $\mathcal{K}_{RV}$ can be seen as semi-direct of product $\mathcal{R}_V$ and $\mathcal{C}$. Some results on this group were obtained by R. Atique in [**2**].

DEFINITION 2.4. Let $f : (K^n, 0) \to (K^p, 0)$ be an analytic map germ, we denote by I_f the ideal generated by components $f_1, ..., f_p$ of f. A germ $l : (K^p, 0) \to (K^s, 0)$

is a reduced defining equation for f if the ideal of the functions which vanish on $(f(K^n),0)$ is generated by I_l, i.e., $I(f(K^n)) = \langle l_1, ..., l_s\rangle$ as an $\mathcal{O}_p$-module.

In [**18**], it's shown that f and $g : (K^n, 0) \to (K^p, 0)$ are $\mathcal{K}$-equivalent if only if there exists $h \in \mathcal{D}_n$ taking I_f to I_g, that is $h^*(I_f) = I_g$, where $h^* : \mathcal{O}_n \to \mathcal{O}_n$ is the ring isomorphism induced by h. It follows that if f and g are $\mathcal{K}$-equivalent then there is diffeomorphism $h : (K^n, 0) \to (K^n, 0)$ such that $h(f^{-1}(0)) = g^{-1}(0)$. We remark that the converse is not true, for example, consider $f(x) = x$ and $g(x) = x^2$. However, we can show the following.

LEMMA 2.5. *Let f, $g : (\mathbb{C}^n, 0) \to (\mathbb{C}^p, 0)$, with I_f and I_g radical ideals. If there exists $h : (\mathbb{C}^n, 0) \to (\mathbb{C}^n, 0)$ such that $h(f^{-1}(0)) = g^{-1}(0)$, then f and g are $\mathcal{K}$-equivalent.*

PROOF. It's enough to show that $h^*(I_g) = I_f$. Let $l \in I_g$ then $l(z) = 0 \, \forall \, z \in g^{-1}(0)$. Taking $x \in f^{-1}(0)$, by hypothesis it follows that $h(x) \in g^{-1}(0)$. Hence,

$$h^*(l)(x) = (l \circ h)(x) = l(h(x)) = 0 \Rightarrow l \circ h \in I(V(I_f)) = \sqrt{I_f} = I_f$$

where the penultimate equality follows from the Hilbert's Nullstellensatz Theorem for germs of analytic functions (see [**21**]), therefore $h^*(I_g) \subseteq I_f$. As h^* is an isomorphism, we can show similarly that $(h^*)^{-1}(I_f) \subseteq I_g$, thus $I_f \subseteq h^*(I_g)$. Therefore $h^*(I_g) = I_f$. □

The following result connects $\mathcal{A}_V$ classification of section to $\mathcal{K}_{RV}$ classification of reduced defining equation of sections.

THEOREM 2.6. *Given g, $g' : (\mathbb{C}^m, 0) \to (\mathbb{C}^n, 0)$, let l, $l' : (\mathbb{C}^n, 0) \to (\mathbb{C}^p, 0)$ be defining equation for g, g' respectively.*
(a) *If g, g' are $\mathcal{A}_V$- equivalent then l, l' are $\mathcal{K}_{RV}$-equivalent.*
(b) *If g, g' are $\mathcal{A}$-finitely determined with $n > 2$ and l, l' are $\mathcal{K}_{RV}$-equivalent then g, g' are $\mathcal{A}_V$-equivalent.*

PROOF. (a) If g, g' are $\mathcal{A}_V$-equivalent then there exists $\phi \in \mathcal{D}_m$ and $\varphi \in \mathcal{D}_n$ such that $g \circ \phi = \varphi \circ g'$ and $\varphi(V) = V$. Hence, $\varphi(\mathrm{Im}(g')) = \mathrm{Im}(g)$,

$$(l \circ \varphi)^{-1}(0) = \varphi^{-1}(l^{-1}(0)) \supseteq \varphi^{-1}(\mathrm{Im}(g)) = \mathrm{Im}(g'), \text{then}$$

$$I((l \circ \varphi)^{-1}(0)) \subseteq I_{l'} \Rightarrow (l \circ \varphi)^{-1}(0) \supseteq l'^{-1}(0)$$

we have

$$(l' \circ \varphi^{-1})^{-1}(0) = \varphi(l'^{-1}(0)) \supseteq \varphi(\mathrm{Im}(g')) = \mathrm{Im}(g), \text{ then}$$

$$I((l' \circ \varphi^{-1})^{-1}(0)) \subseteq I_l \Rightarrow \varphi(l'^{-1}(0)) \supseteq l^{-1}(0) \Rightarrow l'^{-1}(0) \supseteq (l \circ \varphi)^{-1}(0)$$

Therefore

$$(l \circ \varphi)^{-1}(0) = l'^{-1}(0)$$

as I_l and $I_{l'}$ are radical ideals it follows from Lemma 2.5 that l, l' are $\mathcal{K}_{RV}$-equivalent, since $\varphi(V) = V$.

(b) If l, l' are $\mathcal{K}_{RV}$ equivalent then there exists $h \in \mathcal{D}_n$, $h(V) = V$ such that $(l \circ h)^{-1}(0) = l'^{-1}(0)$. Thus, $(h^{-1} \circ g)(\mathbb{C}^m) = g'(\mathbb{C}^m)$, and by uniqueness of the normalisation (see [**16**]) it follows that there exists $\alpha \in \mathcal{D}_m$ such that $h^{-1} \circ g = g' \circ \alpha$, that is, g, g' are $\mathcal{A}_V$-equivalent. □

With the hypothesis added that g, g' are germs of immersion we can connect the $\mathcal{A}_V$-equivalence to the $\mathcal{K}_V$-equivalence.

THEOREM 2.7. *Let g, $g' : (K^m, 0) \to (K^n, 0)$ be germs of immersion. Then g and g' are $\mathcal{A}_V$-equivalent if only if they are $\mathcal{K}_V$-equivalent.*

PROOF. We can suppose that $g(x) = (x, \varphi(x))$ and $g'(x) = (\phi_1(x), \phi_2(x))$ (changing the coordinate only in the source). Supposing g, g' are $\mathcal{K}_V$-equivalent then there exists $H \in \mathcal{D}_{m+n}$ given by $H(x, y) = (h(x), \theta(x, y))$ with $\theta(x, 0) = 0$, $\theta(x, V) \subseteq V$ and

$$H(x, x, \varphi(x)) = (h(x), \theta(x, x, \varphi(x))) = (h(x), \phi_1(h(x)), \phi_2(h(x)))$$

Let $\tilde{\theta} : (K^n, 0) \to (K^n, 0)$ with $\tilde{\theta}(y) = \theta(y_1, ..., y_m, y)$. Then

$$\tilde{\theta} \circ g(x) = \tilde{\theta}(x, \phi(x)) = \theta(x, x, \phi(x)) = (\phi_1(h(x)), \phi_2(h(x))) = g' \circ h(x)$$

and $\tilde{\theta}(V) \subseteq V$, $\tilde{\theta} \in \mathcal{D}_n$, since $H \in \mathcal{D}_n$. Therefore g, g' are $\mathcal{A}_V$-equivalent.

Since $\mathcal{A}_V$ is a subgroup of $\mathcal{K}_V$, it is immediate that if g and g' are $\mathcal{A}_V$-equivalent then they are $\mathcal{K}_V$-equivalent. □

The following example shows that if g or g' is not an immersion then Proposition 2.7 can be false.

EXAMPLE 2.8. Define $g, g' : (K, 0) \to (K^2, 0)$ by $g(t) = (t^2, 0)$, $g'(t) = (t^2, t^3)$ and $V = \{(0, y), y \in K\}$. Taking $H(t, x, y) = (t, x, y - tx)$, $H(t, V) \subseteq K \times V$ and $H(t, t^2, t^3) = (t, t^2, 0)$. Therefore g, g' are $\mathcal{K}_V$-equivalent, but g, g' are not $\mathcal{A}_V$-equivalent, since g, g' are not $\mathcal{A}$-equivalent, since g' is 3-$\mathcal{A}$-determined and g is not $\mathcal{A}$-finitely determined.

LEMMA 2.9. *If $I(V) = \langle \phi_i \rangle_{i=1,...,r}$ as an $\mathcal{O}_n$-module then $I(K^m \times V) = \langle \phi_i \rangle_{i=1,...,r}$ as an $\mathcal{O}_{m+n}$-module.*

PROOF. $P \in I(K^m \times V)$ if and only if $P(x, y) = 0$, $\forall\, (x, y) \in K^m \times V$.

By Hadamard's Lemma (see [**18**]), $P(x, y) - P(0, y) = \sum_{i=1}^{m} q_i(x, y)x_i$ where

$$q_i(x, y) = \int_0^1 \frac{\partial P}{\partial x_i}(tx_1, ..., tx_m, y_1, ..., y_n)dt$$

since $P|_{K^m \times \{y\}} \equiv 0$, we have that $q_i|_{K^m \times \{y\}} \equiv 0\ \forall\, y \in V$.

Therefore, $q_i(K^m \times V) \equiv 0$, i.e., $q_i \in I(K^m \times V)$. From the hypothesis, $P(0, y) = \sum_{i=1}^{r} h_i(y)\phi_i(y)$, with $h_i \in \mathcal{O}_n$, then

$$P(x, y) = \sum_{i=1}^{r} h_i(y)\phi_i(y) + \sum_{i=1}^{m} q_i(x, y)x_i,$$

therefore $I(K^m \times V) \subseteq I(V)_{\mathcal{O}_n} + I(K^m \times V)\, \langle x_1, ..., x_m \rangle$, hence

$$I(K^m \times V) \subseteq I(V)_{\mathcal{O}_{m+n}} + I(K^m \times V)\, \langle x_1, ..., x_m, y_1, ..., y_n \rangle .$$

By Nakayama Lemma (see [**18**]), we have

$$I(K^m \times V) \subseteq I(V)_{\mathcal{O}_{m+n}} = \langle \phi_i \rangle_{i=1,...,r} \text{ as an } \mathcal{O}_{m+n}\text{-module.}$$

As the other inclusion is immediate, the result follows. □

THEOREM 2.10. *Two germs g, $g' : (K^m, 0) \to (K^n, 0)$ are $\mathcal{K}_V$-equivalent if only if there exists $h \in \mathcal{D}_m$ such that $g' \circ h$, g are $\mathcal{C}_V$-equivalent.*

PROOF. If g, g' are $\mathcal{K}_V$-equivalent then there exist germs of diffeomorphism H and h with $H(x,y) = (h(x), \theta(x,y))$, $\theta(x,0) = 0$, $\theta(x,V) \subseteq V$ $\forall x \in K^m$ and $H \circ (id, g) = (h, g' \circ h)$. Let $H' = (h^{-1}, \pi_2)$, where $\pi_2(x,y) = y$ thus, $H' \circ H(id,g) = (id, \theta(id,g)) = (id, g' \circ h)$, that is, g and $g' \circ h$ are $\mathcal{C}_V$-equivalent.

If g, $g' \circ h$ are $\mathcal{C}_V$-equivalent then there exist germs of diffeomorphism H, with $H(x,y) = (x, \theta(x,y))$, $\theta(x,V) \subseteq V$ $\forall x \in K^m$ and $H(id,g) = (id, g' \circ h)$. Let $H'(x,y) = (h(x), y)$, thus $H' \circ H(x,y) = (h(x), \theta(x,y))$. Therefore $H' \circ H(id,g) = (h, \theta(id,g)) = (h, g' \circ h)$, that is, g and g' are $\mathcal{K}_V$-equivalent. □

THEOREM 2.11. *Let g, $g' : (K^m, 0) \to (K^n, 0)$ and $I(V)$ be the ideal generated by $\phi_1, ..., \phi_r$ as an $\mathcal{O}_n$-module, then:*
(a) *If g and g' are $\mathcal{C}_V$-equivalent then $\langle \phi_i \circ g \rangle_{i=1,\ldots,r} = \langle \phi_i \circ g' \rangle_{i=1,\ldots,r}$.*
(b) *If g and g' are $\mathcal{K}_V$-equivalent then $\langle \phi_i \circ g \rangle_{i=1,\ldots,r}$ and $\langle \phi_i \circ g' \rangle_{i=1,\ldots,r}$ are isomorphic. Therefore the germs $F = (\phi_1 \circ g, ..., \phi_r \circ g)$ and $G = (\phi_1 \circ g', ..., \phi_r \circ g')$ are $\mathcal{K}$-equivalent.*

PROOF. (a) It is enough to show that $\langle \phi_i \circ g' \rangle \subseteq \langle \phi_i \circ g \rangle$, since the other inclusion is shown analogously. We show that each $\phi_i \circ g'$ can be written $\phi_i \circ g' = \sum_{j=1}^{r} a_{ij}(\phi_j \circ g)$ with $a_{ij} \in \mathcal{O}_{m+n}$, as g and g' are $\mathcal{C}_V$-equivalent, there exists a germ of diffeomorphism H such that $H(x,y) = (x, \theta(x,y))$, $\theta(x,0) = 0$, $\theta(x,V) \subseteq V$ and $H(x, g(x)) = (x, \theta(x, g(x))) = (x, g'(x))$ as $\theta(x,V) \subseteq V \Rightarrow \phi_i(\theta(x,V)) \equiv 0$ $\Rightarrow \phi_i \circ \theta \in I(K^m \times V)$.

By Lemma 2.9, we can write each $\phi_i \circ \theta$ as

$$\phi_i \circ \theta(x,y) = \sum_{j=1}^{r} a_{ij}(x,y)\phi_j(y) \text{ with } a_{ij} \in \mathcal{O}_{m+n}$$

therefore

$$\phi_i \circ \theta(x, g(x)) = \sum_{j=1}^{r} a_{ij}(x, g(x))\phi_j(g(x)),$$

then $\phi_i \circ \theta(x, g(x)) = \phi_i \circ g'(x)$. It follows that $\phi_i \circ g' = \sum_{j=1}^{r} a_{ij}(\phi_j \circ g)$.

(b) If g and g' are $\mathcal{K}_V$-equivalent then there exists a diffeomorphism h of K^m such that $g' \circ h$ and g are $\mathcal{C}_V$-equivalent, by part (a), $\langle \phi_i \circ g \rangle = \langle \phi_i \circ g' \circ h \rangle$. Thus the isomorphism $h^* : \mathcal{O}_m \to \mathcal{O}_m$ takes $\langle \phi_i \circ g' \rangle$ to $\langle \phi_i \circ g \rangle$. Therefore these ideals are isomorphic. □

The following example shows that the condition $\langle \phi_i \circ g' \rangle$ isomorphic $\langle \phi_i \circ g \rangle$ does not imply that g and g' are $\mathcal{K}_V$-equivalent.

EXAMPLE 2.12. Let $V = \{(x,0), x \in K\} \subseteq K^2$, $I(V) = \langle y \rangle_{\mathcal{O}_2} = \langle \phi \rangle_{\mathcal{O}_2}$. We take g, $g' : (K,0) \to (K^2,0)$ with $g(t) = (t, t^2)$, $g'(t) = (0, t^2)$, so $(\phi \circ g)(t) = t^2 = (\phi \circ g')(t)$, therefore $\langle \phi \circ g \rangle = \langle \phi \circ g' \rangle$. But g and g' are not $\mathcal{K}_V$-equivalent, since they are not $\mathcal{K}$-equivalent.

However, the following result is true.

THEOREM 2.13 ([**14**]). *Let $V = \phi^{-1}(0)$ where $\phi : (K^n, 0) \to (K, 0)$ be a submersion, let g, $g' : (K^m, 0) \to (K^n, 0)$ be embedding and $h : (K^n, 0) \to (K^n, 0)$ be a diffeomorphism such that $h^*(\langle \phi \circ g' \rangle) = \langle \phi \circ g \rangle$, then g and g' are $\mathcal{K}_V$-equivalent.*

THEOREM 2.14. *Let $(V,0) \subset (K^n, 0)$. Two germs l, $l' : (K^n, 0) \to (K^r, 0)$ are $\mathcal{K}_{RV}$-equivalent if and only if there exists germ of diffeomorphism $\phi : (K^n, 0) \to (K^n, 0)$ with $\phi(V) = V$ such that $l' \circ \phi$ and l are $\mathcal{C}$-equivalent.*

PROOF. If l and l' are $\mathcal{K}_{RV}$-equivalent, then there exist germs of diffeomorphism H and ϕ with $H(x,y) = (\phi(x), \theta(x,y))$, $\theta(x,0) = 0$, $\phi(V) = V$ and $H \circ (id, l) = (\phi, l' \circ \phi)$.

Let $H' = (\phi^{-1}, \pi_2)$, where $\pi_2(x,y) = y$. Then $H' \circ H(id,l) = (id, \theta(id,l)) = (id, l' \circ \phi)$, that is, l and $l' \circ \phi$ are $\mathcal{C}$-equivalent and $\phi(V) = V$.

Conversely, if l and $l' \circ \phi$ are $\mathcal{C}$-equivalent and $\phi(V) = V$, then there exists germ of diffeomorphism H with $H(x,y) = (x, \theta(x,y)), H(id,l) = (id, l' \circ \phi)$.

Let $H'(x,y) = (\phi(x), y)$. We have that $H' \circ H(x,y) = (\phi(x), \theta(x,y))$. Therefore $H' \circ H(id,l) = (\phi, \theta(id,l)) = (\phi, l' \circ \phi)$, then l, l' are $\mathcal{K}_{RV}$-equivalent. □

In this work, given an analytic variety germ $(V,0) \subset (K^n,0)$, we will always define a section of V through an equation $h = 0$, with $h : (K^n,0) \to (K^p,0)$. Hence, our main interest is action of the groups $\mathcal{R}_V$ and $\mathcal{K}_{RV}$ on the set $\mathcal{O}(n,p)$ of the germs of analytic maps $h : (K^n,0) \to (K^p,0)$.

As an application of the equivalence studied, we will see in the following proposition that the $\mathcal{A}$ classification of bigerms reduces to $\mathcal{A}_V$ or $\mathcal{K}_{RV}$ classification of sections. This result extends for any dimension of the result of R. Atique [**2**].

A bigerm $G : (K^m,S) \to (K^n,0)$, $S = \{x_1,x_2\}$ and $x_i \in K^m$, $i = 1,2$ is the equivalence classes of the maps that coincide in the neighborhood of S. Without loss of generality we can consider $x_1 = x_2 = 0$ and the germ G as

$$G : \begin{cases} f : (K^m,0) \to (K^n,0) \\ g : (K^m,0) \to (K^n,0) \end{cases}$$

(The equivalence of bigerm is defined by different diffeomorphisms in source and a unique diffeomorphism in the target, see [**2**]).

THEOREM 2.15. *Consider the bigerms*

$$G : \begin{cases} f : (K^m,0) \to (K^n,0) \\ g : (K^m,0) \to (K^n,0) \end{cases} \qquad G' : \begin{cases} f : (K^m,0) \to (K^n,0) \\ g' : (K^m,0) \to (K^n,0) \end{cases}$$

with $n > 2$ and f, g, g' $\mathcal{A}$-finitely determined. Let l and l' be the defining equation for $g(K^m)$ and $g'(K^m)$ respectively and let V be the image of f. The following statements are equivalent
(i) *G, G' are $\mathcal{A}$-equivalent,*
(ii) *g, g' are $\mathcal{A}_V$-equivalent,*
(iii) *l, l' are $\mathcal{K}_{RV}$-equivalent.*

PROOF. If G, G' are $\mathcal{A}$-equivalent then there exist diffeomorphisms ϕ_f, ϕ_g : $(K^m,0) \to (K^m,0)$ and $\varphi : (K^n,0) \to (K^n,0)$ such that $f \circ \phi_f = \varphi \circ f$ and $g \circ \phi_g = \varphi \circ g'$.

Hence $\varphi(\mathrm{Im}(f)) \subseteq \mathrm{Im}(f)$. It follows that $\varphi(V) \subseteq V$, then g, g' are $\mathcal{A}_V$-equivalent.

Assuming that g, g' are $\mathcal{A}_V$- equivalent, there exist diffeomorphisms ϕ_g : $(K^m,0) \to (K^m,0)$ and $\varphi : (K^n,0) \to (K^n,0)$ such that $g \circ \phi_g = \varphi \circ g'$ and $\varphi(\mathrm{Im}(f)) \subseteq \mathrm{Im}(f)$. Therefore $(\varphi \circ f)(K^n)) \subseteq f(K^n)$, by uniqueness of normalisation (see [**16**], [**17**]), there exists ϕ such that $f \circ \phi = \varphi \circ f$, hence G, G' are $\mathcal{A}$-equivalent. It follows from Proposition 2.6 that (ii) is equivalent to (iii). □

3. Finite Determinacy

Results on finite determinacy of groups ($\mathcal{G}_V = \mathcal{R}_V$ or $\mathcal{K}_{RV}$) have been obtained by some authors. The group $\mathcal{R}_V$ was studied in [**5**], [**13**] and [**26**]. There are not many results about group $\mathcal{K}_{RV}$ but there are some results in [**2**].

Our interest is in groups of diffeomorphisms that preserve an analytic variety V. To get these diffeomorphisms the basic technique is the integration of vector fields tangent to V. The main references for studying these vector fields are [**5**], [**8**] and [**30**].

Definition 3.1. ([**5**]) Let $(V, 0) \subset (K^n, 0)$ be the germ of an analytic variety and let ξ be a germ of analytic vector field on $(K^n, 0)$. Then ξ is said to be *logarithmic* for $(V, 0)$ if, when considered as a derivation $\xi : \mathcal{O}_n \to \mathcal{O}_n$, $g \mapsto \xi g$, we have $\xi g \in I(V)$ for all $g \in I(V)$, where $\xi g = dg(\xi)$. The $\mathcal{O}_n$- module of such vector fields is denoted by Θ_V or $Derlog(V)$, that is,

$$\Theta_V = \{\xi \in \mathcal{O}_n^n \ : \ \xi g \in I(V), \forall g \in I(V)\}$$

The main properties of Θ_V are given by the following proposition.

Theorem 3.2 ([**5**]) (i)). *Θ_V is the set of the vector field germs at $0 \in K^n$ which are tangent to the set of regular points of V.*
(ii) *Θ_V is finitely generated as an $\mathcal{O}_n$-module.*
(iii) *If $\xi \in \Theta_V$ and $\xi(0) = 0$, then the flow ϕ_t generated by ξ preserves $(V, 0)$. Thus $\phi_t \in \mathcal{R}_V$ for all t.*

Definition 3.3. A germ $f : (K^n, 0) \to (K^p, 0)$ is k-$\mathcal{G}_V$-finitely determined if all germs g, such that $j^k g(0) = j^k f(0)$ is $\mathcal{G}_V$-equivalent to f. If f is k-$\mathcal{G}_V$-finitely determined for some k, we say that f is $\mathcal{G}_V$-finitely determined.

Definition 3.4. Let $f : (K^n, 0) \to (K^p, 0)$. We denote by Θ_V^0 the subset of Θ_V constituted by vector fields that vanish at origin. The tangent spaces and the extended tangent spaces of the groups $\mathcal{R}_V$ and $\mathcal{K}_{RV}$ are respectively:

$$T\mathcal{R}_V(f) = df(\Theta_V^0), \qquad T\mathcal{R}_{V,e}(f) = df(\Theta_V)$$
$$T\mathcal{K}_{RV}(f) = df(\Theta_V^0) + f^*(m_p)\mathcal{O}_n^p, \quad T\mathcal{K}_{RV,e}(f) = df(\Theta_V) + f^*(m_p)\mathcal{O}_n^p$$

The next proposition gives a necessary condition to $\mathcal{G}_V$-finite determinacy.

Theorem 3.5. *If $f : (K^n, 0) \to (K^p, 0)$ is $\mathcal{G}_V$-finitely determined, then*

$$T\mathcal{G}_V(f) \supset m_n^l \mathcal{O}_n^p,$$

for some positive integer l.

For groups $\mathcal{R}_V$ and $\mathcal{K}_{RV}$, the essential part of proof, as in groups $\mathcal{K}$, $\mathcal{A}$, $\mathcal{L}$ and $\mathcal{R}$, is to note that $J^k\mathcal{G}_V$, the set of k-jets of elements $\mathcal{G}_V$, is a Lie group acting smoothly on $J^k(n, p)$. R. Pellikaan in [**25**] shows that $J^k\mathcal{R}_V$ is a Lie group and as $\mathcal{K}_{RV} = \mathcal{R}_V.\mathcal{C}$ (semi-direct product) it follows that $J^k\mathcal{K}_{RV}$ is Lie group.

The theorems on infinitesimal criterion for $\mathcal{G}_V$-finite determinacy are known, see for instance [**2**], [**6**], [**5**].

Theorem 3.6. *If $T\mathcal{G}_V(f) \supseteq m_n^k \mathcal{O}_n^p$ then f is k-$\mathcal{G}_V$-determined, $\mathcal{G}_V = \mathcal{R}_V$, $\mathcal{K}_{RV}$.*

The following corollary is useful for the classification of germs, since it reduces the estimate of tangent space in the finite dimension case.

COROLLARY 3.7. *If* $T\mathcal{G}_V(f) + m_n^{k+1}\mathcal{O}_n^p \supseteq m_n^k\mathcal{O}_n^p$ *then* f *is* k-$\mathcal{G}_V$*-determined.*

PROOF. This follows Nakayama Lemma, since $T\mathcal{G}_V(f)$ is an $\mathcal{O}_n$-module, for $\mathcal{G}_V = \mathcal{R}_V$, $\mathcal{K}_{RV}$. □

DEFINITION 3.8. (a) We denote by Θ_V^1 the subset of Θ_V constituted by vector fields whose 1-jets are null. If $f : (K^n, 0) \to (K^p, 0)$ we denote

$$T\mathcal{R}_V^1(f) = df(\Theta_V^1)$$
$$T\mathcal{K}_{RV}^1(f) = df(\Theta_V^1) + f^*(m_p)m_n\mathcal{O}_n^p.$$

(b) We denote by $H^k(n,p)$ the vector space of all the maps $K^n \to K^p$, where the components are homogeneous polynomials of degree k.
(c) The standard k–jet space $m_n\mathcal{O}_n/m_n^{k+1}\mathcal{O}_n$ is denoted $J^k(n,1)$.

An efficient method for the classification of singularities finitely determined is the *complete transversal method* described by Bruce, Kirk and du Plessis in [**3**].

THEOREM 3.9 ([**3**]). *Let* $f : (K^n, 0) \to (K^p, 0)$ *be an analytic germ and let* T *be a vector subspace of the set* $H^{k+1}(n,p)$*, such that* $J^{k+1}(T\mathcal{G}_V^1 f) + T \supset H^{k+1}(n,p)$*. Then every* $(k+1)$*-jet* g *with* $j^k g = j^k f$ *lies in the same* $\mathcal{G}_V^1$*-orbit of some* $(k+1)$*-jet of the form* $j^{k+1}f + t$*, for some* $t \in T$*.*

COROLLARY 3.10. *If* $T\mathcal{G}_V^1 f \supseteq m_n^{k+1}\mathcal{O}_n^p$ *then* f *is* k-$\mathcal{G}_V$*-determined.*

LEMMA 3.11. *Let* $f : (K^n, 0) \to (K^p, 0)$ *be an analytic germ and let* $\{\xi_1, ..., \xi_r\}$ *be generators of* Θ_V*. Then* f *is* $\mathcal{K}_{RV}$*-finitely determined if and only if* $Jf(\Theta_V) + f^*(m_p) \supseteq m_n^k$ *for some* k*, where* $Jf(\Theta_V)$ *is the ideal in* $\mathcal{O}_n$ *generated by all the* $(p \times p)$*-minors of the matrix* $(\nabla f_i(\xi_j))_{i=1,\ldots,p}^{j=1,\ldots,r}$ *(if* $p > r$*, then* $Jf(\Theta_V) = 0$*).*

PROOF. The proof is an adaptation of the result correspondent for the group $\mathcal{K}$ ([**15**], Lemma 2.12).

If f is $\mathcal{K}_{RV}$-finitely determined, then by Proposition 3.5, there exists l such that $df(\Theta_V) + f^*(m_p)\mathcal{O}_n^p \supset m_n^l\mathcal{O}_n^p$. Let $u \in m_n^{l\,p} = m_n^k$, $u = u_1.\ldots.u_p$ where $u_i \in m_n^l$. Since

$$(0, \ldots, 0, u_i, 0, \ldots, 0) \in df(\Theta_V) + f^*(m_p)\mathcal{O}_n^p$$

there exists $\eta_i \in \Theta_V$ such that

$$df(\eta_i) = (0, \ldots, 0, u_i, 0, \ldots, 0)\text{mod} f^*(m_p)\mathcal{O}_n^p$$

with $\eta_i = a_1^i\xi_1 + \ldots + a_r^i\xi_r$ and $a_j^i \in \mathcal{O}_n$.

Hence,

$$df\begin{pmatrix} a_1^1\xi_1^1 + \ldots + a_r^1\xi_r^1 & \cdots & a_1^p\xi_1^1 + \ldots + a_r^p\xi_r^1 \\ a_1^1\xi_1^2 + \ldots + a_r^1\xi_r^2 & \cdots & a_1^p\xi_1^2 + \ldots + a_r^p\xi_r^2 \\ \vdots & \vdots & \vdots \\ a_1^1\xi_1^n + \ldots + a_r^1\xi_r^n & \cdots & a_1^p\xi_1^n + \ldots + a_r^p\xi_r^n \end{pmatrix} = \begin{pmatrix} u_1 & 0 & 0 & \cdots & 0 \\ 0 & u_2 & 0 & \cdots & 0 \\ \vdots & \vdots & \vdots & \vdots & \vdots \\ 0 & 0 & 0 & \cdots & u_p \end{pmatrix}$$

$\text{mod} f^*(m_p)\mathcal{O}_n^p$. Thus,

$$df\begin{pmatrix} \xi_1^1 & \xi_2^1 & \cdots & \xi_r^1 \\ \xi_1^2 & \xi_2^2 & \cdots & \xi_r^2 \\ \vdots & \vdots & \vdots & \vdots \\ \xi_1^n & \xi_2^n & \cdots & \xi_r^n \end{pmatrix} \cdot \begin{pmatrix} a_1^1 & a_1^2 & \cdots & a_1^p \\ a_2^1 & a_2^2 & \cdots & a_2^p \\ \vdots & \vdots & \vdots & \vdots \\ a_r^1 & a_r^2 & \cdots & a_r^p \end{pmatrix} = \begin{pmatrix} u_1 & 0 & 0 & \cdots & 0 \\ 0 & u_2 & 0 & \cdots & 0 \\ \vdots & \vdots & \vdots & \vdots & \vdots \\ 0 & 0 & 0 & \cdots & u_p \end{pmatrix}$$

$\mathrm{mod} f^*(m_p)\mathcal{O}_n^p$, otherwise,

$$\begin{pmatrix} \nabla f_1(\xi_1) & \dots & \nabla f_1(\xi_r) \\ \nabla f_2(\xi_1) & \dots & \nabla f_2(\xi_r) \\ \vdots & \vdots & \vdots \\ \nabla f_p(\xi_1) & \dots & \nabla f_p(\xi_r) \end{pmatrix} \cdot \begin{pmatrix} a_1^1 & a_1^2 & \dots & a_1^p \\ a_2^1 & a_2^2 & \dots & a_2^p \\ \vdots & \vdots & \vdots & \vdots \\ a_r^1 & a_r^2 & \dots & a_r^p \end{pmatrix} = \begin{pmatrix} u_1 & 0 & 0 & \dots & 0 \\ 0 & u_2 & 0 & \dots & 0 \\ \vdots & \vdots & \vdots & \vdots & \vdots \\ 0 & 0 & 0 & \dots & u_p \end{pmatrix}$$

$\mathrm{mod} f^*(m_p)\mathcal{O}_n^p$. Taking the determinant of each side of the equation, it follows that $u \in Jf(\Theta_V) + f^*(m_p)$.

Assuming now that $Jf(\Theta_V) + f^*(m_p) \supseteq m_n^k$. It is enough to show that

$$Jf(\Theta_V)\mathcal{O}_n^p \subseteq df(\Theta_V).$$

Let $M = (\nabla f_i(\xi_{j_s}))_{i=1,\dots,p}^{s=1,\dots,p}$ be a $(p \times p)$ sub-matrix of the matrix $(\nabla f_i(\xi_j))_{i=1,\dots,p}^{j=1,\dots,r}$. Let $\{e_1, ..., e_p\}$ be the standard base of K^p, then $(\det M)e_k \in Jf(\Theta_V)\mathcal{O}_n^p$ and

$$(\det M)e_k = \begin{pmatrix} \nabla f_1(\xi_1) & \dots & \nabla f_1(\xi_r) \\ \nabla f_2(\xi_1) & \dots & \nabla f_2(\xi_r) \\ \vdots & \vdots & \vdots \\ \nabla f_p(\xi_1) & \dots & \nabla f_p(\xi_r) \end{pmatrix} \cdot \begin{pmatrix} 0 \\ \vdots \\ cof(\nabla f_k(\xi_{j_1})) \\ \vdots \\ \mathrm{cof}(\nabla f_k(\xi_{j_p})) \\ \vdots \\ 0 \end{pmatrix}$$

where $\mathrm{cof}(\nabla f_k(\xi_{j_s}))$ is the cofactor of $\nabla f_i(\xi_{j_s})$ in the matrix M, then

$$(\det M)e_k = df \begin{pmatrix} \xi_1^1 & \xi_2^1 & \dots & \xi_r^1 \\ \xi_1^2 & \xi_2^2 & \dots & \xi_r^2 \\ \vdots & \vdots & \vdots & \vdots \\ \xi_1^n & \xi_2^n & \dots & \xi_r^n \end{pmatrix} \cdot \begin{pmatrix} 0 \\ \vdots \\ cof(\nabla f_k(\xi_{j_1})) \\ \vdots \\ \mathrm{cof}(\nabla f_k(\xi_{j_p})) \\ \vdots \\ 0 \end{pmatrix},$$

Hence

$$(\det M)e_k = df \begin{pmatrix} \xi_{j_1}^1 \mathrm{cof}(\nabla f_k(\xi_{j_1})) + \dots + \xi_{j_p}^1 \mathrm{cof}(\nabla f_k(\xi_{j_p})) \\ \xi_{j_1}^2 \mathrm{cof}(\nabla f_k(\xi_{j_1})) + \dots + \xi_{j_p}^2 \mathrm{cof}(\nabla f_k(\xi_{j_p})) \\ \vdots \\ \xi_{j_1}^n \mathrm{cof}(\nabla f_k(\xi_{j_1})) + \dots + \xi_{j_p}^n \mathrm{cof}(\nabla f_k(\xi_{j_p})) \end{pmatrix}$$

$$= df(\mathrm{cof}(\nabla f_k(\xi_{j_1}))\xi_{j_1} + \mathrm{cof}(\nabla f_k(\xi_{j_2}))\xi_{j_2} + \dots + \mathrm{cof}(\nabla f_k(\xi_{j_p}))\xi_{j_p})$$

therefore, $(\det M)e_k \in df(\Theta_V)$ and $Jf(\Theta_V)\mathcal{O}_n^p \subseteq df(\Theta_V)$. □

The following proposition is the geometric criterion for the $\mathcal{K}_{RV}$-finitely determinacy.

THEOREM 3.12. *Let $f : (\mathbb{C}^n, 0) \to (\mathbb{C}^p, 0)$ and let $(V, 0) \subseteq (\mathbb{C}^n, 0)$. Then f is $\mathcal{K}_{RV}$-finitely determined if and only if $V(Jf(\Theta_V)) \cap f^{-1}(0) \subseteq \{0\}$.*

PROOF. By Lemma 3.11, f is $\mathcal{K}_{RV}$-finitely determined if and only if $Jf(\Theta_V)+f^*(m_p) \supseteq m_n^k$ for some k. Then, $V(Jf(\Theta_V)+f^*(m_p)) \subseteq V(m_n^k)$. Hence,

$$V(Jf(\Theta_V)) \cap V(f^*(m_p)) \subseteq \{0\} \text{ and } V(Jf(\Theta_V)) \cap f^{-1}(0) \subseteq \{0\}.$$

Going in the other direction, assuming that $V(Jf(\Theta_V)) \cap f^{-1}(0) \subseteq \{0\}$ we get

$$I(V(Jf(\Theta_V)+f^*(m_p))) \supseteq I(\{0\}).$$

Thus, by *Hilbert's Nullstellensatz Theorem* for analytic function germs (see [**21**]),

$$\sqrt{Jf(\Theta_V)+f^*(m_p)} \supseteq m_n,$$

hence $Jf(\Theta_V)+f^*(m_p) \supseteq m_n^k$ for some k. □

The following theorem is the geometric criterion for the $\mathcal{R}_V$-finite determinacy and its proof is a consequence of the proof of Proposition 3.12.

THEOREM 3.13. *Let $(V,0) \subseteq (\mathbb{C}^n,0)$ be an analytic variety germ and let $f : (\mathbb{C}^n,0) \to (\mathbb{C},0)$ be an analytic function germ. Let $V(f) = \{x \in \mathbb{C}^n : \xi f(x) = 0$ for all $\xi \in \Theta_V\}$. Then f is $\mathcal{R}_V$-finitely determined if and only if $V(f) = \{0\}$ or $\emptyset$.*

In the real case, the necessary condition remains true, that is, if f is $\mathcal{G}_V$-finitely determined then: $V(Jf(\Theta_V)) \cap f^{-1}(0) \subseteq \{0\}$ for $\mathcal{G}_V = \mathcal{K}_{RV}$ and the set $\{x \in \mathbb{R}^n : \xi f(x) = 0 \text{ for all } \xi \in \Theta_V\}$ is $\{0\}$ or $\emptyset$ for $\mathcal{G}_V = \mathcal{R}_V$.

THEOREM 3.14. *Let $(V,0) \subseteq (K^n,0)$ be an analytic variety germ and let $f : (K^n,0) \to (K,0)$. If f is $\mathcal{R}_V$-finitely determined then $f^{-1}(0)$ is transverse to V away from 0.*

PROOF. By hypothesis there exists a positive integer l such that $df(\Theta_V) \supseteq m_n^l$, thus we have that $V(df(\Theta_V)) \subseteq \{0\}$. However, $V(df(\Theta_V))$ describes the set of points where $f^{-1}(0)$ is not transverse to V, thus $f^{-1}(0)$ is transverse to V away from 0. □

In the following example we show that the converse of Theorem 3.14 is not true.

EXAMPLE 3.15. Let $V \subseteq (\mathbb{C}^2,0)$ be defined by $\varphi(x,y) = x^3 - y^2 = 0$. We have that Θ_V is generated by $\alpha_1 = (2x, 3y), \alpha_2 = (2y, 3x^2)$. Let $f(x,y) = y^2$, $f^{-1}(0) = x$-axis. It is transverse to V away from 0 but f is not $\mathcal{R}_V$-finitely determined.

Bruce and Roberts, in [**5**], introduced a generalization of Milnor number of a function germ f related to a variety V by $\mu_{BR}(f,V) = \dim \frac{\mathcal{O}_n}{df(\theta_V)}$. We will call it the Bruce-Roberts number of f with respect to V. Like the Milnor number of f, this number shows some geometric properties of f and V. For instance, if one considers the group $\mathcal{R}_V$, then f is finitely determined with respect to the action of $\mathcal{R}_V$ on $\mathcal{O}_n$ if and only if $\mu_{BR}(f,V)$ is finite and, in this case, the codimension of the orbit of f under this action is equal to $\mu_{BR}(f,V)$. We refer to [**5**] for more details.

4. Section of the Singularities S_k

The goal of this section is to classify germs of submersion $h : (\mathbb{C}^3, 0) \to (\mathbb{C}, 0)$ by action of $\mathcal{R}_V$ where V is the image of S_k-singularities classified by Mond in [**22**], $S_k = \{(x, y^2, y^3 + x^{k+1}y)\}$, $k \geq 1$, in this case the sections are space curves $C_k = S_k \cap h^{-1}(0)$.

We have that $\varphi_k(x, y, z) = 2x^{k+1}y^2 + y^3 - z^2 + x^{2(k+1)}y = 0$ is a reduced defining equation for the images of S_k (obtained using Gröbner basis methods) and the germ of the singular set of S_k is a smooth curve given by $\Sigma_k = \{(x, -x^{k+1}, 0)\}$.

Given a finite function germ $f : (C, 0) \to (\mathbb{C}, 0)$ on a complex analytic reduced space curve $(C, 0)$, in [**24**] is shown that

$$\mu(f|_C) = \mu(C) + \deg(f) - 1,$$

where $\mu(C)$ is the Milnor number of the curve, as defined by Buchweitz and Greuel [**7**], $\deg(f)$ is the degree of f (that is, the number of inverse images of a generic value), and $\mu(f|_C)$ is the Milnor number of f, introduced by Goryunov [**19**] for curves in $(\mathbb{C}^3, 0)$ and by Mond and D. van Straten [**23**] for the general case of curves in $(\mathbb{C}^n, 0)$. As Σ_k is a smooth curve it follows that $\mu(h|_{\Sigma_k}) = \deg(h|_{\Sigma_k}) - 1$.

THEOREM 4.1. *Θ_{S_k} is the $\mathcal{O}_3$-module generated by*

$$\eta_1 = (2x, 2(k+1)y, 3(k+1)z) \qquad \eta_2 = (0, 2z, x^{2(k+1)} + 4x^{k+1}y + 3y^2)$$
$$\eta_3 = (x^{k+1} + 3y, -2(k+1)x^k y, 0) \qquad \eta_4 = (z, 0, (k+1)(x^{2k+1}y + x^k y^2))$$

PROOF. If $\varphi_k = 2x^{k+1}y^2 + y^3 - z^2 + x^{2(k+1)}y$ then $S_k = \varphi_k^{-1}(0)$. We have that

$$d\varphi_k = (\varphi_{k_x}, \varphi_{k_y}, \varphi_{k_z}) = (2(k+1)x^k y^2 + 2(k+1)x^{2k+1}y, 4x^{k+1}y + 3y^2 + x^{2(k+1)}, -2z),$$
$$d\varphi_k(\eta_1) = 6(k+1)\varphi_k \text{ and } d\varphi_k(\eta_2) = d\varphi_k(\eta_3) = d\varphi_k(\eta_4) = 0,$$

so η_1, η_2, η_3 and η_4 are in Θ_{S_k}.

Now if $\xi \in \Theta_{S_k}$ then $d\varphi_k(\xi) = \lambda\varphi_k$ for some $\lambda \in \mathcal{O}_3$, then

$$d\varphi_k(\xi - \frac{1}{6(k+1)}\lambda\eta_1) = d\varphi_k(\xi) - \frac{\lambda}{6(k+1)}d\varphi_k(\eta_1) = d\varphi_k(\xi) - \lambda\varphi_k = 0,$$

so we need only to check that η_1, η_2, η_3 and η_4 generate all germs of vector ξ such that $d\varphi_k(\xi) = 0$. Writing $\xi = (\xi_1, \xi_2, \xi_3)$, we have $d\varphi_k(\xi) = 0$ if and only if

$$\xi_1\varphi_{k_x} + \xi_2\varphi_{k_y} - \xi_3\varphi_{k_z} = 0,$$

if and only if (ξ_1, ξ_2, ξ_3) is a syzygy of $(\varphi_{k_x}, \varphi_{k_y}, \varphi_{k_z})$ which is generated by

$$(\varphi_{k_y}, -\varphi_{k_x}, 0) = (y + x^{k+1})\eta_1, \ (\varphi_{k_z}, 0, -\varphi_{k_x}) = -\frac{1}{2}\eta_4 \text{ and } (0, -\varphi_{k_z}, \varphi_{k_y}) = \eta_2,$$

so the result follows. □

We follow the standard classification techniques, namely to classify germs inductively at the jet level until a sufficient jet is obtained (and hence produces a finitely determined germ). For this we use the complete transversal method, Mather's Lemma and the results on finite determinacy. To calculate the Milnor number and the Bruce-Roberts number we use SINGULAR (see [**20**]).

THEOREM 4.2. *Let V be the image of S_k, $k \geq 1$ and $h : (\mathbb{C}^3, 0) \to (\mathbb{C}, 0)$ be a germ of submersion $\mathcal{R}_V$-finitely determined with $\mu(h_{|\Sigma_k}) \leq k$. Then h is $\mathcal{R}_V$-equivalent to one of the following germs*

h	$\mu_{BR}(h, S_k)$	$\mu(C_k)$	$\mu(h_{\vert\Sigma_k})$
x	1	2	0
$y + a_2x^2,\ a_2 \neq 0$	3	5	1
$y + a_3x^3 + a_4x^4,\ a_3 \neq 0$	5	8	2
$y + a_4x^4 + a_5x^5 + a_6x^6,\ a_4 \neq 0$	7	11	3
$\vdots$	$\vdots$	$\vdots$	$\vdots$
$y + a_{k+1}x^{k+1} + a_{k+2}x^{k+2} + \cdots + a_{2k}x^{2k},$ $a_{k+1} \neq 0, 1$	$2k+1$	$3k+2$	k
$y + x^{k+1} + a_lx^l + \cdots + a_{l+k-1}x^{l+k-1},$ $l \geq k+2,\ a_l \neq 0$	$l+k$	$2l+k$	k
$y + a_lx^l + a_{l+1}x^{l+1} + \cdots + a_{l+k-1}x^{l+k-1},$ $l \geq k+2,\ a_l \neq 0$	$l+k$	$l+2k+1$	k
$z + a_2x^2,\ a_2 \neq 0$	4	6	1
$z + a_3x^3 + a_4x^4,\ a_3 \neq 0$	7	10	2
$z + a_4x^4 + a_5x^5 + a_6x^6,\ a_4 \neq 0$	10	14	3
$\vdots$	$\vdots$	$\vdots$	$\vdots$
$z + a_lx^l + a_{l+1}x^{l+1} + \cdots + a_{2l-2}x^{2l-2},$ $a_l \neq 0, l \leq k+1$	$3l-2$	$4l-2$	$l-1$

In what follows we give a sketch of the proof.

With the intention of setting ideas, we start with the case $k = 1$:

Section of S_1.

We start classifying by the 1-jets. By Mather's Lemma and change of scale the orbit in $J^1(3,1)$ of the action of $J^1\mathcal{R}_V$ reduces to: x, y, z and 0.

Case (i): Classification of germs h for which the 1-jet is x.

Consider the germ $h(x,y,z) = x$. Then $T\mathcal{R}_V(h) + \mathcal{M}^2 \supset \mathcal{M}^1$ and x is 1-$\mathcal{R}_V$-determined.

Case (ii): Classification of germs h for which the 1-jet is y.

Consider the germ $h(x,y,z) = y$, by geometric criterion, y is not $\mathcal{R}_V$-finite. Then a complete transversal is given by $h = y + a_lx^l$.

(a): If $l = 2$, $a_2 \neq 0$ and $a_2 \neq 1$ then $h(x,y,z) = y + a_2x^2$ is 2-$\mathcal{R}_V$-determined.

When $a_2 = 1$, $h = y + x^2$ is not $\mathcal{R}_V$-finite, a complete transversal is $y + x^2 + a_lx^l$.

Let $h = y + x^2 + a_lx^l$ with $a_l \neq 0$, then $T\mathcal{R}_V(h) + \mathcal{M}^{l+2} \supseteq \mathcal{M}^{l+1}$ hence $y + x^2 + a_lx^l$ is $(l+1)$-$\mathcal{R}_V$-determined. It shows that in fact it is (l)-$\mathcal{R}_V$-determined for all $a_l \neq 0$.

(b): If $l \neq 2$ and $a_l \neq 0$ then $h = y + a_lx^l$ is l-$\mathcal{R}_V$-determined.

Case (iii): Classification of the germs for which the 1-jet is z.

Consider the germ $h(x,y,z) = z$ which is not $\mathcal{R}_V$-finite.

As $J^2(T\mathcal{R}^1_V(h)) + \mathbb{C}\{x^2, xy, y^2\} \supseteq H^2$, all the 2-jets which 1-jet is z are $\mathcal{R}_V$-equivalent to $z + a_2x^2 + bxy + cy^2$. We have

$$J^2(T\mathcal{R}_V(h)) = \big\langle 2a_2x^2 + 3bxy + 4cy^2 + 3z, 2bxz + 4cyz + 3y^2, 6a_2xy + 3by^2, 2a_2xz + byz\big\rangle,$$

therefore $y^2 \in j^2(T\mathcal{R}_V(h))$, $\forall\, a_2,\, b,\, c$ if $a_2 \neq 0$, $xy \in j^2(T\mathcal{R}_V(h))$, $\forall\, b, c$. Hence, by Mather's Lemma, we have:

For $a_2 \neq 0$, $h = z + a_2x^2 + bxy + cy^2 \sim z + a_2x^2$ is $2-\mathcal{R}_V$-determined.
For $a_2 = 0$, $b \neq 0$, $h = z + bxy + cy^2 \sim z + bxy$ and $\mu(h|_{\Sigma_1}) = 2 > 1$.
For $a_2 = b = 0$, $j^2(h) = z$ and $\mu(h|_{\Sigma_1}) \geq 2$.

Sections of S_k.
We have that orbits in $J^1(3,1)$ of the act of $J^1\mathcal{R}_V$ are: x, y, z and 0.

Case (i): Classification of the germs h for which the 1-jet is x.
Let $h(x,y,z) = x$, so $T\mathcal{R}_V(h) \supset \mathcal{M}^1$ and x is 1-$\mathcal{R}_V$-determined.

Case (ii): Classification of the germs h for which the 1-jet is y.
The generators of the tangent space are: $\eta_1 h = 2(k+1)y$, $\eta_2 h = 2z$, $\eta_3 h = -2(k+1)x^3y$ and $\eta_4 h = 0$ and a l-complete transversal is $y + a_l x^l$.
Let $h = y + a_l x^l$, then $\eta_1 h = 2la_l x^l + 2(k+1)y$, $\eta_2 h = 2z$, $\eta_3 h = la_l x^{k+l} + 3la_l x^{l-1}y - 2(k+1)x^k y$ and $\eta_4 h = la_l x^{l-1}z$.

(a): If $l = k+1$, $a_{k+1} \neq 0$ and $a_{k+1} \neq 1$ then $h = y + a_{k+1}x^{k+1}$ is $\mathcal{R}_V$-finitely determined.
In this case $T\mathcal{R}_V(h) \supseteq \mathcal{M}^{2k+1}$, if $a_{k+1} \neq 0$ and $a_{k+1} \neq 1$, thus $y + a_{k+1}x^{k+1}$ is $(2k+1)$-$\mathcal{R}_V$-determined. Successively using the complete transversal method we obtain $F = y + a_{k+1}x^{k+1} + a_{k+2}x^{k+2} + \ldots + a_{2k}x^{2k} + a_{2k+1}x^{2k+1}$, as $x^{2k+1} \in T\mathcal{R}_V(F)$, then by Mather's Lemma $F \sim y + a_{k+1}x^{k+1} + a_{k+2}x^{k+2} + \ldots + a_{2k}x^{2k}$.
For $a_{k+1} = 1$, let $h = y + x^{k+1}$, then $\eta_1 h = 2(k+1)x^{k+1} + 2(k+1)y$, $\eta_2 h = 2z$, $\eta_3 h = (k+1)x^{2k+1} + (k+1)x^k y$ and $\eta_4 h = (k+1)x^k z$. Hence by geometric criterion, $y + x^{k+1}$ is not $\mathcal{R}_V$-finite. A l-complete transversal for $y + x^{k+1}$ is $y + x^{k+1} + a_l x^l$, $l > (k+1)$.
Let $h = y + x^{k+1} + a_l x^l$ with $a_l \neq 0$. Then $\eta_1 h = 2(k+1)x^{k+1} + 2(k+1)y + 2a_l l x^l$, $\eta_2 h = 2z$, $\eta_3 h = (k+1)x^{2k+1} + (k+1)x^k y + la_l x^{k+l} + 3la_l x^{l-1}y$ and $\eta_4 h = (k+1)x^k z + la_l x^{l-1}z$. Hence $T\mathcal{R}_V(h) + \mathcal{M}^{k+l+1} \supseteq \mathcal{M}^{k+l}$, thus $y + x^{k+1} + a_l x^l$ is $(k+l)$-$\mathcal{R}_V$-determined. Successively using complete transversal method we obtain $F = y + x^{k+1} + a_l x^l + \ldots + a_{k+l-1}x^{k+l-1} + a_{k+l}x^{k+l}$, as $x^{k+l} \in T\mathcal{R}_V(F)$, then by Mather's Lemma $F \sim y + x^{k+1} + a_l x^l + \ldots + a_{k+l-1}x^{k+l-1}$.

(b): If $l \neq (k+1)$ and $a_l \neq 0$ then $h = y + a_l x^l$ is $\mathcal{R}_V$-finitely determined.
If $l > (k+1)$, $T\mathcal{R}_V(h) + \mathcal{M}^{k+l+1} \supseteq \mathcal{M}^{k+l}$ and $y + a_l x^l$ is $(k+l)$-$\mathcal{R}_V$-determined. Successively using the complete transversal method we obtain $F = y + a_l x^l + \ldots + a_{k+l-1}x^{k+l-1} + a_{k+l}x^{k+l}$, as $x^{k+l} \in T\mathcal{R}_V(F)$, then by Mather's Lemma $F \sim y + x^{k+1} + a_l x^l + \ldots + a_{k+l-1}x^{k+l-1}$.
If $l < (k+1)$ then $T\mathcal{R}_V(h) + \mathcal{M}^{2l} \supseteq \mathcal{M}^{2l-1}$ and $y + a_l x^l$ is $(2l-1)$-$\mathcal{R}_V$-determined. Successively using the complete transversal method we obtain $y + a_l x^l + \ldots + a_{2l-2}x^{2l-2} + a_{2l-1}x^{2l-1}$, as $x^{2l-1} \in T\mathcal{R}_V(F)$, then by Mather's Lemma $F \sim y + a_l x^l + \ldots + a_{2l-2}x^{2l-2}$.

Case (iii): Classification of the germs h for which the 1-jet is z.
By geometric criterion, $h = z$ is not $\mathcal{R}_V$-finite. A 2-complete transversal for z is $h = z + a_2x^2 + bxy + cy^2$.

$$\begin{aligned} J^2(T\mathcal{R}_V) = \; & \big\langle 4a_2x^2 + (4+2k)bxy + 4(k+1)cy^2 + 3(k+1)z, \\ & 2bxz + 4cyz + 3y^2, 6a_2xy + 3by^2, 2a_2xz + byz \big\rangle . \end{aligned}$$

We have $j^2x\eta_1h = 3(k+1)xz$, $j^2y\eta_1h = 3(k+1)yz$, $j^2\eta_2h = 2bxz + 4cyz + 3y^2$ e $j^2\eta_3h = 6a_2xy + 3by^2$. Then $a_2xy, y^2 \in j^2(T\mathcal{R}_V(h))$ and by Mather's Lemma:

For $a_2 \neq 0$, $h = z + a_2x^2 + bxy + cy^2 \sim z + a_2x^2$.

For $a_2 = 0$, $b \neq 0$, $h = z + bxy + cy^2 \sim z + bxy$.

For $a_2 = b = 0$, $h = z$.

If $a_2 \neq 0$, $h = z + a_2x^2$ is 2-$\mathcal{R}_V$-determined.

If $a_2 = 0$ and $b \neq 0$ taking $h = z + bxy$. We have $\mu(h|_{\Sigma_k}) = k+1 > k$.

If $b = 0$, let $j^{l-1}h = z$, $l \geq 3$, a l-complete transversal for z is $z + a_lx^l + b_{l-1}x^{l-1}y$ if $3 \leq l \leq (2k+2)$ and $z + a_lx^l$ if $l \geq (2k+3)$.

Then for $h = z + a_lx^l + b_{l-1}x^{l-1}y$, $\eta_1h = 2la_lx^l + 2(l+k)b_{l-1}x^{l-1}y + 3(k+1)z$, $\eta_2h = 2b_{l-1}x^{l-1}z + x^{2k+2} + 4x^{k+1}y + 3y^2$, $\eta_3h = la_lx^{l+k} + 3la_lx^{l-1}y + (l-2k-3)b_{l-1}x^{l+k-1}y + (3l-3)b_{l-1}x^{l-2}y^2$ and $\eta_4h = la_lx^{l-1}z + (l-1)b_{l-1}x^{l-2}yz + (k+1)x^{2k+1}y + (k+1)x^ky^2$

thus, for $a_l \neq 0$, $x^{l-1}y \in T\mathcal{R}_V(h)$ and by Mather's Lemma $h \sim z + a_lx^l$.

Let $h = z + a_lx^l$, with $a_l \neq 0$. If $l > k+1$, $\mu(h|_{\Sigma_k} \geq k+1 > k$, then we do not need to classify. If $l \leq k+1$, h is $(2l-1)$-$\mathcal{R}_V$-determined and $x^{2l-1} \in T\mathcal{R}_V(\tilde{h})$ with $\tilde{h} = z + a_lx^l + a_{l+1}x^{l+1} + \cdots + a_{2l-2}x^{2l-2}$, then $h \sim z + a_lx^l + a_{l+1}x^{l+1} + \cdots + a_{2l-2}x^{2l-2}$.

References

1. V. I. Arnold, *Wave front evolution and equivanant Morse lemma*, Commun. Pure App. Math. **29** (1976), 557–582. MR0436200 (55:9148)
2. R. Wik-Atique, *On the classification of multi-germs of maps from $\mathbb{C}^2$ to $\mathbb{C}^3$ under $\mathcal{A}$-equivalence*, in J.W.Bruce and F.Tari(eds.) *Real and Complex Singularities*, Research Notes in Maths Series, Chapman & Hall / CRC, 119-133. MR1715699 (2000f:58078)
3. J. W. Bruce, N. P. Kirk, and A. A. du Plessis, *Complete transversals and the classification of singularities*, Nonlinearity **10** (1997), 253–275. MR1430752 (97j:58010)
4. J. W. Bruce and P. L. Giblin, *Smooth maps of discriminant varieties*, Proc. London Math. Soc. (3) **50** (1984), 535–551.
5. J. W. Bruce and R. M. Roberts, *Critical points of functions on analytic varieties*, Topology **27** (1988), 57–90. MR935528 (89c:32019)
6. J. W. Bruce and J. M. West, *Functions on a crosscap*, Math. Proc. Cambridge Philos. Soc. **123** (1998), no. 1, 19–39. MR1474861 (98j:58016)
7. R. O. Buchweitz and G. M. Greuel, *The Milnor number and deformations of complex curve singularities.* Invent. Math. **58** (1980), 241–281. MR571575 (81j:14007)
8. J. Damon, *Deformations of sections of singularities and Gorenstein surface singularities*, Amer. J. Math. **109** (1987), 695–722. MR900036 (88i:58017)
9. J. Damon, *$\mathcal{A}$-equivalence and the equivalence of sections of images and discriminants*, Singularity Theory and its Applications: Warwick 1989, Part I, Lecture Notes in Math., **1462** (1991), 93–121, Springer-Verlag, New York. MR1129027 (92m:32057)
10. J. Damon, *The unfolding and determinacy theorems for subgroups of A and K*, Memoirs Amer. Math. Soc. **306** (1984). MR713063 (85h:58024)
11. J. Damon, *Topological triviality and versality for subgroups of A and K*, Memoirs Amer. Math. Soc. **389** (1988). MR961212 (90a:58012)
12. J. Damon, *Topological triviality and versality for subgroups of $\mathcal{A}$ and $\mathcal{K}$: II. Sufficient conditions and applications*, Nonlinearity **5**(1992), 373–412. MR1158379 (93f:58020)
13. A. Dimca, *Function germs defined on isolated hypersurface singularities*, Composition Math. **53** (1984), 245–258. MR766299 (86a:32027)
14. M. C. R. Fuster, S. Mancini, M. A. S. Ruas, *Stability and genericity of composition of maps*, Matemática Contemporânea, **5** (1993), 29–51. MR1305174 (95k:58017)

15. T. Gaffney, *Properties of finitely determined germs*, Thesis, Brandeis University, 1975. MR2625981
16. T. Gaffney, L. Wilson, *Equivalence of generic mappings and C^∞ normalization*, Compositio Mathematica **49** (1983), 291–308. MR707178 (85g:58018)
17. T. Gaffney, L. Wilson, *Equivalence theorems in global singularity theory*, Proc. of Symposia in Pure Math. **40**(1983), 439–447. MR713083 (85b:58017)
18. C. G. Gibson, *Singular points of smooth mappings*. Research Notes in Mathematics **25** London, Pitman. MR533668 (80j:58011)
19. V. V. Goryunov, *Functions on space curves,* J. Lond. Math. Soc. (2) 61 (2000), no. 3, 807–822. MR1766106 (2002f:58068)
20. G. M. Greuel and G. Pfister, SINGULAR 3.1.0, *A computer algebra system for polynomial computations*, http://www.singular.uni-kl.de.
21. R. C. Gunning and H. Rossi, *Analytic functions of several complex variables*, Prentice-Hall, 1965. MR0180696 (31:4927)
22. D. Mond, *On the classification of maps from $\mathbb{R}^2$ to $\mathbb{R}^3$*, Proc. London Math. Soc., **50** (1985), 333–369. MR772717 (86g:58021)
23. Mond, D.and D. van Straten, *Milnor number equals Tjurina number for functions on space curves*. J. Lond. Math. Soc., II. Ser. 63, **1**, (2001) 177–187. MR1802765 (2002e:32040)
24. J. J. Nuño-Ballesteros and J. N. Tomazella, *The Milnor number of a function on a space curve germ,* Bull. Lond. Math. Soc. **40** (2008), no.1, 129–138. MR2409185 (2009d:32032)
25. R. Pellikaan, *Finite determinacy of functions with non-isolated singularities*, Proc. London Math. Soc., (3) **57** (1988), 357–382. MR950595 (89i:32014)
26. S. Izumiya and S. Matsuoka, *Notes on smooth functions germs on varieties*, Proc. Amer. Math. Soc. **97** (1989), 146–150. MR831404 (87g:58013)
27. O. Lyashko, *Classification of critical points of functions on a manifold with singular boundary*, Funct. Anal. Appl. (1984), 187–193.
28. M. A. S. Ruas and J. N. Tomazella, *Topological triviality of families of functions on analytic varieties*, Nagoya Math. Journal, **175** September, (2004). MR2085309 (2005h:32068)
29. M. A. S. Ruas and J. N. Tomazella, *An infinitesimal criterion for topological triviality of families of sections of analytic varieties*, Singularity Theory and Its Applications. Adv. Studies in Pure Math., **43** (2006), 421–436. MR2325149 (2008i:58038)
30. K. Saito, *Theory of logarithmic differential forms and logarithmic vector fields*, J. Fac. Sci. Univ. Tokyo Sect. 1A Math. **27** (1980), 265–291. MR586450 (83h:32023)
31. D. Siersma, *Singularities of functions on boundaries, corners, etc.*, Quart. J. Math. **32** (1981), 119–127. MR606928 (82m:58011)

Departamento de Matemática, Universidade Federal de São Carlos, Rodovia Washington Luis, km 235, Caixa Postal 676, São Carlos, SP, Brazil
E-mail address: `bruna@dm.ufscar.br`

Departamento de Matemática, Universidade Federal de São Carlos, Rodovia Washington Luis, km 235, Caixa Postal 676, São Carlos, SP, Brazil
E-mail address: `tomazella@dm.ufscar.br`

Contemporary Mathematics
Volume **569**, 2012
http://dx.doi.org/10.1090/conm/569/11245

The Artin-Greenberg function of a plane curve singularity

S. Saleh

ABSTRACT. We calculate the Artin-Greenberg function of plane curves. It turns out that this function is a topological invariant of the singularity and that it determines with other invariants the semigroup of the singularity. This generalizes a result of M. Hickel proved for irreducible plane curves.

1. Introduction

Let $\mathbf{K}$ be an algebraically closed field of characteristic zero, and let $\mathbf{A} = \mathbf{K}[[t_1, ..., t_m]]$ be the ring of formal power series in $t_1, \ldots, t_m$ over $\mathbf{K}$. Let $\mathbf{A}[[x_1, ..., x_e]]$ be the ring of formal power series in $x_1, \ldots, x_e$ over $\mathbf{A}$ and let $I = (f_1, ..., f_r)$ be an ideal of $\mathbf{A}$. It has been proved by M. Artin (see [**2**]) and G. Pfister and D. Popescu (see [**11**]) that there exists a function b from $\mathbb{N}$ to $\mathbb{N}$, such that if we denote $t = (t_1, ..., t_m)$, then for all $i \in \mathbb{N}$ and for all $F(t) = (F_1(t), ..., F_e(t)) \in \mathbf{A}^e$, if for all $1 \leq j \leq r$, $f_j(F(t)) \in (t)^{b(i)+1}$ then, there exists $G(t) = (G_1(t), ..., G_e(t)) \in \mathbf{A}^e$ such that $f_j(G(t)) = 0$ for all j, and $G(t) - F(t) \in (t)^{i+1}$ where (t) denotes the maximal ideal in $\mathbf{K}[[t_1, \ldots, t_m]]$. We call the Artin function the minimal function β with this property.

Assume $m = 1$ and let $V = V(I)$ be the analytical variety defined by I. Let $G(t) = (G_1(t), \ldots, G_e(t)) \in \mathbf{A}^e = \mathbf{K}[[t_1]]^e$. If $f_j(G(t)) = 0$ for all $1 \leq j \leq r$ then $G(t)$ is said to be an arc of I in $\mathbf{A}^e$. In this case, the Artin function, which we call equally the Artin-Greenberg function, determines the minimal order of an element in $\mathbf{A}^e$ which allows to have an arc that coincides with this element up to that order. Note that V is non singular if and only if $\beta(i) = i$ for all $i \in \mathbb{N}$. When $V = V(f)$ is a hypersurface singularity, the Artin function has been studied by M. Hickel and M. Lejeune-Jalabert (see [**8**] and [**10**]).

The aim of this paper is to describe the Artin-Greenberg function for plane curve singularities. Let

$$f = y^n + a_1(x)y^{n-1} + \ldots + a_n(x)$$

be a reduced polynomial in $\mathbf{K}[[x]][y]$ and recall, by the Newton-Puiseux Theorem, that:

$$f(t^n, y) = \prod_{i=1}^{n}(y - y_i(t))$$

where $y_1(t), \ldots, y_n(t) \in \mathbf{K}[[t]]$. Assume that f is irreducible, then $y_1(t), \ldots, y_n(t)$ are related by the n-th roots of unity in $\mathbf{K}$. Given a polynomial $h \in \mathbf{K}[[x]][y]$, let

$int(f,h)$ be the order in t of $h(t^n, y_1(t))$. The set

$$\{int(f,h), h \in \mathbf{K}[[x]][y]\}$$

is a semigroup of $\mathbb{N}$, denoted $\Gamma(f)$.

In [**9**], M. Hickel gives explicit formulas of the Artin function, he also shows that this function with the multiplicity of f at the origin determines $\Gamma(f)$.

Assume that f is not irreducible, and let

$$f = \prod_{i=1}^{s} f_i$$

be its decomposition into irreducible polynomials in $\mathbf{K}[[x]][y]$. The set

$$\{(int(f_1,h), \ldots, int(f_s,h)), h \in \mathbf{K}[[x]][y]\}$$

is a semigroup of $\mathbb{N}^s$. It is the semigroup of f, denoted $\Gamma(f)$. Unlike the irreducible case, A. Campillo, F. Delgado, A. Garcia, and S. Gusein-Zade ([**4**], [**6**]) showed that in this case the semigroup is not finitely generated. In [**5**], F. Delgado showed the existence of a finite set of elements in $\Gamma(f)$, the irreducible absolute maximals, which allows with some other invariants of the singularity to describe the semigroup.

In this paper, we calculate the Artin-Greenberg function of such a polynomial and relate it with the semigroup of f. First of all we recall the construction of characteristic exponents of an irreducible polynomial in $\mathbf{K}[[x]][y]$, then we describe the semigroup of f where we refer to Abhyankar-Assi's paper [**1**] to define the tree of contacts of such a polynomial, and give the elements of its set of irreducible absolute maximals. For clarity we compute the Artin-Greenberg function of f when f has two branches, then we show that this function with those of the branches of f and their multiplicities at the origin determine the set of the absolute irreducible maximals and thus the semigroup $\Gamma(f)$. Then we generalize this result to the case of a reducible polynomial with several branches, using for this aim the tree of contacts of f. This results being intrinsic, we have implemented with MAPLE a program that calculates the Artin-Greenberg function of a polynomial with two branches [**12**].

2. Characteristic sequences and semigroup of a reducible polynomial

Let $\mathbf{K}$ be an algebraically closed field of characteristic zero, and let $f = y^n + a_1(x)y^{n-1} + \ldots + a_n(x)$ be a reduced polynomial of $\mathbf{K}[[x]][y]$ of multiplicty n at the origin.
Suppose that f is irreducible in $\mathbf{K}[[x]][y]$. It follows from the Newton-Puiseux theorem that there exists a formal power series $y(t) \in \mathbf{K}[[t]]$ such that $f(t^n, y(t)) = 0$. Moreover,

$$f(t^n, y) = \prod_{w \in U_n} (y - y(wt))$$

where U_n is the set of the n-th roots of unity in $\mathbf{K}$.
Let $(\beta_i)_{0 \leq i \leq g}$ be the set of Puiseux-exponents of f, defined as in [**14**]. Let, as in [**14**], $e_0 = \beta_0 = e_1 = n, e_2 = \gcd(e_1, \beta_1)$, and $e_i = \gcd(\beta_{i-1}, e_{i-1})$ for all $i \geq 3$. We have $e_{g+1} = 1$. Finally let $\bar{\beta}_0 = \beta_0, \bar{\beta}_1 = \beta_1$, and for all $i = 1, \ldots, g-1, \bar{\beta}_{i+1} = \bar{\beta}_i . \dfrac{e_i}{e_{i+1}} + \beta_{i+1} - \beta_i$. We denote by convention $\beta_{g+1} = \bar{\beta}_{g+1} = +\infty$.

Given a nonzero series $u = \sum_p c_p t^p$ in $\mathbf{K}[[t]]$, the t-order of u, denoted $\mathrm{O}_t(u)$, is the minimal integer i such that $u \in (t)^i$, where (t) denotes the maximal ideal of $\mathbf{K}[[t]]$. Given a nonzero element $\phi(t) = (\phi_1(t), \phi_2(t)) \in \mathbf{K}[[t]]^2$, the t-order of ϕ is defined to be $\mathrm{ord}_t(\phi) = \min\{\mathrm{O}_t(\phi_1(t)), \mathrm{O}_t(\phi_2(t))\}$.

Let $\phi(t) = (t^p, Y(t)), \psi(t) = (t^q, Z(t))$ be two elements in $\mathbf{K}[[t]]^2$. The contact between ϕ and ψ is defined to be $c(\phi, \psi) = \dfrac{1}{pq}\mathrm{O}_t(Y(t^q) - Z(t^p))$. The contact of f and ϕ, denoted by $\mathrm{c}(f, \phi)$, is given by

$$c(f, \phi) = \max\{c(f, \psi), \text{ where } \psi = (t^n, y(t)) \text{ is a root of } f(x, y) = 0\}.$$

Let $h = y^m + b_1(x)y^{m-1} + \ldots + b_m(x)$ be an irreducible polynomial in $\mathbf{K}[[x]][y]$, and let $\psi(t) = (t^m, z(t))$ be a root of $h(x, y) = 0$. The contact of f and h, denoted $c(f, h)$, is defined to be $c(f, h) = c(f, \psi)$. We also define the intersection multiplicity between f and h denoted $int(f, h)$ to be $O_t(h(t^n, y_1(t)))$, and we recall that this definition does not depend on the choice of the root y_1 of $f(t^n, y) = 0$. Furthermore, if $c = c(f, h)$, the following holds:

PROPOSITION 2.1 ([**14**]). *i) If* $c < \dfrac{\beta_1}{n}$, *then* $int(f, h) = n.c.m.$

ii) If $\dfrac{\beta_a}{n} \leq c < \dfrac{\beta_{a+1}}{n}$, $0 \leq a \leq g$ *then* $int(f, h) = (\bar{\beta}_a e_a + (nc - \beta_a)e_{a+1}).\dfrac{m}{n}$.

iii) If $c > \dfrac{\beta_a}{n}$, *then* $\dfrac{n}{e_{a+1}} | m$

Let $\Gamma(f) = \{int(f, h) | h \in \mathbf{K}[[x]][y]\}$. It follows by [**13**] that $\Gamma(f)$ is generated by the set $\{\bar{\beta}_0, \bar{\beta}_1, \ldots, \bar{\beta}_g\}$.

Suppose that f is not irreducible and let $f = \prod_{k=1}^s f_k$ be the decomposition of f into irreducible polynomials in $\mathbf{K}[[x]][y]$. Let n^k be the degree in y of $f_k, k = 1, \ldots, s$, and let

$$y^k(t) = \sum_j c_j^k . t^j$$

be a root of $f_k(t^{n^k}, y) = 0$. Let $(\beta_i^k)_{0 \leq i \leq g_k}, (e_i^k)_{0 \leq i \leq g_k+1}, (\bar{\beta}_i^k)_{0 \leq i \leq g_k}$ be the characteristic sequences associated with f_k as defined above.

The tree of contacts of Abhyankar-Assi

Let the notations be as in section 2, we recall in this section the notion of the tree of contacts of f introduced by Abhyankar-Assi (see [**1**]). Let

$$C(f) = \{c(f_k, f_j), 1 \leq k \neq j \leq s\} \cup \{(\frac{\beta_i^k}{n^k})_{i \leq g_k}, k = 1 \ldots, s\}$$

For $M \in C(f)$, we associate with M the following equivalence relation defined as follows:

$$f_k R_M f_j \quad \text{if and only if} \quad c(f_j, f_k) \geq M$$

With the understanding that $c(f_k, f_k) \geq M$ if and only if $M = \dfrac{\beta_i^k}{n^k}$ for some $1 \leq i \leq g_k$. We define the points of the tree of contacts of f at the level M to be the set of equivalence classes of R_M, and we denote this set by $P(M, 1), \ldots, P(M, r)$. Denote by c_{jk} the contact of f_j with f_k, $1 \leq j \neq k \leq s$ and let $h = y^m + b_1(x)y^{m-1} + \ldots + b_m(x)$ be an irreducible polynomial of $\mathbf{K}[[x]][y]$. Consider the tree of contacts of $f.h$ and let $M = \max\{c(f_j, h) | 1 \leq i \leq s\}$. Let l be such that $\max\{c(f_j, h) | 1 \leq$

$j \leq s\} = c(f_l, h)$. We denote by $Q(M)$ the equivalence class of M containing f_l. Note that $c(f_p, h) = M$ for all $f_p \in Q(M)$. Let $c = \min\{c_{jk} | 1 \leq j \neq k \leq s\}$.

PROPOSITION 2.2. If $M \leq c$, then $int(f_k, h) = \frac{n^k}{n^j} int(f_j, h)$, $1 \leq j \neq k \leq s$.

PROOF. Since $M \leq c$, then $c(f_j, h) = M, j = 1, \ldots, s$. Hence, the result is a direct application of the formulas of proposition 2.1. □

PROPOSITION 2.3. If $M > c$, then for all $k \in \{1, \ldots, s\}$:

$$int(f_k, h) = \begin{cases} \frac{n^k}{n^l} int(f_l, h) & \text{if } f_k \in Q(M) \\ \frac{m}{n^l} . int(f_k, f_l) & \text{if } f_k \notin Q(M) \end{cases}$$

PROOF. If $f_k \in Q(M)$ (resp.$f_k \notin Q(M)$), then $c(f_k, h) = M$ (resp.$c(f_k, h) = c(f_k, f_l) < M$). Now use Proposition 2.1. □

DEFINITION 2.4. Let $Q(P) \in T(f)$, and let $f_l \in Q(P)$, we define $Q'(P,l), \bar{Q}'(P,l)$ as follows:

$$\begin{cases} f_k \in Q'(P,l) \text{ if and only if } c(f_k, f_l) > P \\ f_k \in \bar{Q}'(P,l) \text{ if and only if } c(f_k, f_l) < P. \end{cases}$$

Note that, if $f_k \in Q'(P,l)$ then $Q'(P,l) = Q'(P,k)$ and $\bar{Q}'(P,l) = \bar{Q}'(P,k)$. In general

$$Q'(P,l) \cup \bar{Q}'(P,l) \subseteq \{1, \ldots, s\}$$

but we may not have the equality. Set:

$$B = \{P \text{ such that } Q'(P,l) \cup \bar{Q}'(P,l) = \{1, \ldots, s\}\}.$$

Let $\Gamma(f)$ be the semigroup of f. In [**5**], Delgado has shown the existence of a finite set of elements in $\Gamma(f)$, the irreducible absolute maximals $MAI(f)$, which with $\{\Gamma(f_{i_1} \cdot \ldots \cdot f_{i_{s-1}}), 1 \leq i_1 < \ldots < i_{s-1} \leq s\}$ characterizes the elements of $\Gamma(f)$. The result of Delgado has the following interpretation in terms of $T(f)$:

PROPOSITION 2.5 (See [**5**]). *Let $1 \leq j \leq s$ and $1 \leq a \leq g_j$ such that $\frac{\beta^j_a}{n^j} \in B$ and let:*

$$\gamma^{a,j}_k = \begin{cases} \dfrac{n^k}{n^j} \bar{\beta}^j_a = \bar{\beta}^k_a & \textit{if } f_k \in Q'(\frac{\beta^j_a}{n^j}, j) \\ \dfrac{int(f_k, f_j)}{e^j_a} & \textit{otherwise} \end{cases}$$

and $\gamma^{a,j} = (\gamma^{a,j}_1, \ldots, \gamma^{a,j}_s)$ and recall that $c = \min\{c_{jk} = c(f_j, f_k) | 1 \leq j \neq k \leq s\}$.

I) If there exists $i \in \{1, \ldots, s\}$ such that $c < \frac{\beta^i_1}{n^1}$ then:

$$MAI(f) = \bigcup_{j=1}^{s} \{\gamma^{1,j}, \ldots, \gamma^{g_j,j}\}$$

II) Otherwise, let q be the greatest integer such that $\frac{\beta^1_q}{n^1} = \ldots = \frac{\beta^s_q}{n^s} < c$. Let for $j \in 1, \ldots, s$, $q < g_j$:

$$L_j = \begin{cases} q & \textit{if } c \neq \dfrac{\beta^j_{q+1}}{n^j} \\ q+1 & \textit{otherwise.} \end{cases}$$

Set $A = \{j, L_j < g_j\}$. *In this case:*

$$MAI(f) = \{\alpha^1, \ldots, \alpha^q\} \bigcup \bigcup_{j \in A} \{\gamma^{L_j+1,j}, \ldots \gamma^{g_j,j}\}$$

where $\alpha^a = (\bar{\beta}_a^1, \ldots, \bar{\beta}_a^s), a = 1, \ldots, q$.

3. The Artin-Greenberg function

Let the notations be as in section 2. The aim of this part is to give explicit formulas of the Artin-Greenberg function of f in terms of the generators of the semigroup of f.

DEFINITION 3.1. Let $\phi(t) \in \mathbf{K}[[t]]^2$, $\phi(t)$ is said to be congruent to f till the order i, we note $\phi \cong^i f$, if there exists $\psi(t) \in \mathbf{K}[[t]]^2$ such that $f(\psi(t)) = 0$ and $\mathrm{ord}_t(\phi(t) - \psi(t)) \geq i + 1$.

REMARK 3.2. When $f = f_1 \ldots f_s$, $\phi \cong^i f$ if there exists $\psi(t) \in \mathbf{K}[[t]]^2$ such that $f_j(\psi(t)) = 0$ for some $j \in \{1, \ldots, s\}$. In particular, $\phi \not\cong^i f$ if and only if $\phi \not\cong^i f_j$ for all $j \in \{1, \ldots, s\}$.

LEMMA 3.3. $\beta(i) = \max\{\mathrm{O}_t f(\phi(t)) | \phi \not\cong^i f\}$.

PROOF. See [9].

□

COROLLARY 3.4. Let $\beta_j(i)$ be the Artin-Greenberg function of $f_j, j = 1, \ldots, s$. We have $\beta(i) \leq \beta_1(i) + \ldots + \beta_s(i)$.

LEMMA 3.5. Let $i \in \mathbb{N}$, $\phi(t) \in \mathbf{K}[[t]]^2$, if $\phi \not\cong^i f$ then $\mathrm{ord}_t(\phi(t)) \leq i$.

PROOF. See [12]. □

LEMMA 3.6. Let $\phi(t) = (\phi_1(t), \phi_2(t)) \in \mathbf{K}[[t]]^2$ and let $1 \leq j \leq s$.
i) If $i < n^j$ then $\phi \not\cong^i f_j$ if and only if $\mathrm{ord}_t(\phi) \leq i$.
ii) If $i \geq n^j$ then $\phi \not\cong^i f_j$ if and only if

$$\begin{cases} \mathrm{ord}_t(\phi) < \mathrm{O}_t(\phi_1) \\ \text{or } \mathrm{ord}_t(\phi) = \mathrm{O}_t(\phi_1), n^j \text{ doesn't divide } \mathrm{ord}_t(\phi) \\ \text{or } \mathrm{ord}_t(\phi) = \mathrm{O}_t(\phi_1), n^j \text{ divides } \mathrm{ord}_t(\phi) \text{ and } c(f_j, \phi) \leq \dfrac{i}{\mathrm{ord}_t(\phi)} \end{cases}$$

PROOF. See [9].

□

REMARK 3.7. Let $\phi(t) \in \mathbf{K}[[t]]^2$, $\mathrm{ord}_t(\phi) \leq i$. Set $\phi(t) = \bar{\phi}(t^k)$, where $\bar{\phi}$ is a primitive representation of ϕ (a representation $\bar{\phi}$ is said to be primitive if there is no l such that $\bar{\phi}(t) = \Phi(t^l)$). Thus

$$\mathrm{O}_t(f(\phi)) = k. \sum_{j=1}^{s} int(f_j, h)$$

where h is the monic irreducible polynomial of $\mathbf{K}[[x]][y]$ such that $h(\bar{\phi}(t)) = 0$.

We give some notations that we will use in the main theorem, let $j \in \{1,\dots,s\}$, set

$$k_a^j(i) = \begin{cases} [\frac{ie_a^j}{n^j}] & \text{if } [\frac{ie_a^j}{n^j}] \not\equiv 0 \bmod e_a^j \\ [\frac{ie_a^j}{n^j}] - 1 & \text{otherwise} \end{cases}$$

$$\theta^j(i,a) = \begin{cases} [\frac{i}{\beta_a^j}](\bar{\beta}_a^j e_a^j - \beta_a^j.e_{a+1}^j) + ie_{a+1}^j & \text{if } [\frac{i}{\beta_{a+1}^j}] < [\frac{i}{\beta_a^j}] \\ 0 & \text{otherwise} \end{cases}$$

LEMMA 3.8. Let $i \in \mathbb{N}$ and let $1 \le j \le s$. Let $\phi(t) \in \mathbf{K}[[t]]^2$, $\mathrm{ord}_t(\phi(t)) = mk$, and let $\bar{\phi}(t) = \phi(t^k)$ be a primitive representation of $\phi(t)$. Set $\mathrm{ord}_t(\bar{\phi}(t)) = m$, and assume that $km \le i$. Let h be the monic irreducible polynomial of $\mathbf{K}[[x]][y]$ such that $h(\phi(t)) = 0$. Set

$$s_j = max\{0 \le p \le g_j | m \equiv 0(\frac{n^j}{e_{p+1}^j})\}$$

We have the following::

i) If $i < n^j$ then $k.int(f_j, h) \le [\frac{ie_{s_j+1}^j}{n^j}].\bar{\beta}_{s_j+1}^j$.

ii) If $i \ge n^j$ and n^j doesn't divide i then $k.int(f_j, h) \le k_{s_j+1}^j(i).\bar{\beta}_{s_j+1}^j$.

PROOF. i) Note that $c_j \le \frac{\beta_{s_j+1}^j}{n^j}$. Thus, $k.int(f_j,h) \le k.\frac{m}{n^j}.\bar{\beta}_{s_j+1}^j e_{s_j+1}^j$, but by the definition of s_j, $k.\frac{m}{n^j}.e_{s_j+1}^j$ is an integer, which shows that $k.int(f_j,h) \le [\frac{ie_{s_j+1}^j}{n^j}].\bar{\beta}_{s_j+1}^j$.

ii) We have $int(f_j,h) \le \frac{m}{n^j}.\bar{\beta}_{s_j+1}^j.e_{s_j+1}^j$. But $k_{s_j+1}^j(i)$ is the greater integer such that:

1) $\frac{k.m}{n^j}.e_{s_j+1}^j \le k_{s_j+1}^j(i)$

2) $k_{s_j+1}^j(i).\frac{e_{s_j+1}^j}{n^j} \le i$, which shows the result.

□

LEMMA 3.9. Let $i \in \mathbb{N}$ and let $1 \le j \le s$. Let $\phi(t) \in \mathbf{K}[[t]]^2$. Assume that $\mathrm{ord}_t(\phi) = \mathrm{O}_t(\phi_1), n^j$ divides $\mathrm{ord}_t(\phi)$ and that $c(f_j,\phi) \le \frac{i}{\mathrm{ord}_t(\phi)}$. In this case there exists $1 \le a \le h_j$ such that $\mathrm{O}_t(f_j(\phi)) \le \theta^j(i,a)$.

PROOF. Write $\phi(t) = (t^{n^j p}, \phi_2(t)) = (t^{mk}, \theta(t^k))$ and $c_j = c(f_j,\phi)$. Note that $(t^m, \theta(t))$ is primitive.

Let a be the unique integer such that $\beta_a^j \le \frac{i}{p} < \beta_{a+1}^j$, then $[\frac{i}{\beta_{a+1}^j}] + 1 \le p \le [\frac{i}{\beta_a^j}]$.

$\mathrm{O}_t(f_j(\phi)) = \frac{km}{n^j}(\bar{\beta}_{q'}^j e_{q'}^j + (n^j c_j - \beta_{q'}^j)e_{q'+1}^j)$ where $\frac{\beta_{q'}^j}{n^j} \le c_j < \frac{\beta_{q'+1}^1}{n^1}$. But $c_j \le \frac{i}{\mathrm{ord}_t(\phi)}$ then $q' \le a$.

For $q' < a$, $O_t(f_j(\phi)) \leq \frac{km}{n^j}(\bar{\beta}^j_a e^j_a + (\beta^j{}_{a+1} - \beta^j{}_a)e^j_{a+1}) = p.(\bar{\beta}^j_a e^j_a - \beta^j{}_a e^j_{a+1}) + p.\beta^j{}_{a+1}e^j_{a+1} \leq \theta^j(i,a)$.
For $q' = a$, $O_t(f_j(\phi)) \leq \theta^j(i,a)$.

□

4. The Artin-Greenberg function for a reducible polynomial with two branches

Let the notations be as in section 2 and assume that $s = 2$. The aim of this section is to give explicit formulas for $\beta(i)$. Assume without loss of generality that $n^1 \leq n^2$.

REMARK 4.1. Let h be an irreducible polynomial in $\mathbf{K}[[x]][y]$ of degree m in y. Let

$$S = \{\frac{int(f_1,h)}{n^1.m}, \frac{int(f_2,h)}{n^2.m}, \frac{int(f_1,f_2)}{n^1.n^2}\}$$

By prop.2.2. and prop.2.3., two elements of S are equal and the third one is greater or equal to the common value of these two elements.

4.1. Calculation of $\beta(i)$ for $i < n^1$. Let $\phi(t) = (\phi_1(t), \phi_2(t))$ be a primitive element of $\mathbf{K}[[t]]^2$ such that $\mathrm{ord}_t(\phi) = O_t(\phi_1)$. Set $\tilde{m} = \mathrm{ord}_t(\phi(t))$ and let $\phi_1(t) = T^{\tilde{m}}$. Denote by $\tilde{\phi}(T) = (T^{\tilde{m}}, \tilde{\phi}_2(T))$, and let h be the minimal polynomial of $\tilde{\phi}_2(x^{\frac{1}{\tilde{m}}})$ over $\mathbf{K}((x))$. There exist $b_1(x), \ldots, b_{\tilde{m}}(x)$ such that $h(x,y) = y^{\tilde{m}} + b_1(x).y^{\tilde{m}-1} + \ldots + b_{\tilde{m}}(x)$. On the other hand, $O_t(f(\phi(t)) = O_T(f(T^{\tilde{m}}, \tilde{\phi}_2(T)) = int(f,h)$. In particular,

$$\beta(i) = max\{k.int(f,h), h \text{ is a monic irreducible polynomial of degree } m \text{ in } y \text{ of } \mathbf{K}[[x]][y] \text{ such that } k.mult_O(h) \leq i\}.$$

Note that by remark 4.1., we have:

1) If $int(f_1,h) < \frac{m}{n^2}int(f_1,f_2)$, then $k.int(f,h) = k.int(f_1,h)(1 + \frac{n^1}{n^2})$.
2) If $int(f_1,h) > \frac{m}{n^2}int(f_1,f_2)$, then $k.int(f,h) = k.(int(f_1,h) + \frac{m}{n^1}.int(f_1,f_2))$.
3) If $int(f_1,h) = \frac{m}{n^2}int(f_1,f_2)$, then $k.int(f,h) = k.(int(f_2,h) + \frac{m}{n^2}.int(f_1,f_2))$.

Let h be an irreducible monic polynomial of $\mathbf{K}[[x]][y]$ of degree m in y, and let $k \in \mathbb{N}$ be such that $km \leq i$.

◇ If $mult_O(h) < deg_y(h)$ then $k.int(f,h) = k.mult_O(h).n \leq (n^1 + n^2).i$.
◇ If $mult_O(h) = deg_y(h)$, then set:

$$A_1 = \{k.int(f_1,h)(1 + \frac{n^1}{n^2}) | int(f_1,h) < \frac{m}{n^2}int(f_1,f_2)\}.$$
$$A_2 = \{k.(int(f_1,h) + \frac{m}{n^1}.int(f_1,f_2)) | int(f_1,h) > \frac{m}{n^2}int(f_1,f_2)\}.$$
$$A_3 = \{k.(int(f_2,h) + \frac{m}{n^2}.int(f_1,f_2)) | int(f_1,h) = \frac{m}{n^2}int(f_1,f_2)\}.$$

Hence, for $i < n^1$, $\beta(i) = \max(A_1 \cup A_2 \cup A_3 \cup \{(n^1 + n^2).i\})$.

Let q_1 (resp. q_2) be the unique integer such that $\frac{n^1}{e^1_{q_1}} \leq i < \frac{n^1}{e^1_{q_1+1}}$ (resp. $\frac{n^2}{e^2_{q_2}} \leq i < \frac{n^2}{e^2_{q_2+1}}$). With these notations the following theorem holds:

THEOREM 4.2. I) If $c < \frac{\beta_1^1}{n^1}$ or $\frac{\beta_1^2}{n^2}$, then

$$\beta(i) = \max\{(\bar{\beta}_a^1 + \frac{int(f_1, f_2)}{e_a^1}).[\frac{ie_a^1}{n^1}] | a \leq q_1, (\bar{\beta}_a^2 + \frac{int(f_1, f_2)}{e_a^2}).[\frac{ie_a^2}{n^2}] | a \leq q_2\}.$$

II) Otherwise let q be the greatest integer such that $\frac{\beta_q^1}{n^1} = \frac{\beta_q^2}{n^2} < c$. Set for $j = 1, 2$

$$L_j = \begin{cases} q & \text{if } c \neq \frac{\beta_{q+1}^j}{n^j} \\ q+1 & \text{otherwise} \end{cases}$$

i) If $q < q_1$, then

$$\beta(i) = max\left\{(\bar{\beta}_a^1 + \bar{\beta}_a^2).[\frac{ie_a^1}{n^1}] | a \leq q, (\bar{\beta}_a^1 + \frac{int(f_1, f_2)}{e_a^1}).[\frac{ie_a^1}{n^1}] | L_1 < a \leq q_1, \right.$$
$$\left. (\bar{\beta}_a^2 + \frac{int(f_1, f_2)}{e_a^2}).[\frac{ie_a^2}{n^2}] | L_2 < a \leq q_2 \right\}.$$

ii) If $q_1 \leq q$, then $\beta(i) = max\{(\bar{\beta}_a^1 + \bar{\beta}_a^2).[\frac{ie_a^1}{n^1}] | a \leq q_1\}$.

PROOF. We prove II), I) can be proved in a similar way. Let h be a monic irreducible polynomial of $\mathbf{K}[[x]][y]$ of degree m in y, and let $k \in \mathbb{N}$ be such that $km \leq i$. Let $c(f_l, h) = c_l, l = 1, 2$.
Since $\frac{n^1}{e_{q_1}^1} \leq i < \frac{n^1}{e_{q_1+1}^1}$ (resp. $\frac{n^2}{e_{q_2}^2} \leq i < \frac{n^2}{e_{q_2+1}^2}$), then $c_1 \leq \frac{\beta_{q_1}^1}{n^1}$ (resp. $c_2 \leq \frac{\beta_{q_2}^2}{n^2}$).
i) Suppose that $q < q_1$. Clearly, $c_1 \leq \frac{\beta_{s_1+1}^1}{n^1}$ and $c_2 \leq \frac{\beta_{s_2+1}^2}{n^2}$. But, by lemma 3.8., $k.int(f_1, h) \leq [\frac{ie_{s_1+1}^1}{n^1}].\bar{\beta}_{s_1+1}^1$ and $k.int(f_2, h) \leq [\frac{i.e_{s_2+1}^2}{n^2}].\bar{\beta}_{s_2+1}^2$.
Case 1) If $multo(h) < deg_y(h)$, then $k.int(f, h) \leq (n^1 + n^2).i$.
Case 2) If $multo(h) = deg_y(h)$, we have the following cases:
1) $\frac{int(f_1, h)}{m} < \frac{int(f_1, f_2)}{n^2}$:

$\diamond$ If $s_1 + 1 \leq q$, then $int(f_1, h) = \frac{n^1}{n^2} int(f_2, h)$, so

$$k.int(f, h) = k.int(f_1, h)(1 + \frac{n^1}{n^2}) \leq [\frac{i.e_{s_1+1}^1}{n^1}].\bar{\beta}_{s_1+1}^1(1 + \frac{n^1}{n^2}).$$

In order to get the equality, let $\tilde{\phi}(t) = \phi(t) = (t^{[i.\frac{e_{s_1+1}^1}{n^1}].\frac{n^1}{e_{s_1+1}^1}}, y_{<m_{s_1+1}^1}^1(t^{[i.\frac{e_{s_1+1}^1}{n^1}]}))$ (Where if $y^1 = \sum_p c_p t^p$ denotes a root of $f_1(t^{n^1}, y) = 0$ then $y_{<a}^1 = \sum_{p<a} c_p t^p$) and let h be the monic irreducible polynomial such that $h(\phi(t)) = 0$, then

$$k.int(f, h) = [\frac{i.e_{s_1+1}^1}{n^1}].\bar{\beta}_{s_1+1}^1(1 + \frac{n^1}{n^2}).$$

◇ If $s_1+1>q$, then $int(f_2,h)\leq \frac{m}{n^1}.int(f_1,f_2)$, consequently

$$k.int(f,h)\leq [i.\frac{e^1_{s_1+1}}{n^1}].\bar{\beta}^1_{s_1+1}+\frac{km}{n^1}.int(f_1,f_2)$$
$$=[i.\frac{e^1_{s_1+1}}{n^1}].\bar{\beta}^1_{s_1+1}+\frac{km.e^1_{s_1+1}}{n^1}.\frac{int(f_1,f_2)}{e^1_{s_1+1}}$$
$$\leq [i.\frac{e^1_{s_1+1}}{n^1}].(\bar{\beta}^1_{s_1+1}+\frac{int(f_1,f_2)}{e^1_{s_1+1}}).$$

The equality is attained for $\phi(t)=\tilde{\phi}(t)=(t^{[i.\frac{e^1_{s_1+1}}{n^1}].\frac{n^1}{e^1_{s_1+1}}},y^1_{<\beta^1_{s_1+1}}(t^{[i.\frac{e^1_{s_1+1}}{n^1}]}))$.

2) If $\frac{int(f_1,h)}{m}>\frac{int(f_1,f_2)}{n^2}$, then

$$k.int(f,h)=k.(int(f_1,h)+\frac{m}{n^1}.int(f_1,f_2))\leq [\frac{ie^1_{s_1+1}}{n^1}].(\bar{\beta}^1_{s_1+1}+\frac{int(f_1,f_2)}{e^1_{s_1+1}}).$$

The equality is attained for $\phi(t)=\tilde{\phi}(t)=(t^{[\frac{ie^1_{s_1+1}}{n^1}].\frac{n^1}{e^1_{s_1+1}}},y^1_{<\beta^1_{s_1+1}}(t^{[\frac{ie^1_{s_1+1}}{n^1}]}))$.

3) If $\frac{int(f_1,h)}{m}=\frac{int(f_1,f_2)}{n^2}$, then

$$k.int(f,h)=k.(int(f_2,h)+\frac{m}{n^2}int(f_1,f_2))\leq [\frac{i.e^2_{s_2+1}}{n^2}].(\bar{\beta}^2_{s_2+1}+\frac{int(f_1,f_2)}{e^2_{s_2+1}})$$

and the equality is attained for $\phi(t)=\tilde{\phi}(t)=(t^{[\frac{ie^2_{s_2+1}}{n^2}].\frac{n^2}{e^2_{s_2+1}}},y^2_{<\beta^2_{s_2+1}}(t^{[\frac{ie^2_{s_2+1}}{n^2}]}))$.

II.ii) If $q_1\leq q$, the other assertion is clear since $q_1\leq q$ implies that $q_1=q_2$. Moreover, since $c_1\leq \frac{\beta^1_{q_1}}{n^1}$, then we have $c_1=c_2<c$, so $k.int(f,h)=k.int(f_1,h)(1+\frac{n^1}{n^2})$.

□

REMARK 4.3. If f_1 and f_2 are equisingular, and $c>\beta^1_{g_1}$, then

$$\beta(i)=max\{2\bar{\beta}^1_a.[\frac{ie^1_a}{n^1}]|a\leq q_1\}.$$

4.2. Calculation of $\beta(i)$ for $i\geq n^1$.

Case 1: $n^1<n^2$ **and** $n^1\leq i<n^2$**.** Let q_2 be the unique integer such that $\frac{n^2}{e^2_{q_2}}\leq i<\frac{n^2}{e^2_{q_2+1}}$. With these notations the following theorem holds:

THEOREM 4.4. I) If $c<\frac{\beta^1_1}{n^1}$ or $\frac{\beta^2_1}{n^2}$, then

$$\beta(i)=\max\{(n^1+n^2)i,k^1_a(i)(\bar{\beta}^1_a+\frac{int(f_1,f_2)}{e^1_a}),a\leq g_1,$$
$$\theta^1(i,a)+[\frac{i}{\beta^1_a}].\frac{int(f_1,f_2)}{e^1_a}|a\leq g_1,[\frac{ie^2_a}{n^2}](\bar{\beta}^2_a+\frac{int(f_1,f_2)}{e^2_a})|a\leq q_2\}.$$

II) Otherwise let q be the greatest integer such that $\frac{\beta_q^1}{n^1} = \frac{\beta_q^2}{n^2} < c$. Set for $j = 1, 2$

$$L_j = \begin{cases} q & \text{if } c \neq \frac{\beta_{q+1}^j}{n^j} \\ q+1 & \text{otherwise} \end{cases}$$

i) If $q < q_2$, then

$$\beta(i) = max\Big\{(n^1+n^2)i, [\frac{ie_a^2}{n^2}](\bar{\beta}_a^1 + \bar{\beta}_a^2), \theta^1(i,a) + \frac{n^2}{n^1}.\theta^1(i,a) | a \leq q,$$
$$k_a^1(i)(\bar{\beta}_a^1 + \frac{int(f_1,f_2)}{e_a^1}), \theta^1(i,a) + [\frac{i}{\bar{\beta}_a^1}].\frac{int(f_1,f_2)}{e_a^1} | L_1 < a \leq g_1,$$
$$[\frac{ie_a^2}{n^2}](\bar{\beta}_a^2 + \frac{int(f_1,f_2)}{e_a^2}) | L_2 < a \leq q_2\Big\}.$$

ii) If $q_2 \leq q$, then

$$\beta(i) = max\{(n^1+n^2)i, [\frac{ie_a^2}{n^2}](\bar{\beta}_a^1 + \bar{\beta}_a^2), \theta^1(i,a) + \frac{n^2}{n^1}.\theta^1(i,a) | a \leq q_2\}$$

PROOF. We prove II), I) can be proved in a similar way. Let $\phi(t) \in \mathbf{K}[[t]]^2$ be such that $\mathrm{ord}_t(\phi(t)) \leq i$. Let $\phi(t) \in \mathbf{K}[[t]]^2$, be such that $\phi \not\equiv^i f$. In order to calculate $\beta(i)$, we have to calculate $\mathrm{O}_t(f(\phi))$ when one of the following cases holds:
1) $\mathrm{ord}_t(\phi) < \mathrm{O}_t(\phi_1)$.
2) $\mathrm{ord}_t(\phi) = \mathrm{O}_t(\phi_1)$, n^1 doesn't divide $\mathrm{ord}_t(\phi)$.
3) $\mathrm{ord}_t(\phi) = \mathrm{O}_t(\phi_1)$, $n^1 | \mathrm{ord}_t(\phi)$ and $c(f_1, \phi) \leq \frac{i}{\mathrm{ord}_t(\phi)}$.
Set $\phi(t) = \bar{\phi}(t^k)$ where $\bar{\phi}(t)$ is a primitive representation of $\phi(t)$, let h be the monic irreducible polynomial such that $h(\bar{\phi}(t)) = 0$. Let $c(f_l, h) = c_l, l = 1, 2$.
Since $\frac{n^2}{e_{q_2}^2} \leq i < \frac{n^2}{e_{q_2+1}^2}$, then $c_2 \leq \frac{\beta_{q_2}^2}{n^2}$.
Suppose $q_2 > q$. Note that by lemma 3.8., $k.int(f_2,h) \leq [\frac{ie_{s_2+1}^2}{n^2}].\bar{\beta}_{s_2+1}^2$, $k.int(f_1,h) \leq k_{s_1+1}^1(i).\bar{\beta}_{s_1+1}^1$.
case 1) $\mathrm{ord}_t(\phi) < \mathrm{O}_t(\phi_1)$, then $\mathrm{O}_t(f(\phi)) \leq (n^1+n^2).i$. In order to get equality let $\phi(t) = \tilde{\phi}(t) = (0, t^i)$.
case 2) $\mathrm{ord}_t(\phi) = \mathrm{O}_t(\phi_1)$, n^1 doesn't divide $\mathrm{ord}_t(\phi)$. We have three cases:
1) If $int(f_1,h) < \frac{m}{n^2} int(f_1,f_2)$.
⋄ If $s_1 + 1 \leq q$, then

$$\mathrm{O}_t(f(\phi)) \leq k_{s_1+1}^1(i).\bar{\beta}_{s_1+1}^1(1 + \frac{n^1}{n^2}).$$

In order to get equality, let $\phi(t) = \tilde{\phi}(t) = (t^{k_{s_1+1}^1(i).\frac{n^1}{e_{s_1+1}^1}}, y_{<\beta_{s_1+1}^1}^1(t^{k_{s_1+1}^1(i)}))$, so $\mathrm{O}_t(f(\phi)) = k_{s_1+1}^1(i).\bar{\beta}_{s_1+1}^1(1 + \frac{n^1}{n^2})$.

◇ If $s_1+1>q$, then

$$\mathrm{O}_t(f(\phi)) \leq k^1_{s_1+1}(i).(\bar{\beta}^1_{s_1+1}+\frac{int(f_1,f_2)}{e^1_{s_1+1}}).$$

In order to get equality, let $\phi(t)=\tilde{\phi}(t)=(t^{k^1_{s_1+1}(i).\frac{n^1}{e^1_{s_1+1}}}, y^1_{<\beta^1_{s_1+1}}(t^{k^1_{s_1+1}(i)})).$

2) If $int(f_1,h) > \frac{m}{n^2} int(f_1,f_2)$, then

$$\mathrm{O}_t(f(\phi)) = k.(int(f_1,h)+\frac{m}{n^1}.int(f_1,f_2)) \leq k^1_{s_1+1}(i)(\bar{\beta}^1_{s_1+1}(i)+\frac{int(f_1,f_2)}{e^1_{s_1+1}}).$$

The equality is attained for $\phi(t)=\tilde{\phi}(t)=(t^{k^1_{s_1+1}(i).\frac{n^1}{e^1_{s_1+1}}}, y^1_{<\beta^1_{s_1+1}}(t^{k^1_{s_1+1}(i)})).$

3) If $int(f_1,h) = \frac{m}{n^2} int(f_1,f_2)$, then

$$\mathrm{O}_t(f(\phi)) = k.(int(f_2,h)+\frac{m}{n^2}.int(f_1,f_2)) \leq [\frac{i.e^2_{s_2+1}}{n^2}].(\bar{\beta}^2_{s_2+1}+\frac{int(f_1,f_2)}{e^2_{s_2+1}}).$$

The equality is attained for $\phi(t)=\tilde{\phi}(t)=(t^{[\frac{ie^2_{s_2+1}}{n^2}].\frac{n^2}{e^2_{s_2+1}}}, y^2_{<\beta^2_{s_2+1}}(t^{[\frac{ie^2_{s_2+1}}{n^2}]}).$

case 3) $\mathrm{ord}_t(\phi)=\mathrm{O}_t(\phi_1)$, $n^1|\mathrm{ord}_t(\phi)$ and $c(f_1,\phi)=c_1\leq \frac{i}{\mathrm{ord}_t(\phi)}$.

1) If $int(f_1,h) < \frac{n^2}{m}.int(f_1,f_2)$, then $\mathrm{O}_t(f(\phi)) = k.int(f_1,h)(1+\frac{n^1}{n^2})$. But by lemma 3.9, $\mathrm{O}_t(f_1(\phi))\leq \theta^1(i,a)$.
In order to get equality, let

$$\phi(t)=\tilde{\phi}(t)=(t^{n^1\tilde{p}}, y'(t)=\sum_{j,\tilde{p}j<i} c^1_j t^{\tilde{p}j}+Zt^i).$$

Where $\tilde{p}=[\frac{i}{\beta^1_a}]$ and Z is a generic element of $\mathbf{K}$. In these conditions we get $c_1 = \frac{i}{n^1\tilde{p}}$. In this case, $\mathrm{O}_t(f_1(\phi))=\theta^1(i,a)$, and by prop.2.3. $O_t(f_2(\phi))=\frac{n^2}{n^1}.\theta^1(i,a)$.

2) If $int(f_1,h) > \frac{n^2}{m}.int(f_1,f_2)$. Then $\mathrm{O}_t(f(\phi)) = k.(int(f_1,h)+\frac{m}{n^1}.int(f_1,f_2))$.
By the same arguments as the previous case, we get $\mathrm{O}_t(f(\phi)) \leq \theta^1(i,a)+[\frac{i}{\beta^1_a}].\frac{int(f_1,f_2)}{e^1_a}$. The equality is attained for

$$\phi(t)=\tilde{\phi}(t)=(t^{n^1\tilde{p}}, y'(t)=\sum_{j,\tilde{p}j<i} c^1_j t^{\tilde{p}j}+Zt^i).$$

Where $\tilde{p}=[\frac{i}{\beta^1_a}]$ and Z is a generic element of $\mathbf{K}$.

3) If $int(f_1,h) < \frac{n^2}{m}.int(f_1,f_2)$, a same argument than case 2)3) gives the result.

□

Case2: $n^1 < n^2$ and $i \geq n^2$.

THEOREM 4.5. I) If $c < \frac{\beta_1^1}{n^1}$ or $\frac{\beta_1^2}{n^2}$ then

$$\beta(i) = \max\Big\{(n^1+n^2)i, k_a^1(i)(\bar{\beta}_a^1 + \frac{int(f_1,f_2)}{e_a^1}), \theta^1(i,a) + [\frac{i}{\beta_a^1}].\frac{int(f_1,f_2)}{e_a^1} | a \leq g_1,$$
$$k_a^2(i)(\bar{\beta}_a^2 + \frac{int(f_1,f_2)}{e_a^2}), \theta^2(i,a) + [\frac{i}{\beta_a^2}].\frac{int(f_1,f_2)}{e_a^2} | a \leq g_2\Big\}$$

II) Otherwise, let q be the greatest integer such that $\frac{\beta_q^1}{n^1} = \frac{\beta_q^2}{n^2} < c$ and let for $j = 1, 2$

$$L_j = \begin{cases} q & \text{if } c \neq \frac{\beta_{q+1}^j}{n^j} \\ q+1 & \text{otherwise} \end{cases}$$

$$\beta(i) = \max\Big\{(n^1+n^2)i, k_a^1(i)(\bar{\beta}_a^1 + \bar{\beta}_a^2), \theta^1(i,a) + \frac{n^2}{n^1}.\theta^1(i,a), \theta^2(i,a) + \frac{n^1}{n^2}.\theta^2(i,a) | a \leq q,$$
$$k_a^1(i)(\bar{\beta}_a^1 + \frac{int(f_1,f_2)}{e_a^1}), \theta^1(i,a) + [\frac{i}{\beta_a^1}].\frac{int(f_1,f_2)}{e_a^1} | L_1 < a \leq g_1,$$
$$k_a^2(i)(\bar{\beta}_a^2 + \frac{int(f_1,f_2)}{e_a^2}), \theta^2(i,a) + [\frac{i}{\beta_a^2}].\frac{int(f_1,f_2)}{e_a^2} | L_2 < a \leq g_2\Big\}$$

PROOF. Let $\phi(t) \in \mathbf{K}[[t]]^2$, $\text{ord}_t(\phi(t)) \leq i$. In order to calculate $\beta(i)$, we need to calculate $\text{O}_t(f(\phi(t)))$ when ϕ satisfies one of the following conditions:
1) $\text{ord}_t(\phi) < \text{O}_t(\phi_1)$.
2) $\text{ord}_t(\phi) = \text{O}_t(\phi_1)$, n^1 doesn't divide $\text{ord}_t(\phi)$ and

$$\begin{cases} \text{either } n^2 \text{ doesn't divide } \text{ord}_t(\phi) \\ \text{either } n^2 \text{ divides } \text{ord}_t(\phi) \text{ and } c(f_2,\phi) \leq \frac{i}{\text{ord}_t(\phi)}. \end{cases}$$

3) $\text{ord}_t(\phi) = \text{O}_t(\phi_1)$, n^1 divides $\text{ord}_t(\phi)$, $c(f_1,\phi) \leq \frac{i}{\text{ord}_t(\phi)}$ and

$$\begin{cases} \text{either } n^2 \text{ doesn't divide } \text{ord}_t(\phi) \\ \text{either } n^2 \text{ divides } \text{ord}_t(\phi) \text{ and } c(f_2,\phi) \leq \frac{i}{\text{ord}_t(\phi)}. \end{cases}$$

Considering these cases, the proof can be done using lemma 3.8. and lemma 3.9. □

Case3: $n^1 = n^2$. In this case the result is similar to the one of theorem 4.5.

4.3. The Artin function and the semigroup of f. In this part, we give the relation between the Artin-Greenberg function and the semigroup of f. Note that in theorem 4.2. and 4.4., if we consider $(\text{O}_t(f_1(\phi), \text{O}_t(f_2(\phi)))$ as an element of $\mathbb{N}^2$, then by computing

$$\{\beta(\frac{n^l}{e_a^l}), l = 1, 2, 1 \leq a \leq g_l\}$$

we find

$$\begin{cases} \{(\bar\beta^1_a, \frac{int((f_1,f_2)}{e^1_a})|a \leq g_1, \\ (\frac{int((f_1,f_2)}{e^2_a}, \bar\beta^2_a)|a \leq g_2\} & \text{if } c < \frac{\beta^1_1}{n^1} \text{ or } \frac{\beta^2_1}{n^2} \\ \{(\bar\beta^1_a, \bar\beta^2_a)|a \leq q, (\beta^1_a, \frac{int((f_1,f_2)}{e^1_a})|L_1 < a \leq g_1, \\ (\frac{int((f_1,f_2)}{e^2_a}, \bar\beta^2_a)|L_2 < a \leq g_2\} & \text{otherwise} \end{cases}$$

where q is the greatest integer such that $\frac{\beta^1_q}{n^1} = \frac{\beta^2_q}{n^2} < c$ and for $j = 1, 2$

$$L_j = \begin{cases} q & \text{if } c \neq \frac{\beta^j_{q+1}}{n^j} \\ q+1 & \text{otherwise} \end{cases}$$

But this set with the multiplicity at the origin of f_1 and f_2 is equivalent to the set of irreducible absolute maximals of $\Gamma(f)$. In [**9**], M. Hickel, showed that $(\beta_1(i), n^1)$ (resp. $(\beta_2(i), n^2)$) determines $\Gamma(f_1)$ (resp. $\Gamma(f_2)$). Hence, $(\beta_1, n^1), (\beta_2, n^2)$ and β detemine the semigroup of f.

5. The Artin-Greenberg function of a reducible polynomial with many branches

With the same notations as section 2, we generalize the result found for the Artin-Greenberg function of a reducible polynomial with two branches to the polynomial with several branches.
In order to calculate the Artin-Greenberg function when $s \geq 2$, we suppose that the degrees of the components of f are organised in the following way: let $\{N_1, \ldots, N_r\} = \{n^1, \ldots, n^s\}$ be such that: $N^1 < N_2 < \ldots < N_r$. Now let

$$E_1 = \{j, j = 1, \ldots s, f_j \text{ is of degree } N_1 \text{ in } y\}$$

and for all $i \leq r$, let

$$E_i = \{j, j = 1, \ldots s, f_j \text{ is of degree } \leq N_i \text{ in } y\}$$

Let $i \in \mathbb{N}$, $N^u \leq i < N^{u+1}$. Let $q_j, j \notin E_u$ be the unique integer such that $\frac{n^j}{e^j_{q_j}} \leq i < \frac{n^j}{e^j_{q_j+1}}$.
The following theorem states the formula of the Artin function in the general case, we omit the proof, which can be done by generalising the proof of the case of two branches, and using the tree of contacts of f given in section 2.

THEOREM 5.1. I) If there exists $j \in \{1,\dots,s\}$ such that $c < \frac{\beta_1^j}{n^j}$ then

$$\beta(i) = \max\{\sum_{j=1}^{s} n^j.i, k_a^l(i).(\sum_{f_j \in Q'(\frac{\beta_q^l}{n^l},l)} \frac{n^j}{n^l}.\bar{\beta}_a^l + \sum_{f_j \notin Q'(\frac{\beta_q^l}{n^l},l)} \frac{int(f_l,f_j)}{e_a^l})|a \le g_l, l \in E_u;$$
$$[\frac{ie_a^l}{n^l}].(\sum_{f_j \in Q'(\frac{\beta_q^l}{n^l},l)} \frac{n^j}{n^l}.\bar{\beta}_a^l + \sum_{f_j \notin Q'(\frac{\beta_q^l}{n^l},l)} \frac{int(f_l,f_j)}{e_a^l})|a \le q_l, l \notin E_u$$
$$\sum_{f_j \in Q'(\frac{\beta_q^l}{n^l},l)} \frac{n^j}{n^l}.\theta^l(i,a) + \sum_{f_j \notin Q'(\frac{\beta_q^l}{n^l},l)} [\frac{i}{\beta_a^l}]\frac{int(f_j,f_l)}{e_a^l}|a \le g_l, l \in E_u, (\frac{\beta_a^l}{n^l}) \in B\}$$

II) Otherwise, let q be the greatest integer such that $\frac{\beta_q^1}{n^1} = \dots = \frac{\beta_q^s}{n^s} < c$. *If* $q < g_j$, *then let:*

$$L_j = \begin{cases} q & if\ c \ne \dfrac{\beta_{q+1}^j}{n^j} \\ q+1 & otherwise \end{cases}$$

Let $A = \{j, L_j < q_j\}$:
i) If $q < q_v, v \notin E_u$, *then*

$$\beta(i) = \max\left\{\sum_{j=1}^{s} n^j.i, [\frac{ie_a^l}{n^l}].\sum_{j=1}^{s} \bar{\beta}_a^j|a \le q, l \notin E_u; \theta^l(i,a).\sum_{j=1}^{s}\frac{n^j}{n^l}|a \le q, l \in E_u;\right.$$
$$k_a^l(i)\cdot(\sum_{f_j \in Q'(\frac{\beta_q^l}{n^l},l)} \frac{n^j}{n^l}.\bar{\beta}_a^l + \sum_{f_j \notin Q'(\frac{\beta_q^l}{n^l},l)} \frac{int(f_l,f_j)}{e_a^l})|L_l < a \le g_l, l \in E_u \cap A;$$
$$[\frac{ie_a^l}{n^l}]\cdot(\sum_{f_j \in Q'(\frac{\beta_q^l}{n^l},l)} \frac{n^j}{n^l}.\bar{\beta}_a^l + \sum_{f_j \notin Q'(\frac{\beta_q^l}{n^l},l)} \frac{int(f_l,f_j)}{e_a^l})|L_l < a \le q_l, l \notin E_u, l \in A$$
$$\left.\sum_{f_j \in Q'(\frac{\beta_q^l}{n^l},l)} \frac{n^j}{n^l}.\theta^l(i,a) + \sum_{f_j \notin Q'(\frac{\beta_q^l}{n^l},l)} [\frac{i}{\beta_a^l}]\frac{int(f_j,f_l)}{e_a^l}|L_l < a \le g_l, l \in E_u \cap A, (\frac{\beta_a^l}{n^l}) \in B\right\}$$

ii) If there exists $v \notin E_u$ such that $q_v \le q$, then

$$\beta(i) = \max\{\sum_{j=1}^{s} n^j.i; [\frac{ie_a^l}{n^l}].\sum_{j=1}^{s}\bar{\beta}_a^j|a \le q_l, l \notin E_u; \theta^l(i,a).\sum_{j=1}^{s}\frac{n^j}{n^l}|a \le q_l, l \in E_u(\frac{\beta_a^l}{n^l}) \in B\}.$$

REMARK 5.2. In theorem 5.1., if we consider $(O_t(f_1(\phi(t)),\dots,O_t(f_s(\phi(t)))$ instead of $O_t(f_1(\phi)) + \dots + O_t(f_s(\phi))$ and for $i \in \{\frac{n^j}{d_a^j}, j = 1,\dots,s, a = 1,\dots,g_l\}$, we get the set of the irreducible absolute maximals of $\Gamma(f)$ as described in prop.2.5. without the multiplicity at the origin of f. Hence, $\beta(i)$ and $(\beta_j(i), mult_0(f_j))$ determine the set of the irreducible absolute maximals of $\Gamma(f)$.

References

1. [AAs]S. S. Abhyankar, A. Assi. Jacobian of meromorphic curves, Proc. Mathematical Sciences, Springer India, vol.109, no. 2 (May 1999)117-163. MR1687728 (2000i:14086)
2. [Ar]M. Artin. Algebraic approximation of structures over complete local rings, Inst. Hautes Etudes Sci. Pub. Math., no. 36 (1969), 23-58. MR0268188 (42:3087)
3. [B]V.A.S. Bayer. Semigroup of two algebroid plane curves, Manuscripta matematica, vol. 59 (1987) 347-374.
4. [CDG] A.Campillo, F.Delgado, S.M.Gusein-Zade. On generators of the semigroup of a curve singularity, J. London Math. Soc.(2) 60(1999), 420-430. MR1724869 (2000m:14002)
5. [D] F. Delgado de la Mata. The semigroup of values of a curve singularity with several branches, Manuscripta Mathematica, tome 59 (1987), 347-374 . MR909850 (89f:14022)
6. [Ga] A.Garcia.- Semigroups associated to singular points of plane curves, J.Reine. Angew. Math., 336(1981), 165-184. MR671326 (84h:14037)
7. [Gr]M.Greenberg. Rational points in Henselian discrete valuation rings, Inst. Hautes Etudes Sci. Pub. Math., 31 (1966), 59-64. MR0207700 (34:7515)
8. [H1]M.Hickel. Fonction de Artin et germes de courbes tracées sur un germe d'espaces analytiques, Amer. J. Math., 115 (1993), 1299-1334. MR1254735 (95c:32006)
9. [H2] M. Hickel. Calcul de la fonction d'Artin-Greenberg d'une branche de courbe plane, Pacific Journal of Math., vol 213, no. 1 (2004), 37-46. MR2040249 (2004m:14051)
10. [LJ]M. Lejeune Jalabert. Courbes tracées sur un germe d'hypersurfaces, Amer. J. of Math., 112 (1990), pp 525-568. MR1064990 (92a:14002)
11. [PPo]G.Pfister, D.Popescu. Die strange approximationeigenschaft lokaler ringe, Invent. Math., 30 (1975), 145-174. MR0379490 (52:395)
12. [S]S.Saleh. Calcul de la fonction d'Artin d'une singularité plane, Thèse de doctorat de l'université d'Angers (2010), (http://tel.archives-ouvertes.fr/docs/00/54/36/87/PDF/thesesaleh.pdf.)
13. [W] C.T.C. Wall. Singular points of plane curves, Cambridge University Press (2004). MR2107253 (2005i:14031)
14. [Z]O. Zariski. Le problème des modules pour les branches planes, Lecture Notes (ed. F. Kmety and M. Merle), École Polytechnique (1973). MR0414561 (54:2662)

Université d'Angers, Département de mathématiques Larema, 49045, Angers Cedex 01, France

E-mail address: saharsaleh84@hotmail.com.

Contemporary Mathematics
Volume **569**, 2012
http://dx.doi.org/10.1090/conm/569/11244

Singularities with critical locus a complete intersection and transversal type A_1

Mamuka Shubladze

ABSTRACT. In this paper we study germs of holomorphic functions $f : (\mathbb{C}^m, 0) \to (\mathbb{C}, 0)$ with the following two properties:

(i) the critical locus Σ of f is an isolated complete intersection singularity (icis);

(ii) the transversal singularity of f in points of $\Sigma \backslash \{0\}$ is of type A_1.

We first compute the homology of the Milnor fibre and then show that the homotopy type of the Milnor fibre F of f is a bouquet of spheres.

1. Introduction

Let O be the ring of holomorphic germs $f : (\mathbb{C}^m, 0) \to (\mathbb{C}, 0)$. Let $I \subset O$ be a reduced ideal defining an icis Σ of arbitrary dimension k. As usual $J(f)$ denote the jacobian ideal of f, namely:

$$J(f) = \Big(\frac{\partial f}{\partial x_1}, \dots, \frac{\partial f}{\partial x_m}\Big).$$

We consider, as in [**Pe-1**], [**Pe-2**], the group D_I of local analytic isomorphisms $\varphi : (\mathbb{C}^m, 0) \to (\mathbb{C}^m, 0)$ such that $\varphi^*(I) = I$.

Let $f \in O$ be a germ whose critical set contains Σ. Then by [**Pe-1**], [**Pe-2**], $f \in I^2$. The group D_I acts an I^2, and the extended codimension of the orbit of f with respect to this action is

$$c_e(f) = \dim \frac{I^2}{I^2 \cap J(f)}$$

we shall focus our attention on germs $f \in I^2$ with $c_e(f) < \infty$. We are interested in the topology of Milnor fibre of f. We know that if the dimension k of the singular locus Σ is 1 then the Milnor fibre is equivalent to a bouquet of n dimensional spheres or a bouquet of one $n-1$ dimensional sphere and several n dimensional spheres [**Si**, Theorem 5.7].

2010 *Mathematics Subject Classification.* Primary 32S05, 14B05, 32S25, 32S50, 32S55, 57R45.

This work was done while the author was a guest at the Max-Planck-Institut für Mathematik. He thanks the institute and its staff members for the support and the good mathematical atmosphere. The author would like to express his gratitude to Professor D. Siersma for helpful discussions and suggestions.

If $k = m - 1$ i.e. codim $\Sigma = 1$, then again F is homotopy equivalent to a bouquet of a circle and several n-dimensional spheres [**Sh-1**], [**Sh-2**], [**Ne-1**]. If $k = 2$ the bouquet theorem is also valid: the Milnor fibre F is homotopy equivalent to a bouquet of spheres [**Za**], [**Ne-2**].

We consider case when $k \geq 3$ and give the properties in which case we can prove the

THEOREM. *The Milnor fibre F of $f = (\mathbb{C}^m, 0) \to (\mathbb{C}, 0)$ is homotopy equivalent to a bouquet of spheres $F \simeq S^n \vee S^{m-1} \vee S^{m-1} \vee \cdots \vee S^{m-1}$, where $n = m - k$.*

2. Non-Isolated Singularities with Transversal Type A_1

Let as above $I \subset O$ be a reduced ideal defining an icis (isolated complete intersection singularity) Σ of dimension k and suppose that $I = (g_1, \ldots, g_n)$ with $n = m - k$. We shall assume that $n \geq 2$ and $k \geq 3$; the cases $k = 1$, $k = 2$ and $n = 1$ are studied in [**Si**], [**Ne-2**], [**Za**] and respectively [**Sh-1**] and [**Ne-1**]. Let $f \in O$ be a germ whose critical set contains Σ. It follows that $f \in I^2$, and we have decomposition

$$f = \sum_{i,j=1}^{n} h_{ij}\, g_i\, g_j$$

with $h_{ij} = h_{ji}$ [**Pe-1**], [**Pe-2**]. Moreover, the class of $h_{ij} + I$ in O/I is uniquely determined by f [**Za**].

The $D(k,p)$ singularity was introduced in [**Pe-1**] and [**Pe-2**]. Their local equations, in a suitable coordinate system x_{ij} $(1 \leq i \leq j \leq p)$, $z_1, \ldots, z_q, y_1, \ldots, y_n$, is

$$f(x, y, z) = \sum_{1 \leq i \leq j \leq p} x_{ij}\, y_i\, y_j + \sum_{l=p+1}^{n} y_l^2.$$

Note also the singular locus of a $D(k,p)$ singularity is smooth and of dimension $k = \frac{1}{2}p(p+1) + q$, while $m = k + n$. A $D(k,0)$ singularity in [**Pe-1**], [**Pe-2**] is also called $A(k)$, where

$A(k) := D(k,0) : \sum\limits_{l=1}^{n} y_l^2.$

We note also:

$D(k,1) : x y_1^2 + \sum\limits_{l=2}^{n} y_l^2.$

REMARK 2.1. As in [**Sh-3**], see also [**Za**], it is easy to see that following are valid

(1) A singular point $z \in \Sigma$ is a singular point of type $D(k,0)$ if the matrix $(h_{ij}(z))$ has rank n.

(2) A singular point $z \in \Sigma$ is a singular point of type $D(k,1)$ if the matrix rank $(h_{ij}(z)) = n - 1$ and $\operatorname{grad}_z(\det(h_{ij}(z)))|_\Sigma \neq 0$.

Let D be defined as in [**Za**] by $D(z) = \det(h_{ij}(z))_{ij}$ and if $D(0) = 0$ then the ideal $I + D = (g_1, \ldots, g_n, D)$ defines a complete intersection Δ in $(\mathbb{C}^m, 0)$, which depends only on f [**Za**]. Let us denote by Δ the zero set of $I + (D)$.

The following result is similar to [**Si**], [**Sh-1**] criterion of finite codimension.

THEOREM 2.2. *Let $f \in I^2$, $f = \sum\limits_{ij=1}^{n} h_{ij}\, g_i\, g_j$ and Σ is an icis, and Δ is an icis too. Then*

(a) *the critical locus of* f *is* Σ *and the germ of* f *at every point of* $\Sigma\backslash\{0\}$ *outside* Δ *is equivalent to a* $D(k,0)$ *singularity, and at points of* Δ *is equivalent to a* $D(k,1)$.

(b) $c_e(f) < +\infty$.

PROOF. (a) If $z \in \Delta$ and $z \neq 0$ then $\operatorname{rank}((h_{ij})_{ij}) = n-1$ since Σ is icis of dimension $k = m-n$. Since $D = \det((h_{ij})_{ij})$ is isolated singularity on Σ so grad $D|_\Sigma \neq 0$ at the point of $\Delta\backslash\{0\}$, which means that f at z is of type $D(k,1)$ by the remark of 2.2. Let now $z \overline{\in} \Delta$ and $z \neq 0$. Then we have $\det(h_{ij})_{ij}) \neq 0$ at this point z, so $\operatorname{rank}((h_{ij})_{ij}) = n$ and using Remark 2.1 f at this point z has a $D(k,0)$ singularity.

(b) Let f be some representative of the germ of the given singularity. In the domain where it is given we define a sheaf of O modules as follows

$$\mathcal{F}(u) = I^2/\tau_e(f),$$

with $\tau_e(f) = I^2 \cap J(f)$, where I^2 and $\tau_e(f)$ are considered as modules over the holomorphic functions on u. It is clear that $\mathcal{F}$ is coherent. We will use the fact: $\mathcal{F}$ is concentrated in a point $\Leftrightarrow \dim \Gamma(\mathcal{F}) < \infty$. For $z \in \mathbb{C}^m\backslash\Sigma$, f is regular at z and we have $\dim \mathcal{F}_z = 0$ since $I^2_z \cong O_z$ and $(\tau_e(f))_z \cong O_z$. If $z \in \Sigma\backslash\{0\}$ then as we proved above f is of type $D(k,p)$, $p \leq 1$ at z and we have $\dim \mathcal{F}_z = 0$ since $c_e(D(k,p)) = 0$. So $\mathcal{F}$ is concentrated at 0, hence $c_e(f) < \infty$. □

3. The Deformation of Nonisolated Singularities

In the case of an ordinary isolated singularity it is useful to consider a generic approximation with only ordinary Morse point [**Br**]. At every Morse point one can study its local Milnor fibration, with Milnor fibre homotopy equivalent to one n-sphere S^n ("the vanishing cycle"). The Milnor fibre of the original f then has the homotopy type of the wedge of those spheres.

We want to mimic the constructions in our case.

Let Σ be a k-dimensional complete intersection defined by the ideal $I = (g_1, \dots, g_n)$. Then $f = \sum_{i,j=1}^{n} h_{ij}\, g_i\, g_j$. Assume that $D(z) = \det((h_{ij})_{i,j})$ is an isolated singularity and $I + (D)$ is isolated complete intersection.

Let $G : (\mathbb{C}^m \times \mathbb{C}^r, 0) \to (\mathbb{C}^m \times \mathbb{C}^r, 0)$ be a versal deformation of $(\Sigma, 0)$ with $G(z,v) = (G_1(z,v), \dots, G_n(z,v), v)$ and $G_i(z,0) = g_i(z)$ [**Loo**]. Consider the deformation

$$f_s : (\mathbb{C}^m \times S, 0) = \mathbb{C}^m \times \mathbb{C}^r \times \mathbb{C}^n \times \mathbb{C}^{n(n+1)/2} \times \mathbb{C}^{m-n}, 0) \to \mathbb{C}$$

given by

$$\begin{aligned} f_s(z) &= f(z, v, u, a, b) \\ &= \sum_{i,j=1}^{n} \Big(h_{ij}(z) + a_{ij} + \sum_{t=1}^{m} b_{tj}\, x_t\, \delta_{ij}\Big) \cdot (G_i(z,v) - u_i)(G_j(z,v) - u_j), \end{aligned}$$

where $a_{ij} = a_{ji}$, and S is the space of parameters (a, b, u, v). In the case $k = \dim \Sigma = 1, 2$ or $m-1$, there exists a dense subset U in S such that for every $s \in U$, the germ of f_s at the points of Σ_s is of type $D(k,0)$ or $D(k,1)$. Moreover, the set of points of Σ_s where f_s is of type $D(k,1)$ is exactly Δ_s and this set is a Milnor fibre of the icis Δ. For an arbitrary k, we know at least two cases when such deformation exists: i) the germ f at any point $z \in \Sigma\backslash\{0\}$ is of type $D(k,0)$, ii) the matrix $(h_{ij}(0))_{ij}$ has rank $n-1$ (see [**Si**], [**Sh-1**], [**Za**]). From this page assume the existence of such a deformation for arbitrary k. Then following are valid [**Za**], [**Bo**], [**Ne-2**].

LEMMA 3.1. *There exist an ε-ball B_ε with center $D \in \mathbb{C}^m$, a proper analytic set $(A,0) \subset (S,0)$, and a neighborhood U of $0 \in S$, such that for any $s \in U \backslash A$ has the following:*

(a) $\Sigma_s = \{z \in B_z : G_i(x,v) = u_i,\ i = 1,\dots,n\}$ *is the Milnor fibre of* Σ.

(b) *The zero set of $D_s(z) = \det(h_{ij}(z) + a_{ij} + \Sigma_t\, b_{tj}\, z_t\, \delta_{ij})$ intersets Σ_s transversally; hence $\Delta_s = D_s^{-1}(0) \cap \Sigma_s$ is smooth. In particular Δ_s is (diffeomorphic to) the Milnor fibre of Δ.*

(c) *The singularities of f_s in $B_\varepsilon \backslash \Sigma_s$ are of type A_1.*

(d) *The germ of f_s at any point of $\Sigma_s \backslash \Delta_s$ is of type $D(k,0)$ and at any point of Δ_s is of type $D(k,1)$.*

(e) *Fix ε sufficiently small and δ sufficiently small with respect to ε. If U is sufficiently small with respect to ε and δ, then $f_s^{-1}(t)$ (as a stratified set) intersects ∂B_ε transversally for any $s \in U$ and $t \in \Lambda = \{|t| \leq \delta\}$. In particular, the topological type of the smooth fibres of the maps*

$$f_s : X_s = f_s^{-1}(\Lambda) \cap B_\varepsilon \to \Lambda \quad (s \in U)$$

is independent of the parameter $s \in U$. (In fact, even the fibrations $f_s : f_s^{-1}(\partial\Lambda) \cap B_\varepsilon \to \partial\Lambda$ are equivalent to the fibration $f : f^{-1}(\partial\Lambda) \cap B_\varepsilon \to \partial\Lambda$. In particular, the corresponding fibres are homotopically equivalent).

(f) *The spaces X_s $(s \in U)$ are contractible.*

4. The Topology of Milnor Fibre

Let f_s be a deformation of f described above, having only $D(k,0)$ and $D(k,1)$ singularities on Σ_S, and suppose that the number of A_1 (Morse) points is σ. The critical set of f consists of

(a) a manifold Σ_0 which is the Milnor fibre Σ_s of a k-dimensional isolated complete intersection singularity Σ. The local singularities of f on Σ_0 are $D(k,0)$ and $D(k,1)$.

(b) $\Sigma_1 = \{b_1\}, \dots, \Sigma_\sigma = \{b_\sigma\}$, where the local singularity of f is isolated of type A_1.

Define $B_1, B_2, \dots, B_\sigma$ as disjoint $2m$ dimensional balls in the space $\mathbb{C}^m$ centred at points $b_1, \dots, b_\sigma$ and $D_1, D_2, \dots, D_\sigma$ disjoint two dimensional disks at the points $f_s(b_1), \dots, f_s(b_\sigma)$. Choose them such that $f_s : B_i \cap {f_s}^{-1}(D_i) \to D_i$ define a locally trivial Milnor fibration and the following transversality condition holds: $f_s^{-1}(t) \cap \partial B_i$, $\forall t \in D_i$, $i = 1, \dots, \sigma$.

The situation at the points of $b_1, \dots, b_\sigma$ is well known; we consider the situation along Σ_s.

First we define B^0 a tubular neighborhood

$$B^0 = \left\{ z \in B_\varepsilon : \sum_{i=1}^{n} \left| G_i(z,v) - u \right|^2 \leq \rho \right\} \quad \text{of} \quad \Sigma_s \cap B_\varepsilon,$$

which is diffeomorphic for sufficiently small ρ to the product $(\Sigma_s \cap B_\varepsilon) \times Q^n$, where Q^n is a closed n-ball in $\mathbb{C}^n$ with center at the origin [**Si**].

Let us denote $X_{t,s} = f_s^{-1}(t) \cap B_\varepsilon$ and $F^0 = B^0 \cap X_{t,s}$, then for a sufficiently small t we have

$$H_{*-1}(X_{s,t}) = H_*(X_s, X_{s,t}) = \begin{cases} H_*(B^0, F^0) & \text{if } * \neq m, \\ H_m(B^0, F^0) \oplus \mathbb{Z}^\sigma & \text{if } * = m, \end{cases}$$

[**Si**].

First compute the homology of the point (B^0, F^0). Following [**Si**], [**Za**] we shall consider in B^0 coordinates $(w_1, \dots, w_n, w_{m-k+1}, \dots, w_m)$ such that $(w_1, \dots, w_n) \in Q^n$ are the functions defined by $w_i(z) = G_i(x, v) - u_i$ and $w_{m-k+1}, \dots, w_m \in \Sigma_s$ (recall that $\dim \Sigma = k$ and $m = n + k$). Then $(w_1, \dots, w_n)$ are holomorphic functions on B^0 and $(w_{m-k+1}, \dots, w_m)$ are real differentiable [**Si**]. Now consider the projection $\pi : (w_{m-k+1}, \dots, w_m) : (B^0, F^0) \to \Sigma_s$. From this page assume that $n \neq 1$. Then in a similar way to [**Si**], [**Za**], [**Sh-3**] we can prove

LEMMA 4.1. *If ρ and tubular neighborhoods $U_1 \subset U_2 \subset \Sigma_s$ of $\Delta_s \subseteq \Sigma_s$ are sufficiently small then*

(a) *$\pi : (B^0 \backslash \pi^{-1}(U_1), F^0 \backslash \pi^{-1}(U_1)) \to \Sigma_s \backslash U_1$ is a locally trivial fibration with fibre equal to the pair ($\mathbb{C}^{m-k}$, Milnor fibre of $x_1^2 + \dots + x_n^2$),*

(b) *the map given by the composition $\pi^{-1}(U_2) \to U_2 \to \Delta_s$ is a fibration of the pair $(\pi^{-1}(U_2), F^0 \cap \pi^{-1}(U_2))$ with fibre equal to the pair ($\mathbb{C}^{n+k}$, Milnor fibre of $x_{n+1}x_1^2 + x_2^2 + \dots + x_n^2$).*

For a subset $W \subseteq \Sigma_s$ we shall denote by F_W the following set: $F_W = \pi^1(W) \cap F^0$.

The following statements hold

LEMMA 4.2. *$H_q(F_{\Sigma_s \backslash U_1}) = 0$ for $q = n-2$ and $q = n$. Moreover $H_{n-1}(F_{\Sigma_s \backslash U_1}) = \mathbb{Z}_2$, $H_{m-1}(F_{\Sigma_s \backslash U_1}) = \mathbb{Z}^{\mu_\Delta + \mu_\Sigma}$, $H_1(F_{\Sigma_s \backslash U_1}) = 0$, $q \geq n-2$.*

PROOF. The case $k = 1$ was studied in [**Si**] and $k = 2$ in [**Ne-2**], so assume $k \geq 3$. As in [**Ne-2**] we may assume that $n > k+2$, because if w is a new variable, then the Milnor fiber F_w of $f(z) + w^2$ is the suspension of the Milnor fibre F of f, in particular $H_*(F) = H_{*+1}(F_w)$.

Consider the fibration $\pi : F_{\Sigma_s \backslash U_1} \to \Sigma_s \backslash U_1$.

The base space $\Sigma_s \backslash U_1$ is homotopy equivalent to $\Sigma_s \backslash U_1 \simeq \Sigma_s \underset{\Delta_s}{\sqcup} U_1 \times S^1 \simeq S^1 \vee S^k \vee \dots \vee S^k$, a bouquet of the circle S^1 and k-dimensional spheres. The number of k-dimensional spheres μ is equal to $\mu_\Sigma + \mu_\Delta$ [**Za**], [**Sh-4**]. The homotopy type of fibre of π is S^{n-1} but unfortunately we cannot use the Gysin exact sequence for this fibration π because π is not orientable. But the total space $F_s^0 \backslash F_{U_1}$ is homotopy equivalent to $E' \underset{S^{n-1}}{\cup} E''$, where $E' \to S^1$ and $E'' \to \overset{\mu}{\underset{k=1}{\vee}} S_i^k$ are fibre bundles with fibre S^{n-1} and in $E' \underset{S^{n-1}}{\cup} E''$ a fibre of E' is identified with a fiber of E''. For the fibration $E' \to S^1$ which is nonorientable and its monodromy is equal -1 [**Ne-2**] we may use the Wang exact sequence. We obtain

$$\to H_q(S^{n-1}) \to H_q(E') \to H_{q-1}(S^{n-1}) \to H_{q-1}(S^{n-1}) \to \cdots$$

Finally, we obtain short exact sequence

$$0 \longrightarrow H_n(E') \longrightarrow \mathbb{Z} \xrightarrow{\ \alpha\ } \mathbb{Z} \longrightarrow H_{n-1}(E') \longrightarrow 0,$$

α is multiplication by 2. Therefore $H_{n-1}(E') = \mathbb{Z}_2$, $H_q(E') = 0$, $q \neq 0$, $q \neq n-1$. □

On the other hand, we have an orientable fibration $E'' \to \overset{\mu}{\underset{i=1}{\vee}} S_i^k$, because $k \geq 3$. Hence we may use the Gysin exact sequence to obtain

$$\to H_q(E'') \to H_q(\overset{\mu}{\underset{i=1}{\vee}} S_i^k) \to H_{q-n}(\overset{\mu}{\underset{i=1}{\vee}} S_i^k) \to H_{q-1}(E'') \to \cdots$$

Since $n \geq k+2$ and $k \geq 3$ we obtain $H_{n-2}(E'') = H_n(E'') = 0$ and $H_{n-1}(E'') \simeq \mathbb{Z}$, $H_{m-1}(E'') \simeq \mathbb{Z}^{\mu_\Delta + \mu_\Sigma}$.

The total space $F_{\Sigma_s \backslash U_1} = E' \cup E''$, where $E' \cap E'' \simeq S^{n-1}$. Using the Mayer-Vietoris theorem we obtain

$$\to H_q(E' \cap E'') \to H_q(E') \oplus H_q(E'') \to H_q(E' \cup E'') \to H_{q-1}(E' \cup E'') \to \cdots$$

After short computations we obtain a short exact sequence

$$0 \to H_n(E) \to \mathbb{Z} \to \mathbb{Z}_2 \oplus \mathbb{Z} \to H_{n-1}(E) \to 0.$$

Therefore $H_q(E) = 0$, $H_{n-1}(E) = \mathbb{Z}_2$ and $H_{n-2}(E) = 0$.

Similarly we obtain $H_{m-1}(E) = \mathbb{Z}^{\mu_\Delta + \mu_\Sigma}$ and $H_q(E) = 0$, $q \geq n-2$.

LEMMA 4.3. *$H_{n-2}(F_{U_2 \backslash U_1}) = 0$, $H_{n-1}(F_{U_2 \backslash U_1}) = \mathbb{Z}_2$, $H_{m-2}(F_{U_2 \backslash U_1}) = \mathbb{Z}_2^{\mu_\Delta}$ and $H_q(F_{U_2 \backslash U_1}) = 0$ if $q \geq n-2$ and $q \neq n-1, m-2$.*

PROOF. We have a fibration $\pi : F_{U_2 \backslash U_1} \to U_2 \backslash U_1$ with fibre S^{n-1}. Since $U_2 \backslash U_1$ is homotopy equivalent to $S^1 \times \Delta_s$ using the homotopy exact sequence of fibration π we obtain $H_{n-2}(F_{U_2 \backslash U_1}) = 0$. Because π is not orientable $H_{n-1}(F_{U_2 \backslash U_1}) = \mathbb{Z}_2$. As in [**Ne-2**], since the base space has a product structure, one can write $F_{U_2 \backslash U_1}$ as a fibre bundle over Δ_s with fibre Z, where fibre Z is the total space of a fibre bundle with base S^1 and fibre S^{n-1}. Using the Wang exact sequence we obtain $H_{n-1}(Z) = \mathbb{Z}_2$, $H_q(Z) = 0$, $q \neq 0, n-1$. Because Δ_s is simply connected, it follows from the Serre spectral sequence [**Me**] $H_*(\Delta_s; H_*(Z)) \Rightarrow H_*(F_{U_2 \backslash U_1})$ that $H_{m-2}(F_{U_2 \backslash U_1}) = \mathbb{Z}_2^{\mu_\Delta}$ and $H_q(F_{U_2 \backslash U_1}) = 0$ if $q \geq n-2$, $q \neq n-1, m-2$. □

LEMMA 4.4. *$H_{n-1}(F_{U_2}) = 0$, $H_n(F_{U_2}) = \mathbb{Z}$ and $H_{m-1}(F_{U_2}) = \mathbb{Z}^{\mu_\Delta}$.*

PROOF. This follows from the fibration $F_{U_2} \to \Delta_s$ (cf. Lemma 3.1 (b)), whose fibre has the homotopy type of S^n. □

COROLLARY 4.5.

$$H_q(F^0, F_{U_2}) = \begin{cases} \mathbb{Z}, & \textit{if} \quad q = 0, \\ \mathbb{Z}^{\mu_\Delta + \mu_\Sigma}, & \textit{if} \quad q = m-1, \\ 0, & \textit{otherwise.} \end{cases}$$

PROOF. Using the long exact sequence for the pair $(F_{\Sigma_s \backslash U_1}, F_{U_2 \backslash U_1})$ we obtain

$$\to H_q(F_{U_2 \backslash U_1}) \to H_q(F_{\Sigma_s \backslash U_1}) \to H_q(F_{\Sigma_s \backslash U_1}, F_{U_2 \backslash U_1}) \to \\ \to H_{q-1}(F_{U_2 \backslash U_1}) \to \cdots$$

Since $F_{U_2 \backslash U_1} \hookrightarrow F_{\Sigma_s \backslash U_1}$ is an inclusion using excision $H_q(F_{\Sigma_s \backslash U_1}, F_{U_2 \backslash U_1}) = H_q(F^0, F_{U_2})$, and Lemma 4.2, 4.3 we obtain $H_q(F_s^0, F_{U_2}) = 0$ if $q \neq 0, n, m-1$. For the n-dimensional homology group we have exact sequence

$$0 \to H_n(F^0, F_{U_2}) \to \mathbb{Z}_2 \hookrightarrow \mathbb{Z}_2 \to H_{n-1}(F^0, F_{U_2}) \to 0.$$

So $H_n(F^0, F_{U_2}) = 0$ and $H_{n-1}(F^0, F_{U_2}) = 0$. For the $m-1$ dimensional homology group we have the following exact sequence

$$0 \to \mathbb{Z}^{\mu_\Delta + \mu_\Sigma} \to H_{m-1}(F^0, F_{U_2}) \to \mathbb{Z}_2^{\mu_\Delta} \to 0.$$

As we know we have fibrations

$$\begin{array}{ccc} F_{\Sigma_s\backslash U_1} & \xleftarrow[i_1]{} & F_{U_2\backslash U_1} \\ \downarrow & & \downarrow \\ \Sigma_s \underset{\Delta_s}{\sqcup} S^1 \times \Delta_s & \xleftarrow[i_2]{} & S^1 \times \Delta_s \end{array}$$

Let $b_1, \ldots, b_{\mu_\Sigma}$ generators of $H_{m-1}(F_{\Sigma_s\backslash U_1})$. Take into account $\Delta_s \simeq \bigvee_{i=1}^{\mu_\Delta} S_i^{k-1}$. Let $f_{i,\pm}$ be the map

$$D^k_{i,\pm} = \left[0, \frac{1}{2}\right] \times S_i^{k-1}/\{1\} \times S_i^{k-1} \to S^1 \times \Delta_s/\{1\} \times \Delta_s \to$$
$$\to \Sigma_s \underset{\Delta_s}{\sqcup} (S^1 \times \Delta_s), \quad i = 1, \ldots, \mu_0.$$

The pullback of the fibration $F_{\Sigma_s\backslash U_1} \to \Sigma_s \underset{\Delta_s}{\sqcup} (S^1 \times \Delta_s)$ along $f_{i,+}$ is trivial. Therefore we have following diagram

$$\begin{array}{ccc} (D^k_{i,+} \times S^{n-1}, S_i^{k-1} \times S^{n-1}) & \xrightarrow{\widetilde{f}_{i,+}} & (F_{\Sigma_s\backslash U_1}, F_{U_2\backslash U_1}) \\ \downarrow & & \downarrow \\ (D^k_{i,+}, S_i^{k-1}) & \xrightarrow{f_{i,+}} & (\Sigma_s \underset{\Delta_s}{\sqcup} (S^1 \times \Delta_s), S^1 \times \Delta_s) \end{array} .$$

Let $a_i \in H_{n-1}(F_{\Sigma_s\backslash U_1}, F_{U_2\backslash U_1})$ be the image of a generator of $H_{m-1}(D^k_{i,+} \times S^{n-1}, S_i^{k-1} \times S^{n-1}) \cong \mathbb{Z}$ under $(\widetilde{f}_{i,+})$. There is a homotopy between $f_{i,+}$ and $f_{i,-}$ (as a map of pairs), namely

$$D_i^k \times [0,1] = ([0,1] \times S_i^{k-1}/\{0\} \times S_i^{k-1}) \times [0,1] \to \Sigma_s \underset{\Delta_s}{\sqcup} (S^1 \times \Delta_s).$$

$$([t,x], S) \mapsto \begin{cases} f_{i,+}([1-2s)t, x), & 0 \le s \le \frac{1}{2}, \\ f_{i,-}([2s-1)t, x), & \frac{1}{2} \le s \le 1. \end{cases}$$

Therefore $(\widetilde{f}_{i,+})_*$ and $(\widetilde{f}_{i,-})_*$ define the same element a_i. Hence $2a_i$ as an element $H_{m-1}(F_{\Sigma_s\backslash U_1})$ is represented by $\widetilde{f}_{i,+} \cup \widetilde{f}_{i,-}$, which means that

$$H_{m-1}(F_{\Sigma_s\backslash U_1}, F_{U_2\backslash U_1}) = H_{m-1}(F^0, F_{U_2}) = \mathbb{Z}^{\mu_\Delta + \mu_\Sigma}. \qquad \square$$

COROLLARY 4.6.

$$H_q(F^0) = \begin{cases} \mathbb{Z}, & \textit{if} \quad q = q = n, \quad q = 0, \\ \mathbb{Z}^{2\mu_\Delta + \mu_\Sigma}, & \textit{if} \quad q = m-1, \\ 0, & \textit{otherwise.} \end{cases}$$

PROOF. Use the long exact sequence of the pair (F^0, F_{U_2}) and above lemmas.

Now we consider the pair (B^0, F^0) and the corresponding exact sequence in homology to obtain $H_*(B^0, F^0) = H_{*-1}(F^0)$. As we mentioned in the beginning of this section for the Milnor fibre $F = X_{t,s}$ the homology group is equal

$$H_{*-1}(F) = \begin{cases} H_{*-1}(F^0) & \text{if} \quad * \neq m, \\ H_{m-1}(F^0) \oplus \mathbb{Z}^\sigma & \text{if} \quad * = m. \end{cases}$$

Therefore finally we obtain

$$H_*(F) = \begin{cases} \mathbb{Z} & \text{if } \quad * = 0, n, \\ \mathbb{Z}^{2\mu_\Delta + \mu_\Sigma + \sigma} & \text{if } \quad * = m-1, \\ 0 & \text{otherwise.} \end{cases}$$

□

We show now that our Milnor fibre is homotopy equivalent to a wedge of spheres $S^n \vee S^{m-1} \vee \cdots \vee S^{m-1}$.

LEMMA 4.7. *Let X be a $(n-2)$-connected CW complex of dimension $n \geq 3$ with given homology $H_n(X, \mathbb{Z}) = \mathbb{Z}$, $H_{m-1}(X, \mathbb{Z}) = \mathbb{Z}^\mu$, $\widetilde{H}_k(X, \mathbb{Z}) = 0$ if $k \neq n, m-1$. Then we have a homotopy equivalence*

$$X \simeq S^n \vee S^{m-1} \vee \cdots \vee S^{m-1}.$$

PROOF. For $n \geq 3$ we have that X is simply connected. According to Hurewicz's theorem $\pi_n(X) \simeq H_n(X) = \mathbb{Z}$. We may attach an n-cell e_n corresponding to a generator φ of $\pi_{n-1}(X)$. Let $\widetilde{X} = X \underset{\Phi}{\cup} e_n$. So we have $\pi_{n-1}(\widetilde{X}) = 0$ and $\pi_k(\widetilde{X}) = \pi_k(X) = 0$, $k \leq n-2$.

Moreover we can prove that $\widetilde{X}$ is homotopy equivalent to a bouquet of μ copies of an $m-1$ dimensional sphere (see [**Si**], Prop. 6.1).

Consider the following Hurewicz diagram

$$\begin{array}{ccc}
0 & & \\
\downarrow & & \downarrow \\
\mathbb{Z}^\mu = H_{m-1}(X) & \longleftarrow & \pi_{m-1}(X) \\
\downarrow \alpha_1 & & \downarrow \alpha_2 \\
H_{m-1}(\widetilde{X}) & \xleftarrow{\cong} & \pi_{m-1}(\widetilde{X}) \\
\downarrow \beta_1 & & \downarrow \beta_2 \\
H_{m-1}(\widetilde{X}, X) & \longleftarrow & \pi_{m-1}(\widetilde{X}, X) \\
\downarrow \delta_1 & & \downarrow \gamma_2 \\
H_{m-2}(X) & \longleftarrow & \pi_{m-2}(X) \\
\downarrow & & \downarrow
\end{array}$$

This implies $\beta_2 = 0$ so α_2 is surjective. □

Let now $Y = S^n \vee S^{m-1} \vee \cdots \vee S^{m-1}$, and $\widetilde{Y} = D^{n+1} \vee S^{m-1} \vee \cdots \vee S^{m-1}$, where $\partial D^{n+1} = S^n$. Define $h : Y \to X$ and $\widetilde{h} : \widetilde{Y} \to \widetilde{X}$ as follows

$$h \mid S^n = \text{generator of} \quad \pi_n(X),$$
$$h \mid S^{m-1} = \text{lifted generator of} \quad \pi_{m-1}(\widetilde{X}),$$
$$h \mid D = e_n.$$

It is obvious that $H_q(X) = H_q(Y)$, if $q \neq m-1$. For $q = m-1$ consider

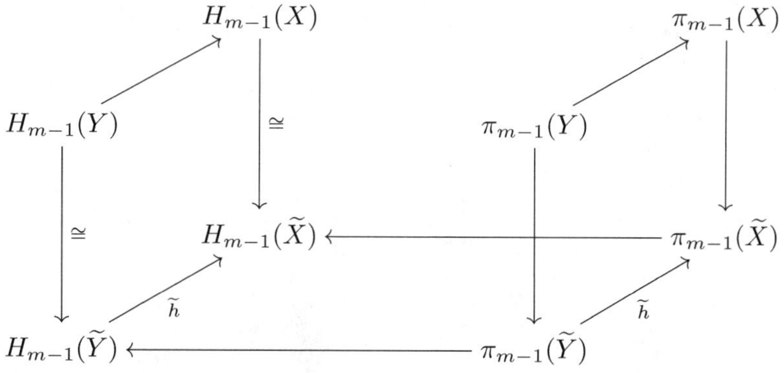

The following maps are isomorphisms

$$\begin{aligned}
\widetilde{h} : \pi_{m-1}(\widetilde{Y}) \to \pi_{m-1}(\widetilde{X}) &\quad \text{by construction,}\\
\pi_{m-1}(\widetilde{X}) \to H_{m-1}(\widetilde{X}) &\quad \text{by Hurewicz's theorem,}\\
\pi_{m-1}(\widetilde{Y}) \to H_{m-1}(\widetilde{Y}) &\quad \text{by Hurewicz's theorem,}\\
H_{m-1}(Y) \to H_{m-1}(\widetilde{Y}) &\quad \text{by exactness,}\\
H_{m-1}(X) \to H_{m-1}(\widetilde{X}) &\quad \text{by exactness.}
\end{aligned}$$

It follows that h is a homotopy equivalence, because $H_*(Y) \cong H_*(X)$, X and Y are simply connected, as a consequence of Whitehead's theorem.

THEOREM 4.8 (Main Theorem). *Let $\Sigma = \{g_1 = \cdots = g_n = 0\}$ be an isolated complete intersection with $n \neq 1$, and $f : (\mathbb{C}^m, 0) \to (\mathbb{C}, 0)$ be a holomorphic function with singular locus $\Sigma(f) = \Sigma$, i.e. $f = \sum_{i,j=1}^{n} h_{ij} g_i g_j$ with $D = \det((h_{ij})_{ij})$ isolated singularity at the origin and $(g_1, \ldots, g_n, D)$ icis, and deformation f_s described above exists. Then the Milnor fibre of f is homotopy equivalent to a bouquet of $\mu_{m-1}(f) = 2\mu_\Delta + \mu_\Sigma + \sigma$ copies of S^{m-1} and one copy of S^n, where μ_Σ (respectively μ_Δ) is the Milnor number of Σ (respectively of Δ), and σ is the number of Morse points which occur in a special deformation of f.*

PROOF. We know that Milnor fibre F is $n-2$ connected (see [**Ka-Na**]). If above $n \geq 3$ so F is simply connected and we can apply Lemma 4.7 and find $F \simeq S^n \vee S^{m-1} \vee \cdots \vee S^{m-1}$. For $n = 2$ as in [**Si**] we can prove $\pi_1(F) = H_1(F) = 0$. So in this case also F is simply connected, and using Corollary 4.6, this finishes the proof of the main theorem. □

REMARK 4.9. The case when $n = 1$ is special, and it must be considered separately, no matter which kind of dimensional non-isolated singularity we want to study. In this case $f = g^2 h$, and it is easy to see π is fibration over $\Sigma_s \setminus \Delta_s$, not over Σ_s as in any other case. For this reason the number of spheres in the bouquet changes. More precisely, we can prove the Milnor fibre of f is homotopy equivalent to a bouquet of $\mu_{m-1}(f) = 2(\mu_\Delta + \mu_\Sigma) + \sigma$ copies of S^{m-1} and one copy of S^n, where μ_Σ (respectively μ_Δ) is the Milnor number of Σ (respectively of Δ), and σ is the number of Morse points which occur in a special deformation of f. See [**Sh-2**].

References

Bo. J. Bobadilla, Approximations of non-isolated singularities of finite codimension with respect to an isolated complete intersection singularity. *Bull. London Math. Soc.* **35** (2003), no. 6, 812–816. MR2000028 (2004j:32030)

Br. E. Brieskorn, Die Monodromie der isolierten Singularitaten von Hyperflachen. (German) *Manuscripta Math.* **2** 1970 103–161. MR0267607 (42:2509)

Go-Gu. M . Golubitsky, V. Guillemin, Stable mappings and their singularities. *Graduate Texts in Mathematics*, Vol. 14. *Springer-Verlag, New York-Heidelberg*, 1973. MR0341518 (49:6269)

Ka-Na. M. Kato, Y. Matsumoto, On the connectivity of the Milnor fiber of a holomorphic function at a critical point. *Manifolds—Tokyo* 1973 (*Proc. Internat. Conf., Tokyo*, 1973), 131–136. *Univ. Tokyo Press, Tokyo*, 1975. MR0372880 (51:9084)

Loo. E. J. N. Looijenga, Isolated singular points on complete intersections. *London Mathematical Society Lecture Note Series,* 77. *Cambridge University Press, Cambridge*, 1984. MR747303 (86a:32021)

Me. J. McCleary, A user's guide to spectral sequences. Second edition. *Cambridge Studies in Advanced Mathematics*, 58. *Cambridge University Press, Cambridge*, 2001. MR1793722 (2002c:55027)

Ne-1. A. Nemethi, The Milnor fiber and the zeta function of the singularities of type $f = P(h, g)$. *Compositio Math.* **79** (1991), no. 1, 63–97. MR1112280 (93a:32058)

Ne-2. A. Nemethi, Hypersurface singularities with 2-dimensional critical locus. *J. London Math. Soc.* **(2) 59** (1999), no. 3, 922–938. MR1709088 (2001a:32037)

Pe-1. G. R. Pellikaan, Hypersurface singularities and resolutions of Jacobi modules. *Dissertation, Rijksuniversiteit te Utrecht, Utrecht,* 1985. *With a Dutch summary. Drukkerij Elinkwijk B. V., Utrecht*, 1985. MR816348 (87j:32024)

Pe-2. G. R. Pellikaan, Finite determinacy of functions with nonisolated singularities. *Proc. London Math. Soc.* **(3) 57** (1988), no. 2, 357–382. MR950595 (89i:32014)

Sh-1. M. Shubladze, Flat isolated singularities of analytic functions of three complex variables. (Russian) *Soobshch. Akad. Nauk Gruzin. SSR* **128** (1987), no. 2, 241–244. MR945129 (90b:32022)

Sh-2. M. Shubladze, Isolated hypersurface singularities of the transversal type A_1. *Bull. Georgian Acad. Sci.* **153** (1996), no. 1, 7–10. MR1651960 (2000a:32057)

Sh-3. M. Shubladze, Nonisolated singularities of codimension 1. Complex analysis. *J. Math. Sci.* (N. Y.) **118** (2003), no. 5, 5467–5505. MR2048103 (2005d:32049)

Sh-4. M. Shubladze, On the homotopy type of the complement to the Milnor fiber of an isolated singularity. (Russian) *Soobshch. Akad. Nauk Gruzin. SSR* **126** (1987), no. 3, 489–492. MR921219 (89c:32028)

Si. D. Siersma, Singularities with critical locus a 1-dimensional complete intersection and transversal type A_1. *Topology Appl.* **27** (1987), no. 1, 51–73. MR910494 (88m:32015)

Za. A. Zaharia, Topological properties of certain singularities with critical locus a 2-dimensional complete intersection. *Topology Appl.* **60** (1994), no. 2, 153–171. MR1302470 (95k:32039)

Tbilisi State University, 0128 Chavchavadze av. 1, Tbilisi, Georgia
E-mail address: mamushubladze@yahoo.com